D1672826

Differential Geometry and Topology

CONTEMPORARY SOVIET MATHEMATICS

Series Editor: Revaz Gamkrelidze, *Steklov Institute, Moscow, USSR*

COHOMOLOGY OF INFINITE-DIMENSIONAL LIE ALGEBRAS
D. B. Fuks

DIFFERENTIAL GEOMETRY AND TOPOLOGY
A. T. Fomenko

LINEAR DIFFERENTIAL EQUATIONS OF PRINCIPAL TYPE
Yu. V. Egorov

THEORY OF SOLITONS: The Inverse Scattering Method
S. Novikov, S. V. Manakov, L. P. Pitaevskii, and V. E. Zakharov

TOPICS IN MODERN MATHEMATICS: Petrovskii Seminar No. 5
Edited by O. A. Oleinik

Differential Geometry and Topology

A. T. Fomenko

Moscow State University
Moscow, USSR

Translated from Russian by
D. A. Leites

CONSULTANTS BUREAU • NEW YORK AND LONDON

Library of Congress Cataloging in Publication Data

Fomenko, A. T.
Differential geometry and topology.

(Contemporary Soviet mathematics)
Bibliography: p.
Includes index.
1. Geometry, Differential. 2. Topology. I. Title. II. Series.
QA641.F65 1987 516.3′6 87-2444
ISBN 0-306-10995-6

This translation is published under an agreement with the Copyright Agency of the USSR (VAAP).

© 1987 Consultants Bureau, New York
A Division of Plenum Publishing Corporation
233 Spring Street, New York, N.Y. 10013

All rights reserved

No part of this book may be reproduced, stored in a retrieval system, or transmitted in any form or by any means, electronic, mechanical, photocopying, microfilming, recording, or otherwise, without written permission from the Publisher

Printed in the United States of America

PREFACE

Differential geometry and topology are two of the youngest but most developed branches of modern mathematics. They arose at the juncture of several scientific trends (among them classical analysis, algebra, geometry, mechanics, and theoretical physics), growing rapidly into a multibranched tree whose fruits proved valuable not only for their intrinsic contribution to mathematics but also for their manifold applications. Some of these applications are mentioned in this book. With such a lot of "parents," modern differential geometry and topology naturally inherited many of their features; being at the same time young areas of mathematics, they possess vivid individuality, the main characteristics being, perhaps, their universality and the synthetic character of the methods and concepts employed in their study. Here geometric concepts and descriptiveness, the language of algebra and functional and differential methods, and so on, are interlinked. This synthetic character of posing problems and finding their solution is, to a certain extent, in tune with the natural sciences of the Renaissance, when mathematics, mechanics, and astronomy were considered as the unique system of knowledge of the laws of the Universe. Though not claiming to be that all-encompassing, modern geometry enables us, nevertheless, to solve many applied problems of fundamental importance.

The aim of this book is to give a short review of some geometric and differential methods used widely in theoretical investigations and in various applications.

With this goal in mind, we begin with the description of an important class of mathematical objects — the so-called CW-complexes that arise in certain concrete problems, for instance in the study of level surfaces of smooth functions on manifolds. In the study of CW-complexes we often encounter the following problem: Do two complexes coincide? This leads us to the search for invariants of CW-complexes that are preserved with respect to homotopic equivalence. Homology and cohomology groups are some of these invariants. We are interested mainly in methods of their computations regardless of their theoretical value. In this book we generally adhere to the following principle: We illustrate the practical applications of some methods, presenting their formal theoretical extension to general cases (which are often technically cumbersome) in a more condensed form and, in certain instances, referring the reader to the specialized literature. For instance, the method of computation of (co)homology using spectral sequences is illustrated by a number of meaningful examples important in further applications. Similarly, we illustrate the main consequences of the root theory of semisimple Lie algebras with the model example of the Lie group of nondegenerate matrices with determinant 1 and that of classical matrix Lie groups and Lie algebras. We do not concern ourselves with extending nontrivial questions and proofs to the general case.

The synthetic character of the architecture of the book is due to the fact that the questions considered here are situated at the juncture of several mathematical disciplines. In this book the following themes are interlinked: complexes, homology, bundles, theory of critical points of smooth functions on manifolds, bordisms, topology of three-dimensional manifolds, Lie groups and Lie algebras, root theory of semisimple Lie algebras, symplectic geometry, Hamiltonian systems, and problems of integration of mechanical systems (for example, equations of motion of a multidimensional solid body with a fixed point in the absence of gravity).

The choice and organization of the material in this book are based on special lectures given by the author at the Mechanical and Mathematical Faculty of Moscow State University and on the required course in geometry and topology offered to students of mathematics.

Some of the material covered in this book is supported by material from two textbooks (B. A. Dubrovin, S. P. Novikov, and A. T. Fomenko, Modern Geometry, Nauka, Moscow, 1979, and A. S. Mishchenko and A. T. Fomenko, Course in Differential Geometry and Topology, Moscow University Press, Moscow, 1980). Though the author did not attempt to make the book self-contained, many parts of it may be read without recourse to the literature.

This book is intended for undergraduate and postgraduate students in mathematics and mechanics and also for specialists in related disciplines who are interested in the applications of modern geometry to the solution of applied problems.

A. T. Fomenko

CONTENTS

Chapter 1

CW-COMPLEXES AND BUNDLES. HOMOLOGY, COHOMOLOGY, AND HOW TO COMPUTE THEM

Chapter 2

CRITICAL POINTS OF SMOOTH FUNCTIONS ON MANIFOLDS

Chapter 3

TOPOLOGY OF THREE-DIMENSIONAL MANIFOLDS

Chapter 4

SYMMETRIC SPACES

Chapter 5

SYMPLECTIC GEOMETRY

Chapter 6

GEOMETRY AND MECHANICS

Chapter 1

CW-COMPLEXES AND BUNDLES. HOMOLOGY, COHOMOLOGY, AND HOW TO COMPUTE THEM

1. CW-COMPLEXES AND THEIR SIMPLEST PROPERTIES

1.1. Preliminaries

Many problems of mechanics and theoretical physics make it necessary to study the properties of smooth manifolds; this is why smooth manifolds often appear even at the first stages of studying any concrete applied problem. The main properties of manifolds and related structures are described in [1]; therefore we will sometimes use this well-developed apparatus, referring the reader also to [2], where other refined properties of manifolds are studied. At the same time, besides manifolds, objects of another kind often arise even at the first stages of the study. They are the so-called CW-complexes arranged locally not as rigidly as smooth manifolds (therefore lacking many of their properties) but sometimes providing a more flexible apparatus enabling one to determine some invariants of the problem in question. The simplest examples are the level surfaces of a smooth function defined on a manifold, for example, level surfaces of the potential function or of the full energy. These surfaces arise in physics as "surfaces" along which the trajectories of mechanical systems with constant energy (important examples will be given below) move, but these "surfaces" may also have singular points, i.e., they are not manifolds. Nevertheless, they are CW-complexes which give a natural extension of the class of manifolds (as we will see, any manifold is a CW-complex). First, let us given an informal description of a CW-complex. Like manifolds, they are "glued" from open balls; but, in the case of manifolds, these balls are glued by homeomorphisms (on the common part), while in the case of CW-complexes each succeeding ball is glued by its boundary to the already constructed part of the complex with respect to a continuous mapping of a boundary of the ball, and this mapping must not be a homeomorphism of the boundary with its image; i.e., on the boundary, the glueings of different degrees of complexity are admitted. It is a weakening of the requirements of glueing "elementary bricks," i.e., balls that enhance steeply the supply of objects to construct. At the same time, the obtained class is meaningful but not all-embracing and, therefore, has many rich properties. Let us now introduce precise definitions.

Definition 1.1. The topological space X is a CW-complex σ_i^k if it is presentable as the union of nonintersecting sets called cells, where k is the dimension of the cell and i is its number, i.e., $X=\bigcup_{k=0}^{\infty}\bigcup_{i\in I_k}\sigma_i^k$, where I_k is the set of indices. For any cell σ_i^k there is fixed a continuous mapping $\chi_i^k\colon D^k\to X$ of a closed k-dimensional ball D^k in the space X, called a characteristic map, with the following properties:

1) the restriction of χ_i^k onto the open ball Int D^k (with the closure D^k) is a homeomorphism of this open ball onto σ_i^k;

2) the boundary of each cell, i.e., the set $\partial\sigma_i^k = \bar{\sigma}_i^k \setminus \sigma_i^k$ (where $\bar{\sigma}_i^k$ stands for the closure of σ_i^k in X), is contained in the union of a finite number of cells of lesser degrees;

3) the set $Y\subset X$ is closed if and only if the pullback $(\chi_i^k)^{-1}(Y)\subset D^k$ is closed in D^k for any σ_i^k.

We shall mainly consider the finite CW-complexes, i.e., those consisting of a finite number of cells. The subset $Y\subset X$ is a subcomplex if Y is closed in X and Y is a CW-complex, all its cells and characteristic maps being at the same time cells and characteristic maps in X. Among the set of different subcomplexes in X, there are naturally distinguished n-dimensional skeletons of X, which will be denoted by X^n, i.e., subcomplexes consisting of the union of all cells of dimension $\leq n$. Let us give the simplest examples of CW-complexes.

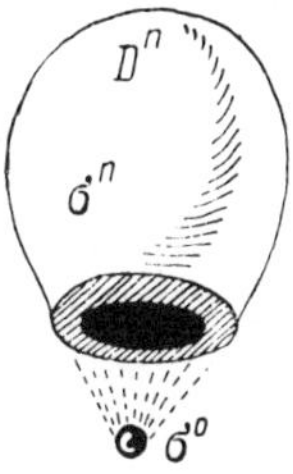

Fig. 1

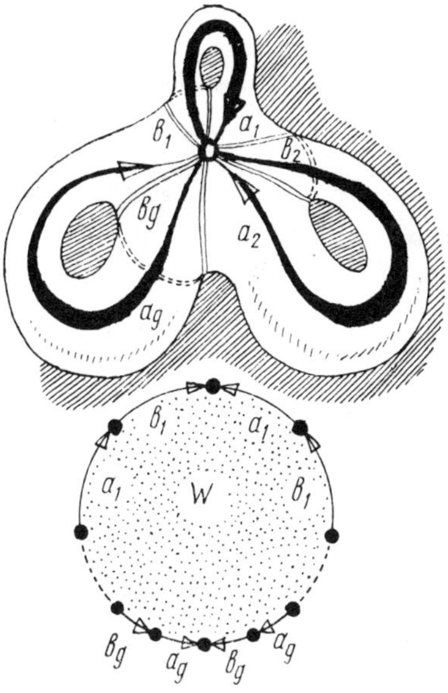

Fig. 2

1.2. Examples of CW-Complexes

Example 1. The standard n-dimensional sphere S^n may be presented in the form of a union of two cells: zero-dimensional and n-dimensional $S^n = \sigma^0 \cup \sigma^n$, so that the characteristic map $\chi^n : D^n \to S^n$ maps all of the boundary of the ball into one point, i.e., the zero-dimensional cell; see Fig. 1. Surely, the presentation of a sphere in the form of a CW-complex is not unique; we have produced the simplest decomposition.

Example 2. Let M^2 be a two-dimensional smooth, compact, connected closed (i.e., without boundary) manifold. Then, as is known from the classification theorem of two-dimensional surfaces (see, e.g., [1], Sec. 5), M^2 may be presented in the form of a fundamental two-dimensional polyhedron W with an identification on its boundary. For example, if M^2 is orientable, then W may be conditionally written in the form (see Fig. 2) $W = a_1 b_1 a_1^{-1} b_1^{-1} \ldots a_g b_g a_g^{-1} b_g^{-1}$. To recover M^2, we must identify the same letters of the word W, their orientation being taken into account. It turns out that this presentation of M^2 enables us to consider M^2 as a CW-complex consisting of a zero-dimensional cell, a two-dimensional one,

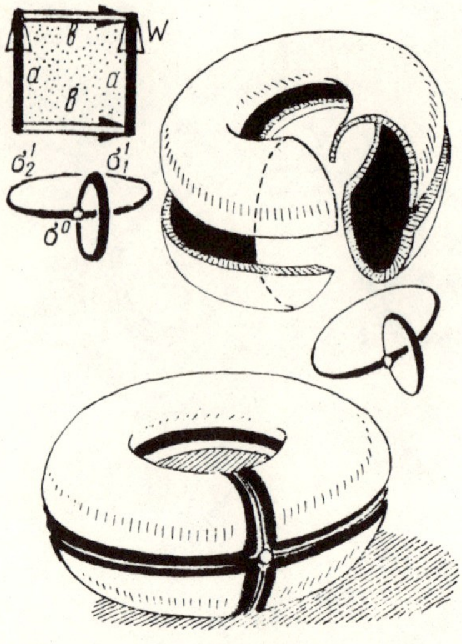

Fig. 3

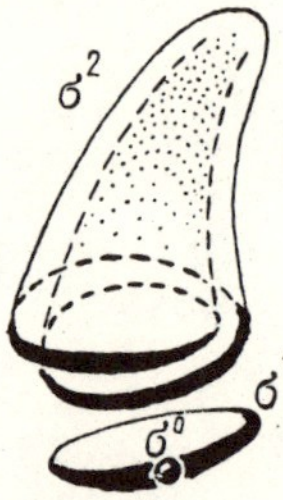

Fig. 4

and 2g one-dimensional ones, where g is a genus of a surface, i.e., the number of handles which, being attached to the sphere, result in the orientable surface. In fact, since by the classification theorem of surfaces all vertices of the fundamental polyhedron are glued into one point, it is natural to take this point as the unique zero-dimensional cell σ^0. The boundary of the polyhedron is turned, after the necessary identifications (see Fig. 2), into a set of 2g circles attached to one point, namely to σ^0. The two-dimensional cell is the interior of the two-dimensional closed polyhedron W attached to the set of one-dimensional cells with respect to the characteristic map defined explicitly by the above formula. This process is shown in Fig. 3 for the torus, i.e., for g = 1. Since the torus is homeomorphic to a sphere with a handle, then we have actually described the process of attachment of W to the set of one-dimensional cells for any g. Thus,

$$M^2 = \sigma^0 \cup \left(\bigcup_{k=1}^{2g} \sigma_k^1\right) \cup \sigma^2.$$

There is a similar decomposition in the nonorientable case (prove!), i.e., when $W = c_1^2 c_2^2 \ldots c_p^2$, where p is the number of "Moebius bands." They, being attached to the sphere,

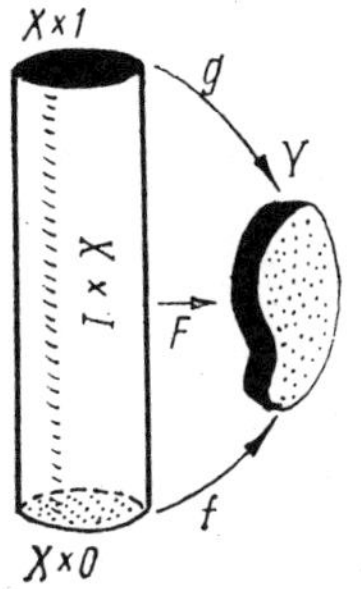

Fig. 5

result in an arbitrary nonorientable manifold:

$$M^2 = \sigma^0 \cup \left(\bigcup_{k=1}^{p} \sigma_k^1\right) \cup \sigma^2.$$

Example 3. Let X be a real projective space $\mathbf{R}P^n = \{(\lambda x_1, \ldots, \lambda x_{n+1});\ \lambda \neq 0 \text{ and } \sum_{i=1}^{n+1} x_i^2 \neq 0\}$; then $\mathbf{R}P^n = \sigma^0 \cup \sigma^1 \cup \ldots \cup \sigma^n$, where $\sigma^{n-k} = \{(0, \ldots, 0, \lambda x_{k+1}, \ldots, \lambda x_{n+1})\} = \mathbf{R}P^{n-k} \setminus \mathbf{R}P^{n-k-1}$. For $\mathbf{R}P^2$ this decomposition is conditionally depicted in Fig. 4. Thus, the closure $\bar{\sigma}^s$ of each cell σ^s is homeomorphic to the projective space $\mathbf{R}P^s$. The complex projective space is similarly split into the union of cells: $\mathbf{C}P^n = \sigma^0 \cup \sigma^2 \cup \sigma^4 \cup \ldots \cup \sigma^{2n}$.

Though we have defined CW-complexes constructively, without reference to a concrete realization, it is worth mentioning the following useful fact (though we will not apply it in what follows): Any finite CW-complex may be embedded in a finite-dimensional Euclidean space (prove!). This statement is analogous to the corresponding embedding theorem for smooth manifolds and is proved by the same scheme. In what follows, if there is no mention to the contrary, the term "complex" stands for the "finite CW-complex."

We shall often encounter continuous mappings. Therefore, recall the important notion of a homotopy.

Definition 1.2. Let f, g: $X \to Y$ be two continuous mappings of topological spaces X, Y. These mappings are called homotopic (and we write $f \approx g$) if there is a continuous mapping $F:X \times I \to Y$ of the direct product $X \times I$ (where $I = [0, 1]$ is the closed segment) in Y such that $F|_{X\times 0} = f$, $F|_{X\times 1} = g$ (see Fig. 5).

In other words, f and g are homotopic if there is a family of $\varphi_t:X \to Y$ continuous in the totality of variables $t \in I$, $x \in X$ such that $\varphi_0(x) \equiv f(x)$, $\varphi_1(x) \equiv g(x)$.

Spaces X and Y are called homotopically equivalent if there are mappings $f:X \to Y$ and $g:Y \to X$ such that $gf \approx 1_X$ and $fg \approx 1_Y$, where 1_X is the identity mapping of X onto itself.

2. SINGULAR HOMOLOGY GROUPS

2.1. Singular Simplexes, the Boundary Operator, and Homology Groups

In various mathematical and applied problems the question often arises: Do two CW-complexes "coincide," for example, are they homeomorphic or homotopically equivalent? This question is quite complicated; however, to answer it, sometimes it suffices to compute several algebraic characteristics related naturally to complexes and preserved with respect to homotopic equivalence. Namely, if these characteristics are different, then complexes are by no means homotopically equivalent (and, moreover, homeomorphic). The coincidence of these algebraic characteristics (among them being homology and cohomology groups) does not mean that complexes "coincide." We will only describe the so-called singular and cell (co)homology, since they will be enough to solve the problems considered in this book.

The standard simplex Δ^k of dimension k is the set of points $x = (x_0, \ldots, x_k) \in R^{k+1}$ defined as follows: $x_0 \geqslant 0, \ldots, x_k \geqslant 0$; $x_0 + x_1 + \ldots + x_k = 1$.

Definition 2.1. A singular simplex f^k of dimension k of the complex X is a continuous mapping of the standard simplex Δ^k in the space X. Furthermore, an integral k-dimensional singular chain c of X (or just chain) is a formal linear combination of singular simplexes f^k of X with integer coefficients only a finite number of which are nonzero. We will denote chains by $c = \sum_{(i)} a_i f_i^k$.

The set of all k-dimensional chains of the complex is evidently transformed into the Abelian group by addition. It is a free Abelian group denoted by $C_k(X)$. It turns out that these groups are a good material to construct the algebraic invariants which enables us to sometimes distinguish complexes among each other. To construct these invariants, it is necessary to define a natural homomorphism $\partial_k: C_k(X) \to C_{k-1}(X)$, usually called the boundary homomorphism or the boundary operator. Since all groups $C_k(X)$ are free, to define ∂_k it suffices to define it on each singular simplex f^k corresponding to a single generator or free group.

<u>Definition 2.2</u>. The boundary operator ∂_k is the mapping $\partial_k(f^k) = \sum_{i=0}^{k} (-1)^i f_i^{k-1}$, where $f_i^{k-1} = f^k|_{\Delta_i^{k-1}}$ is the restriction of a continuous mapping f^k onto the i-th face of the standard simplex Δ^k.

Recall that each face Δ_i^{k-1} of the k-dimensional simplex is a (k − 1)-dimensional simplex itself and $\Delta_i^{k-1} = (x_0, \ldots, x_{i-1}, x_{i+1}, \ldots, x_k) \subset \Delta^k$. The definition of ∂_k implies that $\partial_k \circ \partial_{k+1} = 0$, meaning that Ker $\partial_k \supset$ Im ∂_k, where Ker and Im stand for the kernel and the image of ∂^k. It is just this simple property that enables us to define homology groups that play such an important role in geometry and topology.

<u>Definition 2.3</u>. The group of singular homology $H_k(X)$ of dimension k of X is the quotient group $H_k(X) = \text{Ker}\,\partial_k / \text{Im}\,\partial_{k+1}$. The elements of the group $B_k(X) = \text{Im}\,\partial_{k+1}$ in the group $C_k(X)$ are sometimes called k-dimensional boundaries and the group $B_k(X)$ is called the group of k-dimensional boundaries. Elements of the subgroup $Z_k(X) = \text{Ker}\,\partial_k$ in the group $C_k(X)$ will be called k-dimensional cycles and those in the group $Z_k(X)$ will be called the group of k-dimensional cycles.

In these terms the homology group is sometimes defined as the quotient group of the group of cycles by the subgroup of the boundaries: $H_k = Z_k/B_k$. The definitions of the boundary operator, the cycle, and the boundary are formal analogs of the intuitive notions of the "bounded surface" and its "boundary," related to smooth compact manifolds. In particular, a closed compact smooth orientable submanifold Z in X may be considered as a "cycle," whereas the boundary B of the manifold with boundary as a "boundary" (see Fig. 6). Sometimes one says that the cycle z_k is homologic to zero in X

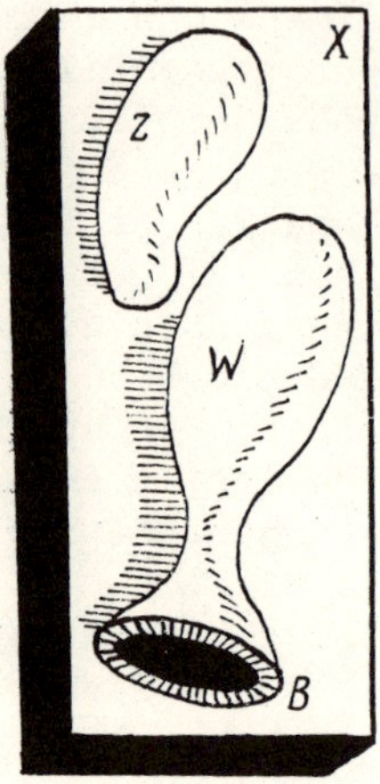

Fig. 6

(notation $z_k \sim 0$) if $z_k \in B_k(X)$, i.e., if there is a (k + 1)-dimensional chain c_{k+1} such that $\partial_k c_{k+1} = z_k$. Two cycles z_k and z_k' are called homologic in X (notation $z_k \sim z_k'$) if the cycle $z_k - z_k'$ is homologic to zero in X. It is convenient to imagine groups $H_k(X)$ in the case where they have a finite number of generators, as follows: since by definition all groups $H_k(X)$ are Abelian, then finitely generated Abelian groups $H_k(X)$ admit of the following presentation: $\mathbf{Z} \oplus \ldots \oplus \oplus \mathbf{Z} \oplus \mathbf{Z}_{p_1} \oplus \ldots \oplus \mathbf{Z}_{p_N}$, where $\mathbf{Z}$ is the free Abelian group and $\mathbf{Z}_{p_i}$ is a finite cyclic group of order p_i. The number of free summands in the decomposition of $H_k(X)$ (for Abelian groups this number is uniquely defined) is called the k-th Betti number or the rank of $H_k(X)$. If there are no finite cyclic groups in the decomposition of $H_k(X)$ [i.e., $H_k(X)$ is a free Abelian group], then one says that $H_k(X)$ is torsion-free. For example, if the space X consists of one point x, then it is easy to show that $H_0(x) = \mathbf{Z}$ and $H_k(x) = 0$ for $k \neq 0$. In fact, in each dimension k there is only one singular simplex $f^k: \Delta^k \to x$, i.e., $C_k(X) = \mathbf{Z}$, hence $\partial f^k = \left(\sum_i (-1)^i\right) f^{k-1}$, i.e., $\partial f^k = 0$ for k = 0 and odd k, and $\partial f^k = f^{k-1}$ for even k. Thus, $Z_k = B_k$ for odd k and $Z_k = B_k = 0$ for even k, completing the computation of H_k.

For any space X we have by definition $H_k(X) = 0$ for $k < 0$. If X is piecewise connected, then $H_0(x) = \mathbf{Z}$ (verify!). If the space X is not piecewise connected, then $H_0(X) = \bigoplus_{i \in I} \mathbf{Z}$, where I is the set of components of X.

2.2. Chain Complexes, Chain Homotopy, and Homotopic Invariance of Homology Groups

The collection of groups $\{C_k(X)\}$ and related homomorphisms $\{\partial_k\}$ is naturally organized in the following sequence: $\ldots \to C_k \xrightarrow{\partial_k} C_{k-1} \to \ldots \to C_1 \xrightarrow{\partial_1} C_0 \xrightarrow{\varepsilon} \mathbf{Z} \to 0$, where $\partial_k \circ \partial_{k+1} \equiv 0$ and ε is an epimorphism (see below). This sequence is sometimes referred to as a chain complex, and groups $\mathrm{Ker}\, \partial_k / \mathrm{Im}\, \partial_{k+1}$ are called k-dimensional homology groups of the chain complex. The homomorphism ε is defined by the formula $\varepsilon\left(\sum_i a_i f_i^0\right) = \sum_i a_i$ and $\mathrm{Im}\, \varepsilon \in \mathbf{Z}$. If X is a piecewise connected space, then $\mathrm{Ker}\, \varepsilon = \mathrm{Im}\, \partial_1$. Chain complexes are very convenient in the study of properties of homology groups. Let X and Y be two spaces and $g: X \to Y$ a continuous mapping. Definitions 2.1 and 2.2 imply that g induces a collection of homomorphisms $g_k: C_k(X) \to C_k(Y)$ of corresponding groups. It is convenient to arrange this collection of homomorphisms into a table (diagram), where $e: \mathbf{Z} \to \mathbf{Z}$ is the identity mapping:

$$\begin{array}{ccccccccc}
\ldots & \longrightarrow & C_1(X) & \xrightarrow{\partial_1} & C_0(X) & \longrightarrow & \mathbf{Z} & \longrightarrow & 0 \\
 & & \downarrow{\scriptstyle g_1} & & \downarrow{\scriptstyle g_0} & & \downarrow{\scriptstyle e} & & \\
\ldots & \longrightarrow & C_1(Y) & \xrightarrow{\partial_1} & C_0(Y) & \longrightarrow & \mathbf{Z} & \longrightarrow & 0
\end{array}$$

Its important properties are the relations $g_{k-1}\partial_k = \partial_k' g_k$ for any k and $e\varepsilon = \varepsilon' g_0$ (this follows from the definition of the boundary operator); in other words, the diagram commutes. In this case, the mapping of chain complexes $\alpha = \{g_k\}: C = \{C_k(X)\} \to C' = \{C_k(Y)\}$ is called a chain homomorphism. This implies that homomorphisms g_k induce homomorphisms $g_{k*}: \mathrm{Ker}\, \partial_k / \mathrm{Im}\, \partial_{k+1} \to \mathrm{Ker}\, \partial_k' / \mathrm{Im}\, \partial_{k+1}'$; i.e., they induce homomorphisms of homology groups $g_{k*}: H_k(X) \to H_k(Y)$.

<u>Lemma 2.1</u>. Homomorphisms $g_* = \{g_{k*}\}$ have the following properties:

1) if $g: X \to Y$ and $h: Y \to Z$ are continuous mappings, then $(h \circ g)_* = h_* \circ g_*$;

2) if $1_X: X \to X$ is the identity mapping, then $(1_X)_*: H_k(X) \to H_k(X)$ is the identity mapping for any k.

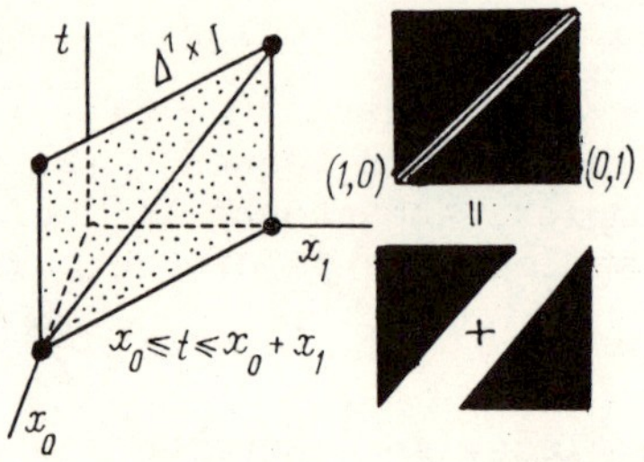

Fig. 7 Fig. 8

The proof follows from the definition of the corresponding homomorphisms.

Corollary 2.1. Groups of singular homology are topological invariants, i.e., if X and Y are homeomorphic, then their singular homology is isomorphic.

Proof. If $g: X \to Y$ is a homeomorphism, then so is $g^{-1}: Y \to X$; now apply Lemma 2.1.

It turns out that a stronger statement holds: the homology of homotopically equivalent spaces is isomorphic (the homotopic invariance of homology). To prove this, we will need the notion of a "chain homotopy" of chain complexes. Let C and C' be chain complexes and $\varphi = \{\varphi_k\}: C \to C'$ and $\psi = \{\psi_k\}: C \to C'$ be two chain homomorphisms.

Definition 2.4. Let us say that D is a chain homotopy of the complex C into the complex C', if $D = \{D_k\}$, where $D_k: C_k \to C'_{k+1}$ are homomorphisms such that $D_{k-1} \circ \partial_k + \partial'_{k+1} \circ D_k = \varphi_k - \psi_k$ for each k. In algebraic terms we have the following table (diagram):

$$\cdots \longrightarrow C_{k+1} \xrightarrow{\partial_{k+1}} C_k \xrightarrow{\partial_k} C_{k-1} \to \cdots$$
$$ \xrightarrow{D_k} \downarrow \varphi_k - \psi_k \xrightarrow{D_{k-1}}$$
$$\cdots \longrightarrow C'_{k+1} \xrightarrow{\partial'_{k+1}} C'_k \xrightarrow{\partial'_k} C'_{k-1} \to \cdots$$

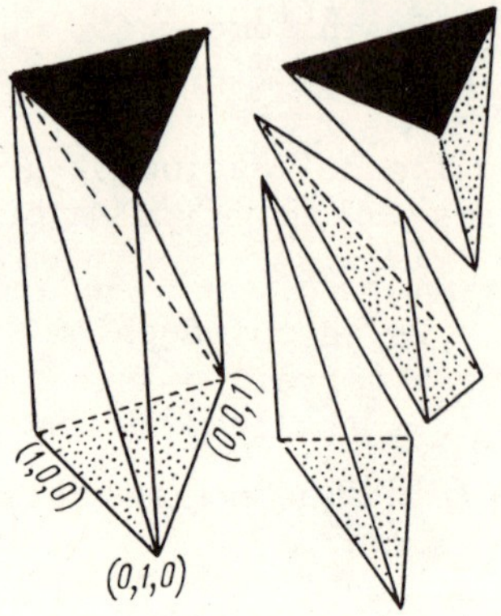

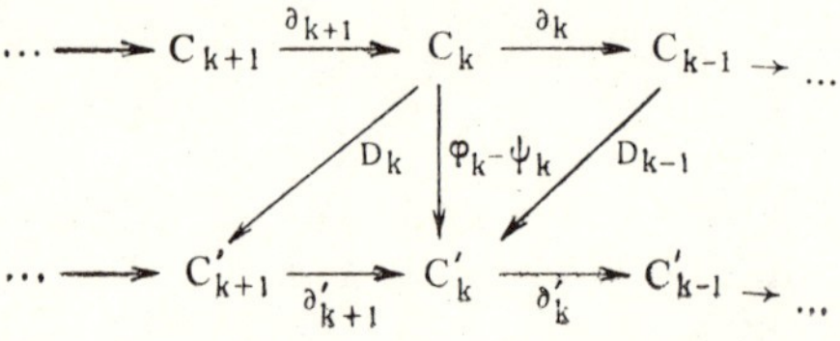

It is natural to say that two chain homomorphisms φ and ψ related by a chain homotopy are homotopic (or chain homotopic). The definition of homology groups implies that homotopic chain homomorphisms induce the same mappings of homology. In fact, if $z \in C_k$ is a cycle, i.e., $\partial_k z_k = 0$, then the definition of a chain homotopy implies

$$\varphi_k(z) - \psi_k(z) = D_{k-1}\partial_k(z) + \partial'_{k+1} D_k(z) = \partial'_{k+1}(D_k(z)),$$

i.e., images of this cycle under φ_k and ψ_k differ only by a boundary as required.

Theorem 2.1. Let $f, g: X \to Y$ be homotopic continuous mappings of spaces X and Y. Then the induced homomorphisms $f_*, g_*: H_k(X) \to H_k(Y)$ coincide for all k. In particular, homotopically equivalent spaces have isomorphic homology groups.

Proof. Since f and g are homotopic, there exists a continuous mapping $F: X \times I \to Y$, where $I = [0, 1]$, such that $F|X\times 0 = f$, $F|X\times 1 = g$. Over each k-dimensional simplex Δ^k let us take a (k + 1)-dimensional cylinder $\Delta^k \times I$ (the direct product of the simplex by the segment) and split it into the union of (k + 1)-dimensional simplexes $\Delta^k \times I = \bigcup_{i=0}^{k} \Delta_i^{k+1}$. The simplex Δ_i^{k+1} is defined in the space of variables $(x_0, \ldots, x_k, t)$ as follows: $\Delta_i^{k+1} = \{(x_0, \ldots, x_k, t): x_0 + \ldots + x_{i-1} \leq t \leq x_0 + \ldots + x_i\}$. Figures 7 and 8 depict the corresponding splitting of cylinders $\Delta^1 \times I$ and $\Delta^2 \times I$ (i.e., when $k = 1, 2$). Let $\varphi: \Delta^k \to X$ be an arbitrary singular simplex of the space X. Define the mapping $\varphi \times 1_I: \Delta^k \times I \to X \times I$ of the cylinder $\Delta^k \times I$ onto $X \times I$. Since on $X \times I$ the mapping F is defined, then the composition $F \circ (\varphi \times 1_I): \Delta^k \times I \to Y$ is well defined. The restriction of this mapping onto (k + 1)-dimensional simplexes Δ_i^{k+1} define k + 1 singular simplexes of the space Y. Denote by $D_k(\varphi)$ the sum of these simplexes. We have assigned to each simplex φ the simplex $D_k(\varphi)$ and therefore the homomorphism $D_k: C_k(X) \to C_{k+1}(Y)$. It remains to note that the collection of homomorphisms $\{D_k\}$ defines the chain homotopy that connects two chain mappings $\varphi = \{f_k\}$ and $\psi = \{g_k\}$ induced by f and g. In fact, from a geometrical point of view, the relation $\partial'_{k+1} D_k = -D_{k-1}\partial_k + f_k - g_k$ means just that the complete boundary of the "singular pyramid" $\Delta^k \times I$ consists of three components, the lateral boundary $-D_{k-1}\partial_k$

and two "bases": f_k and $-g_k$ (see Fig. 8). Thus, homotopic mappings f and g induce chain homotopic mappings $\{f_k\}$ and $\{g_k\}$ and, hence, the same mappings of homology. If spaces X and Y are homotopically equivalent, then there are mappings $p:X \to Y$ and $h:Y \to X$ such that $ph \sim 1_Y$ and $hp \sim 1_X$, whence (and by Lemma 2.1) we obtain the second heading of the theorem. It is just the homotopic invariance of homology groups that is responsible for the important role that they play in geometry and topology. In particular, these groups are invariants of spaces (with respect to homotopic equivalence). If the homologies of two spaces are different, then these spaces cannot be homotopically equivalent.

3. THE EXACT HOMOTOPIC SEQUENCE OF A PAIR

3.1. The Construction of the Exact Sequence

In computation of homology the so-called "exact sequence of a pair" is extremely useful. Suppose X and Y are topological spaces and Y is the closed subspace in X. Clearly, $C_k(Y) \subset C_k(X)$, and we may consider the group of relative chains $C_k(X, Y) = C_k(X)/C_k(Y)$. Since the boundary operator acts as $\partial_k:C_k(X) \to C_{k-1}(X)$, $\partial_k:C_k(Y) \to C_{k-1}(Y)$, it induces an operator $C_k(X, Y) \to C_{k-1}(X, Y)$ which will be denoted for simplicity by the same symbol ∂_k. In analogy with constructions of Sec. 2, we may define groups $\operatorname{Ker} \partial_k = Z_k(X, Y) \supset B_k(X, Y) = \operatorname{Im} \partial_k$ (of relative cycles and relative boundaries) and consider the quotient group $H_k(X, Y) = Z_k(X, Y)/B_k(X, Y)$, called the group of relative k-dimensional homology of the space X modulo the subspace Y. In accordance with the considerations of Sec. 2, we verify that groups of relative homology are also topologically and homotopically invariant (prove!). Now let us pass on to the construction of the new operator $\partial:H_k(X, Y) \to H_{k-1}(Y)$. Let $z_k \in C_k(X, Y)$ be a relative cycle and $\tilde{z}_k \in C_k(X)$ its arbitrary representative (in the coset). Since $\partial_k z_k = 0$, then $\partial_k \tilde{z}_k \in C_{k-1}(Y)$. This absolute cycle we will denote by $\partial z_k \in Z_{k-1}(Y)$. The homology class of this cycle does not depend on the choice of a representative of the homology class of z_k. This defines a homomorphism (operator) $\partial:H_k(X, Y) \to H_{k-1}(Y)$ which will be called the boundary operator (see Fig. 9). Furthermore, denote by $i:Y \to X$ the inclusion; it induces the homomorphism $i_*:H_k(Y) \to H_k(X)$. Since any absolute cycle may be considered as a relative one (modulo the subspace Y), one more natural mapping $j_*:H_k(X) \to H_k(X, Y)$ arises.

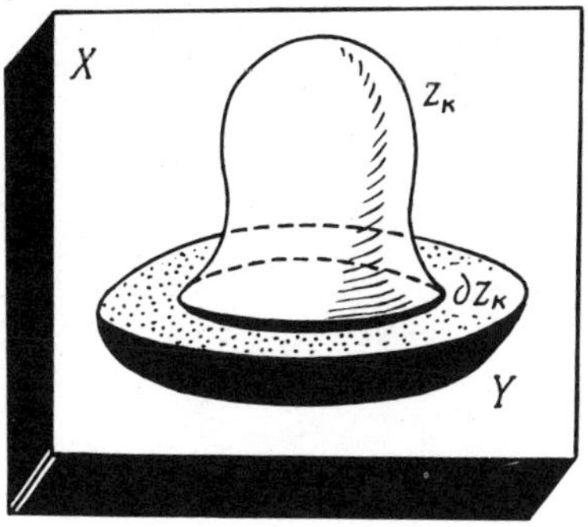

Fig. 9

Theorem 3.1. The following sequence of groups and homomorphisms is exact, i.e., the image of each outcoming homomorphism coincides with the kernel of each entering homomorphism:

$$\cdots \to H_{k+1}(X, Y) \xrightarrow{\partial} H_k(Y) \xrightarrow{i_*} H_k(X) \xrightarrow{j_*} H(X, Y) \xrightarrow{\partial} H_{k-1}(Y) \to \cdots$$

Proof. Let us verify, for example, the exactness in the term $H_k(X, Y)$. If $\partial\alpha = 0$, where $\alpha \in H_k(X, Y)$, then ∂z_k is homologic to zero in Y [where z_k is a representative of α, i.e., $z_k \in Z_k(X, Y)$]. But then, adding to z_k a k-dimensional chain that carries this homology in the subspace Y, we obtain a k-dimensional chain which is now a cycle from the point of view of the enveloping space X, i.e., we have presented the initial element α as the image of an element β, i.e., $\alpha = j_*\beta$, where $\beta \in H_k(X)$ (see Fig. 10). Thus, Ker $\partial \subset$ $\subset$ Im j_*. The converse inclusion Ker $\partial \subset$ Im j_* follows from the fact that any absolute cycle in X considered as a relative one has a zero boundary in Y. The exactness of the sequence in other terms is verified by similar constructions.

In what follows it is useful to know the following simple properties of exact sequences (proofs will be left to the reader as an easy exercise).

1) The sequence $0 \to A \to 0$ is exact if and only if $A = 0$.

2) The sequence $0 \to A \xrightarrow{\alpha} B \to 0$ is exact if and only if groups A and B are isomorphic and the homomorphism α is an isomorphism.

3) The sequence $0 \to A \xrightarrow{i} B \xrightarrow{\pi} C \to 0$ is exact if and only if the group A is a subgroup of B and the homomorphism $i: A \to B$ is an inclusion (monomorphism), $C = B/A$ (the quotient group), and $\pi: B \to B/A = C$ is the natural projection onto the quotient group.

The following property will not be used explicitly, but its proof is a useful exercise and, therefore, we will formulate it. Suppose in the commuting diagram

$$\begin{array}{ccccccccc} A_1 & \to & A_2 & \to & A_3 & \to & A_4 & \to & A_5 \\ \downarrow{\scriptstyle\alpha_1} & & \downarrow{\scriptstyle\alpha_2} & & \downarrow{\scriptstyle\alpha_3} & & \downarrow{\scriptstyle\alpha_4} & & \downarrow{\scriptstyle\alpha_5} \\ B_1 & \to & B_2 & \to & B_3 & \to & B_4 & \to & B_5 \end{array}$$

both lines are exact sequences, α_2, α_4 are isomorphisms, α_1 is an epimorphism, and α_5 is a monomorphism. Then α_3 is an isomorphism.

3.2. The Reduction of the Relative Homology to the Absolute One

If X and Y are piecewise connected, then $H_0(X, Y) = 0$. It turns out that the relative homology reduces, in essence, to the absolute one. We will say that the pair of spaces (X, Y) is a cell pair if X and Y are CW-complexes and the closed subspace Y is a CW-subcomplex in X. As earlier, we will restrict ourselves to finite complexes. Let us denote by X/Y the space obtained from X after the retraction of the subspace Y into one point.

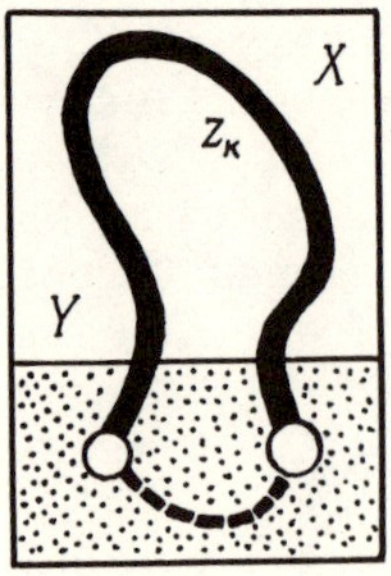

Fig. 10

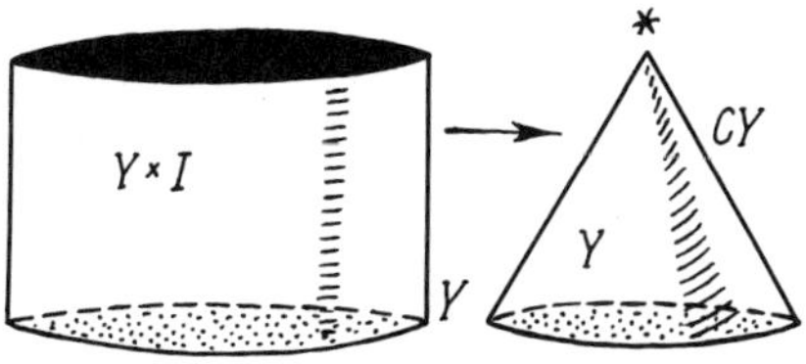

Fig. 11

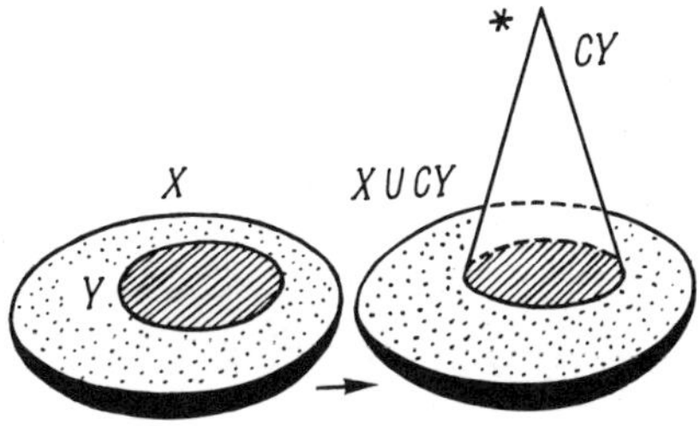

Fig. 12

Theorem 3.2. Let (X, Y) be a cell pair. Then $H_k(X, Y) = H_k(X/Y)$ for $k \neq 0$.

Proof. Let us denote by CY the cone over Y, i.e., the space obtained from the cylinder Y × I by retraction of the upper base of the cone into one point (see Fig. 11). Let us construct the new space $X \cup CY$, i.e., let us identify the subspace Y in X with the base Y in the cone CY (see Fig. 12). Since X, Y are finite complexes, then, retracting the cone CY in itself into one point, we obtain the homotopical equivalence $X \cup CY \approx X/Y$ (see Fig. 13). Thus, to prove the theorem, it suffices to verify that $H_k(X, Y) = H_k(X \cup CY)$ for $k > 0$. The exact sequence of the pair $(X \cup CY, *)$, where $*$ is the vertex of the cone, implies that $H_k(X \cup CY, *) = H_k(X \cup CY)$ for $k > 0$; hence we must prove that $H_k(X, Y) = H_k(X \cup CY, *)$ for $k > 0$. Before proving this statement, we will need one technical argument connected with the so-called barycentric subdivision. Let Δ^k be the standard simplex. Its barycentric subdivision $\beta\Delta^k$ is defined inductively as follows. If $k = 1$, then $\beta\Delta^1$ is obtained by adding the new vertex in the middle of the segment Δ^1. If $k = 2$, then $\beta\Delta^2$ is obtained by connecting the center of the triangle Δ^2 with its vertices

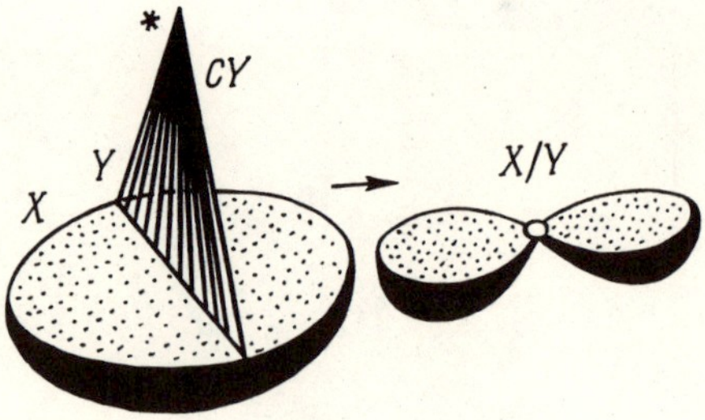

Fig. 13

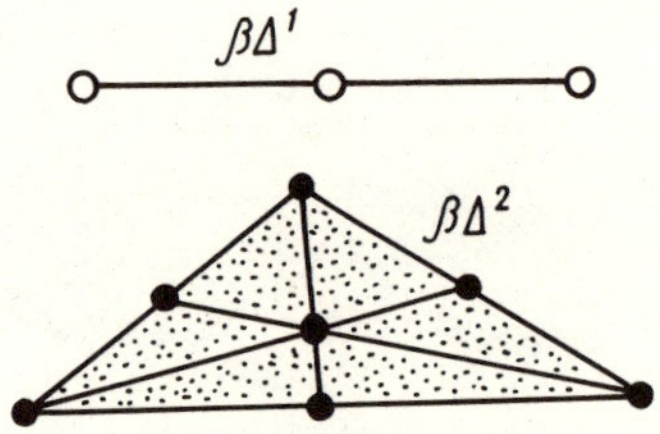

Fig. 14

and the midpoints of its sides (see Fig. 14). Finally, for an arbitrary k the subdivision $\beta\Delta^k$ is obtained, as follows: in Δ^k, we fix its center and divide the simplex Δ^k into pyramids with the vertex in this center, the bases being simplexes of barycentric subdivisions of the face of Δ^k. Now let $f:\Delta^k \to X$ be an arbitrary singular simplex of the space X. Denote by βf the chain equal to the sum of all singular simplexes obtained by restriction of the initial mapping f onto k-dimensional simplexes of the barycentric subdivision $\beta\Delta^k$ of the initial simplex Δ^k. Assigning to each singular simplex f [i.e., to an elementary chain in $C_k(X)$] the singular chain βf, we obtain the homomorphism $\beta_k:C_k(X) \to C_k(X)$. We claim that the collection of mappings $\beta = \{\beta_k\}$ defines the chain homomorphism β of the chain complex $C(X)$ into itself and this mapping is chain homotopic to the identity mapping. The fact that the mapping β is a chain one follows from the fact that the restriction of the formal sum of singular simplexes onto a barycyclic subdivision equals the formal sum of these restrictions. Now let us construct explicitly a chain homotopy E connecting the identity mapping with

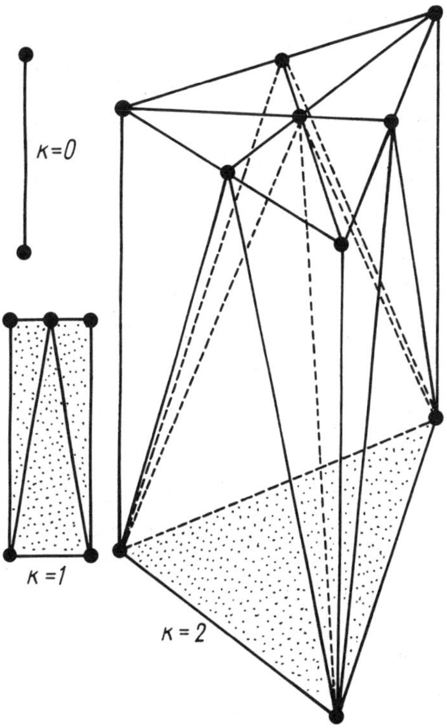

Fig. 15

β. For this, let us define for each $k \geqq 0$ the division of the prism $\Delta^k \times I$ into the sum of simplexes as follows. The lower base of the prism (cylinder) $\Delta^k \times I$ coincides with the initial simplex Δ^k. The upper base of the prism will be considered in the form of barycyclic subdivisions of the standard simplex Δ^k. The direct product $\Delta_i^{k-1} \times I \subset \Delta^k \times I$, where Δ_i^{k-1} is a face of Δ^k, will be decomposed into simplexes as was done in Theorem 2.1 at the division of $\Delta^{k-1} \times I$. All this defines a division of $\Delta^k \times I$. At the last stage, we divide this prism into the union of pyramids whose bases are simplexes in the above-mentioned division of faces of the prism, and the vertex is the center of the upper base of the prism $\Delta^k \times I$. On Fig. 15 this division is shown for $k = 0, 1, 2$.

In other words, this recursive process of division of $\Delta^k \times I$ may be described as follows. If the division is already defined for $q < k$, then on a part of the boundary of

$\Delta^k \times I$, namely on the union (on the "cup") $(\Delta^k \times 0) \cup (\partial\Delta^k \times \times I) \subset \Delta^k \times I$, the division must be performed so that $\Delta^k \times 0$ is the standard simplex and $\partial\Delta^k \times I$ is divided accordingly with the recursive process for $q < k$. After this we divide the whole prism into k-dimensional simplexes whose bases are $(k - 1)$-dimensional simplexes of the above-mentioned division of the part of the boundary (of the "cup") and the vertex is the center of the upper face. Clearly, in this process the upper side is affected by a barycyclic subdivision.

Now let $f:\Delta^k \to X$ be a singular simplex. Denote by $D_k f \in \in C_{k+1}(X)$ the $(k + 1)$-dimensional chain which is the sum of all $(k + 1)$-dimensional singular simplexes obtained at the restriction of the mapping $\Phi:\Delta^k \times I \to Y$, where $\Phi(x, t) = f(x)$, onto simplexes of the above-constructed division of $\Delta^k \times I$. We obtain homomorphisms $D_k:C_k(X) \to C_{k+1}(X)$ which define the required chain homotopy connecting the identity mapping of the complex $C(X)$ onto itself with the barycyclic mapping β (see its construction above). To prove that $\{D_k\}$ define chain homotopy, let us reproduce the considerations from the proof of Theorem 2.1.

Now let us proceed to the proof of our theorem. Let us consider the inclusion $(X, Y) \to (X \cup CY, CY)$. It induces the homomorphism $\alpha:H_k(X, Y) \to H_k(X \cup CY, CY) = H_k(X \cup CY, *)$. Here we have made use of the fact that $CY \approx *$, i.e., the cone is retractable along itself into a point. Let us prove that α is an epimorphism. Let $z \in Z_k(X \cup CY, CY)$ be a cycle. We must find in the group $Z_k(X, Y)$ a cycle such that the above mapping transforms it into a cycle homological to $z \in Z_k(X \cup \cup CY, CY)$. Let us present the cone CY as the union of two subsets: the cone A, which is part of CY, consisting of points such that $t \geq \frac{1}{2}$; and the "truncated cone" B, which is a part of CY, consisting of points such that $t \leq \frac{1}{2}$ (see Fig. 16). Singular simplexes that constitute the cycle z refined with respect to barycyclic subdivisions give us each time (see above) cycles homological to z but consisting of ever smaller simplexes. Since the number of simplexes is finite and X, Y are compact, there is a sufficiently small barycyclic subdivision such that if a singular simplex of this subdivision intersects with A, then it completely belongs to CY (here we consider the image of the standard simplex under the continuous mapping that defines the singular simplex). This cycle z' (which is a barycyclic subdivision of z) is homologic to z. Let us discard all simplexes of z' that intersect with A. Since this operation is performed within the

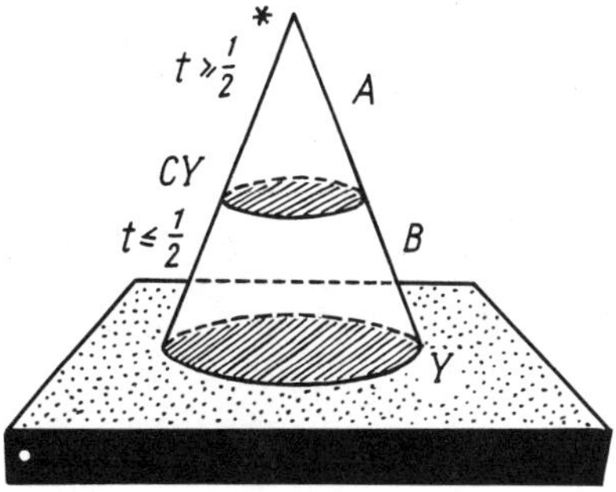

Fig. 16

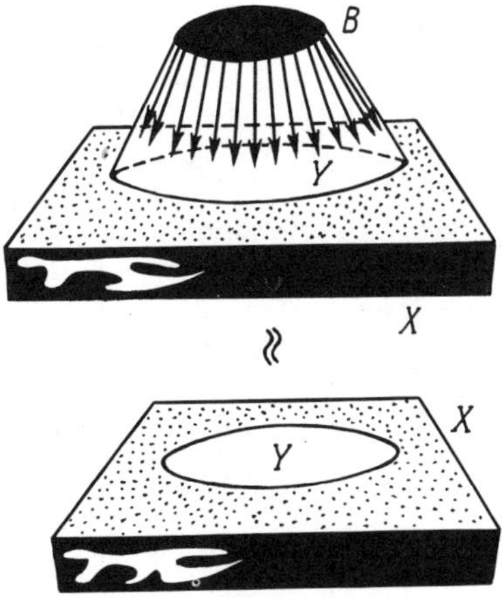

Fig. 17

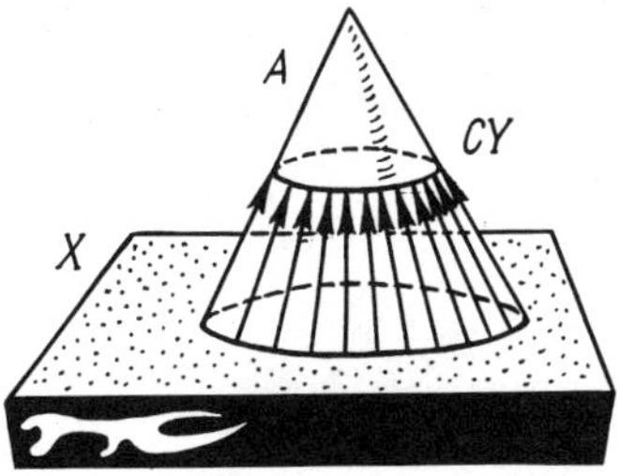

Fig. 18

limits of CY, then from the point of view of relative homology the new cycle remains in the same homology class as z' (hence as z). Thus, $z'' \in H_k(X \cup B, B)$. At the same time, due to the homotopic invariance of homology groups, there is an isomorphism $H_k(X, Y) = H_k(X \cup B, B)$ since $(X \cup B, B) \approx (X, Y)$ (see Fig. 17). Thus we have produced explicitly a cycle $z'' \in H_k(X, Y)$ which is mapped by the homomorphism α onto the cycle $z \in H_k(X \cup CY, CY)$. Thus the epimorphicity of α is proved. The monomorphicity is proved by a similar scheme making use of the pair $(X \cup CY, B) \approx (X \cup CY, CY)$ (see Fig. 18). The theorem is proved.

4. CELL HOMOLOGY

4.1. Computation of the Singular Homology of a Sphere

We have described above one of the methods of defining homology groups using singular simplexes. However, for CW-complexes, it is possible to define the so-called cell homology which coincides, as it turns out, with singular homology, but having the following important advantage: cell homology is a good deal easier to compute than the singular one. We have already seen that even in the simplest case when the space X consists of one point the computation of singular homology requires several considerations (though elementary ones). If the space X is more complicated, then, as even simple examples show, the computation of its singular homology becomes much more difficult. In practice, cell homology is most often used because of its ease of computation and better descriptiveness. Let us introduce this homology bearing in mind the notion of singular homology that enables us not only to prove the existence of groups of singular and cell homology of a complex but to produce a clear connection between them. Let us begin with the simplest example: the computation of singular homology of a sphere. Since the zero-dimensional sphere S^0 consists of a pair of points, then, by Sec. 3, we have $H_k(S^0) = 0$ for $k \neq 0$ and $H_0(S^0) = \mathbf{Z} \oplus \mathbf{Z}$.

<u>Lemma 4.1</u>. The homology groups of the n-dimensional sphere S^n, where $n > 0$, are as follows: $H_k(S^n) = 0$ for $k \neq 0, n$ and $H_k(S^n) = \mathbf{Z}$ for $k = 0, n$.

<u>Proof</u>. Consider the pair (D^1, S^0), where $S^0 = \partial D^1$ (the geometric boundary of a one-dimensional disk, i.e., the segment). The beginning of the exact sequence of the pair $(D^1,$

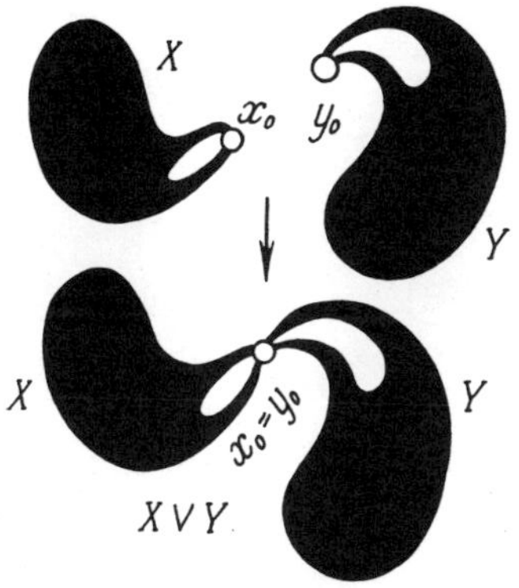

Fig. 19

S^0), i.e., $H_1(D^1) \to H_1(D^1, S^0) \to H_0(S^0) \to H_0(D^1) \to 0$, is of the following form: $0 \to H_1(S^1) \to \mathbf{Z} \oplus \mathbf{Z} \to \mathbf{Z} \to 0$. Here we made use of the fact that D^1 is homotopically equivalent to a point and, therefore, $H_1(D^1) = 0$, $H_0(D^1) = \mathbf{Z}$. Furthermore, $H_1(D^1, S^0) = H_1(D^1/S^0) = H^1(S^1)$ (by Theorem 3.2), since $D^1/S^0 = S^1$. Since the sequence is exact, then $H_1(S^1) = (\mathbf{Z} \oplus \mathbf{Z})/\mathbf{Z} = \mathbf{Z}$. In dimensions greater than one (i.e., $k > 1$), we have $H_k(D^1) \to H_k(D^1, S^0) \to H_{k-1}(S^0)$. Since groups $H_k(D^1)$ and $H_{k-1}(S^0)$ are trivial (see above), then $H_k(D^1, S^0) = H_k(S^1) = 0$ for $k > 0$ by the exactness of the sequence. Suppose the lemma is proved for all spheres S^i of dimension $i \leq n-1$. Consider again the exact sequence of the pair (D^n, S^{n-1})

$$H_k(D^n) \to H_k(D^n, S^{n-1}) \to H_{k-1}(S^{n-1}) \to H_{k-1}(D^n).$$

Since $H_i(D^n) = 0$ for $i > 0$ and $H_k(D^n, S^{n-1}) = H_k(S^n)$, then for $k > 1$ the sequence has the following form: $0 \to H_k(S^n) \to H_{k-1}(S^{n-1}) \to 0$; whence, by exactness, we get $H_k(S^n) = H_{k-1}(S^{n-1})$. For $k = 1$, the beginning of the sequence is as follows:

$$H_1(D^n) \to H_1(D^n, S^{n-1}) \to H_0(S^{n-1}) \to H_0(D^n),$$

i.e., $0 \to H_1(S^n) \to \mathbf{Z} \to \mathbf{Z}$. Since the last arrow stands for the identity mapping of zero-dimensional homology, then the exactness implies $H_1(S^n) = 0$. The inductive hypothesis completes the proof of the theorem.

Recall the definition of the bouquet of two topological spaces. Suppose X and Y are spaces with fixed points x_0 and

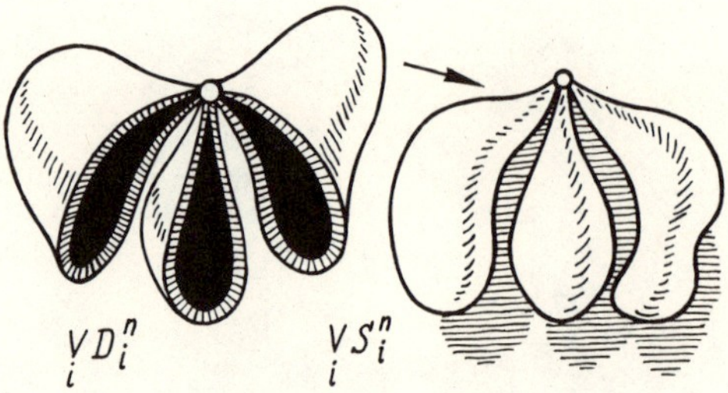

Fig. 20

y_0, respectively. Let us construct the new space $X \vee Y$ identifying x_0 and y_0 (see Fig. 19). Other points of X and Y are not identified.

Lemma 4.2. Let $X = \bigvee_i S_i^n$ be a bouquet of n-dimensional spheres, numbered by the index i, where $1 \leq i \leq N$. If $n > 0$, $k > 0$, then the isomorphism $H_k(X) = \bigoplus_i H_k(S_i^n) = \mathbf{Z} \oplus \ldots \oplus \mathbf{Z}$ (N times) holds.

The proof is practically identical to the inductive proof of Lemma 4.1. We must only consider the exact sequence of the pair $(\bigvee_i D_i^n, \bigvee_i \partial D_i^n)$ taking into account that $(\bigvee_i D_i^n)/(\bigvee_i \partial D_i^n) = \bigvee_i S_i^n$ (see Fig. 20).

Note the following fact, useful in what follows. As a generator of the group $\mathbf{Z} = H_n(D^n, S^{n-1})$, we may take the homology class of the simplest singular chain $1 \cdot f$, where $f: \Delta^n \to D^n$ is a homeomorphism of the simplex onto a disk (ball). Hence the orientation of the sphere $S^n = D^n/S^{n-1}$ may be defined by fixing 1 as a generator in the group $\mathbf{Z} = H_n(S^n)$. The change in orientation of the sphere is equivalent to the replacement $1 \to -1$ in $H_n(S^n)$.

4.2. Groups of Cell Chains

Suppose X is a finite CW-complex. Let us try to compute its singular homology in terms of its cells and their characteristic mappings, i.e., in terms in which this complex is defined (see Definition 1.1). The set of all k-dimensional cells of X will be denoted by X_k. Let X^k be a k-dimensional skeleton of X. Let us assume that the orientations of all cells of the complex are fixed. Let us enumerate all k-dimensional cells and let A_k be the set of indices that are in one-to-one correspondence with the set X_k. Then by Lemma 4.2 we get

$$H_i(X^k, X^{k-1}) = H_i\Big(\bigvee_{\alpha\in A_k} S_\alpha^k\Big) = \begin{cases} 0 & \text{for } i \neq k \\ P_k(X) & \text{for } i = k. \end{cases}$$

Here $P_k(X)$ stands for the free Abelian group whose generators are in one-to-one correspondence with A_k. Since elements of this group are naturally identified with linear combinations of the form $\sum_\alpha a_\alpha \sigma_\alpha^k$, where σ_α^k is a k-dimensional cell of X (recall that the complex is finite; therefore so are all these linear combinations), then $P_k(X)$ is a finitely generated group. The group $P_k(X)$ is called the group of cell k-dimensional chains of X. Groups $P_k(X)$ and $C_k(X)$ are not isomorphic, generally.

Before we proceed, let us consider the so-called exact homologic sequence of a triple — a variant of the exact sequence of a pair. Let (X, Y, Z) be a triple of spaces where Y and Z are closed in X. Consider two embeddings $(Y, Z) \to (X, Z)$ and $(X, Z) \to (X, Y)$. Let $\partial: H_k(X, Y) \to H_{k-1}(Y, Z)$ be a boundary homomorphism generated by $\partial: H_k(X, Y) \to H_{k-1}(Y)$ defined above and by the fact that each absolute cycle from $H_{k-1}(Y)$ may be considered as a relative one mod Z, i.e., to consider it as an element of $H_{k-1}(Y, Z)$. With these remarks taken into account, we obtain the following exact sequence of a triple:

$$\cdots \to H_k(X, Y) \xrightarrow{\partial} H_{k-1}(Y, Z) \to H_{k-1}(X, Z) \to H_{k-1}(X, Y) \xrightarrow{\partial} \cdots$$

The proof of its exactness is left to the reader (see the proof of Theorem 3.1).

4.3. Groups of Cell Homology

Returning to groups $P_k(X) = H_k(X^k/X^{k-1})$ and $P_{k-1}(X) = H_{k-1}(X^k/X^{k-2})$, we consider the exact sequence of the triple (X^k, X^{k-1}, X^{k-2}). At this moment we need from this sequence only the homomorphism $H_k(X^k, X^{k-1}) \to H_{k-1}(X^{k-1}, X^{k-2})$ which in our new notation is written as $P_k(X) \to P_{k-1}(X)$. Denoting this homomorphism by ∂_k, we obtain the chain complex $\{P_k(X),$

$\partial_k\}$, namely $\ldots \to P_k(X) \xrightarrow{\partial_k} P_{k-1}(X) \to \ldots$. As for any chain complex, the homology groups of this complex are defined, i.e., groups $\mathrm{Ker}\, \partial_k / \mathrm{Im}\, \partial_{k+1}$, usually called the cell homology of X.

It turns out there there is a canonical isomorphism between the homology of the described chain complex and the singular homology of X. This is just a main statement of this section that enables us to reduce the computation of the singular homology of X to the computation of the homology of a much simpler chain complex whose structure will be described below. It so happens that this reduction is so effective that the majority of concrete computations of homology is based on this theorem. In particular, it immediately implies that groups of singular homology of finite complexes are finitely generated (as Abelian groups). Having at our disposal only the definition of singular homology, it is quite difficult to discern this fact, since functional spaces $C_k(X)$ of singular chains are, generally speaking, "infinite-dimensional" [unlike finitely generated groups $P_k(X)$].

4.4. Theorem on Coincidence of Singular and Cell Homology of a Finite Complex

Theorem 4.1. For a finite CW-complex X, groups of singular homology $H_k(X)$ and homology groups of the chain complex $\{P_k(X), \partial_k\}$, i.e., groups $\mathrm{Ker}\, \partial_k / \mathrm{Im}\, \partial_{k+1}$ (groups of cell homology), are isomorphic.

First let us prove the following lemma.

Lemma 4.3. For $k > 1$, there is an isomorphism $H_k(X^{k+1}, X^{k-2}) = H_k(X)$.

Proof. Consider the triple of complexes $(X^{k+1}, X^{k-2}, X^{k-3})$ and the corresponding exact sequence

$$H_k(X^{k-2}, X^{k-3}) \to H_k(X^{k+1}, X^{k-3}) \to H_k(X^{k+1}, X^{k-2}) \to H_{k-1}(X^{k-2}, X^{k-3}).$$

Extreme terms of the sequence are zero, i.e., $H_k(X^{k+1}, X^{k-3}) = H_k(X^{k+1}, X^{k-2})$. Let us repeat the same considerations for the triple of complexes $(X^{k+1}, X^{k-3}, X^{k-4})$; then the exact sequence gives that $H_k(X^{k+1}, X^{k-3}) = H_k(X^{k+1}, X^{k-4})$. Pro-

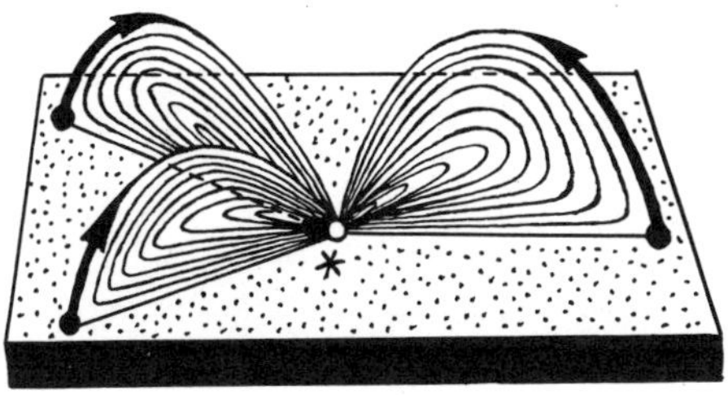

Fig. 21

gressing further in dimension, we obtain the following chain of isomorphisms:

$$H_k(X^{k+1}, X^{k-2}) = H_k(X^{k+1}, X^{k-3}) = H_k(X^{k+1}, X^{k-4}) = \cdots$$
$$= H_k(X^{k+1}, X^0) = H_k(X^{k+1})$$

for $k > 1$. If the skeleton X^0 consists only of one point, then the equality holds for $k = 1$. The case is that any finite connected CW-complex is homotopically equivalent to a finite complex such that its zero-dimensional skeleton consists only of one point. For this it suffices to consider all zero-dimensional cells of the initial complex and to connect each of them by a continuous path that belongs to X^1 with the fixed vertex (zero-dimensional cell $*$). Furthermore, let us accomplish a homotopy depicted in Fig. 21 which retracts all zero-dimensional cells into one. Thus the equality proved above may be considered fulfilled for any $k \geq 1$.

Lemma 4.4. We have $H_k(X) = H_k(X^{k+1})$.

Proof. Let us show that for any $i < k + 1$ the equality $H_i(X^{k+1}) = H_i(X^{k+2})$ holds. In fact, consider the exact sequence of the pair (X^{k+2}, X^{k+1}): $0 = H_{i+1}(X^{k+2}, X^{k+1}) \to \to H_i(X^{k+1}) \to H_i(X^{k+2}) \to H_i(X^{k+2}, X^{k+1}) = 0$. This implies the necessary equality, in particular $H_k(X^{k+1}) = H_k(X^{k+2})$.

Returning to the proof of Lemma 4.3, we obtain

$$H_k(X^{k+1}, X^{k-2}) = H_k(X^{k+1}) = H_k(X^{k+2}) = \cdots = H_k(X),$$

as required.

Lemma 4.5. There exists the isomorphism Ker $\partial_k/\text{Im }\partial_{k+1} = H_k(X^{k+1}, X^{k-2})$, where ∂_k is a homomorphism that defines the complex of groups of cell chains (see above).

Proof. Consider the following commutative diagram:

$$\begin{array}{c} H_k(X^{k-1}, X^{k-2}) = 0 \\ \downarrow \\ P_{k+1}(X) = H_{k+1}(X^{k+1}, X^k) \xrightarrow{\partial} H_k(X^k, X^{k-2}) \rightarrow H_k(X^{k+1}, X^{k-2}) \rightarrow \\ H_k(X^{k+1}, X^k) = 0 \\ \downarrow \\ H_k(X^k, X^{k-1}) = P_k(X) \\ \downarrow \\ H_{k-1}(X^{k-1}, X^{k-2}) = P_{k-1}(X) \end{array}$$

In this diagram the row is a segment of the exact sequence of the triple (X^{k+1}, X^k, X^{k-2}), the column is a segment of the exact sequence of the triple (X^k, X^{k-1}, X^{k-2}), and homomorphisms i_*, j_* are induced by the corresponding embeddings of the pair i, j. The commutativity of the diagram means that $\partial_{k+1} = j_*\partial$. Recall that, by the definition of groups of cell chains, we have $P_\alpha(X) = H_\alpha(X^\alpha, X^{\alpha-1})$. Since the row and the column are parts of exact sequences, then i_* is an epimorphism and j_* is a monomorphism. This implies that

$$H_k(X^{k+1}, X^{k-2}) = H_k(X^k, X^{k-2})/\text{Ker } i_* = H_k(X^k, X^{k-2})/\text{Im } \partial.$$

Since j_* is a monomorphism, then

$$H_k(X^k, X^{k-2})/\text{Im } \partial = j_* H_k(X^k, X^{k-2})/j_* \text{ Im } \partial = \text{Im } j_*/\text{Im } j_* \partial$$

$$= \text{Ker } \partial_k/\text{Im } j_* \partial = \text{Ker } \partial_k/\text{Im } \partial_{k+1}.$$

Here we have used the following identities: Im j_* = Ker ∂_k (by the exactness of the sequence), and $j_*\partial = \partial_{k+1}$ (by the commutativity of a diagram). Thus, we have obtained $H_k(X^{k+1}, X^{k-2}) = \text{Ker } \partial_k/\text{Im } \partial_{k+1}$. The lemma is proved.

Proof of Theorem 4.1. Lemmas 4.3 and 4.5 imply $H_k(X) = \text{Ker } \partial_k/\text{Im } \partial_{k+1} = H_k(X^{k+1}, X^{k-2})$, as required. Note that in the proof of Lemma 4.3 in the case where $k = 1$, it was not necessary to make use of the homotopic equivalence of X and the complex that has only one zero-dimensional cell. It suffices in the proof of Lemma 4.5, in the case where k = 1, to consider the following diagram:

$$\begin{array}{ccccc} & & H_1(X^0)=0 & & \\ & & \downarrow & & \\ P_2(X)=H_2(X^2,X^1) & \xrightarrow{\partial} & H_1(X^1) \to H_1(X^2) \to H_1(X^2,X^1)=0 & & \\ & \overset{\partial_2}{\searrow} & \downarrow & & \\ & & H_1(X^1,X^0)=P_1(X) & & \\ & & \downarrow & & \\ & & H_0(X^0)=P_0(X) & & \end{array}$$

The case k = 0 is trivial, since

$$P_1(X)=H_1(X^1,X^0)\xrightarrow{\partial_0} H_0(X^0)\to H_0(X^1)\to H_0(X^1,X^0)=0.$$

Theorem 4.1 is completely proved.

4.5. A Geometric Definition of Groups of Cell Homology

It only suffices to clarify the geometrical meaning of the operator ∂_k in the chain complex $\{P_k(X), \partial_k\}$. For this, consider two cells σ^k and σ^{k-1} in X. Let us assume that the orientations of these cells are fixed and that the characteristic mappings $\chi^k: D^k \to X$ and $\chi^{k-1}: D^{k-1} \to X$ agree with these orientations. Consider the following continuous mapping $\partial D^k = S^{k-1} \xrightarrow{\chi^k} X^{k-1}/X^{k-2}$, i.e., let us map the boundary of a cell onto the quotient space of the (k − 1)-dimensional skeleton with respect to the (k − 2)-dimensional skeleton. As was mentioned above, this quotient space is homeomorphic to the bouquet of (k − 1)-dimensional spheres. Since in the (k − 1)-dimensional skeleton X^{k-1}, we have distinguished the cell σ^{k-1}, then its image under the factorization $X^{k-1} \xrightarrow{\alpha} X^{k-1}/X^{k-2}$ is a sphere S^{k-1} of the bouquet X^{k-1}/X^{k-2}. Thus, in the bouquet of spheres, a sphere is distinguished that enables us to define the natural projection of all the bouquet onto this sphere. Namely, the fixed sphere is stable while other spheres are mapped into the fixed point of the bouquet (see Fig. 22). The constructed mapping of the boundary of the ball D^k in the union of spheres with the subsequent projection onto the fixed sphere defines a continuous mapping $S^{k-1} \to S^{k-1}$,

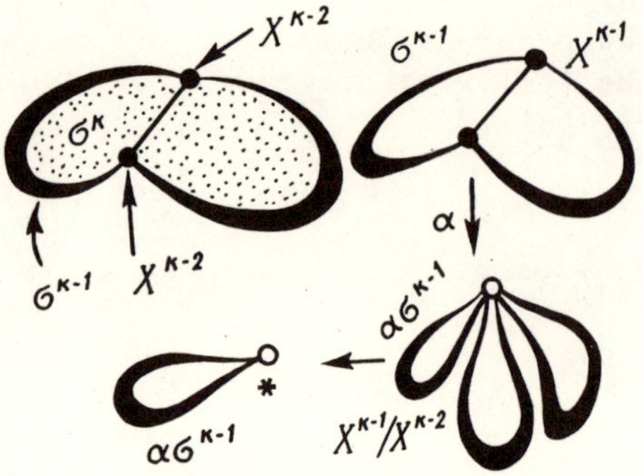

Fig. 22

in particular, defines the homomorphism of $\mathbf{Z} = H_{k-1}(S^{k-1})$ in itself. Each homomorphism $\mathbf{Z} \to \mathbf{Z}$ is uniquely defined by an integer number m which is the image of the unit of $\mathbf{Z}$. This number is called the degree of the mapping. When the constructed mapping is smooth (e.g., as it is in many of the concrete examples considered in what follows). The number m coincides with the usual degree of a smooth mapping defined for mappings of orientable closed manifolds of the same dimension (see [1, 2]).

We have assigned to each pair of cells σ^k and σ^{k-1} an integer number called the incidence coefficient and usually denoted as $[\sigma^k:\sigma^{k-1}]$. From its definition it follows that this number depends on the chosen orientations of cells; the change of one of the orientations changes the sign of $[\sigma^k:\sigma^{k-1}]$.

Theorem 4.2. Let σ^k be an arbitrary generator of $P_k(X) = H_k(X^k, X^{k-1})$. Then the action of the boundary operator ∂ on this generator is defined by the formula $\partial\sigma^k = \Sigma[\sigma^k:\sigma^{k-1}] \times \sigma^{k-1}$, where the sum encompasses all $(k-1)$-dimensional cells σ^{k-1} of X.

This statement gives us a clear geometric interpretation of the boundary operator ∂ defined above in algebraic terms for groups of cell chains. If the cell σ^{k-1} does not intersect with the closure of σ^k, then $[\sigma^k:\sigma^{k-1}] = 0$.

Proof. Consider two triples $(D^k, S^{k-1}, \varnothing)$ and (X^k, X^{k-1}, X^{k-2}) and the continuous mapping $(D^k, S^{k-1}, \varnothing) \to (X^k,$

X^{k-1}, X^{k-2}), where $D^k \to X$ is a characteristic mapping σ^k and $S^{k-1} \to X^{k-1}$ is the restriction of this mapping onto the boundary of the ball. The exact sequences of these triples may be naturally arranged in the following commutative diagram:

$$\begin{array}{ccccccc}
0 & & \mathbf{Z} & & \mathbf{Z} & & \\
\| & & \| & & \| & & \\
H_k(D^k) & \to & H_k(D^k, S^{k-1}) & \xrightarrow{j} & H_{k-1}(S^{k-1}) & \to & 0 \\
 & & {\scriptstyle i_*}\downarrow & & {\scriptstyle \varphi_*}\downarrow & & \\
 & & H_k(X^k, X^{k-1}) & \xrightarrow{\partial} & H_{k-1}(X^{k-1}, X^{k-2}) & & \\
 & & \| & & \| & & \\
 & & P_k(X) & & P_{k-1}(X) & &
\end{array}$$

Consider the element $1 \in \mathbf{Z} = H_k(D^k, S^{k-1})$. Under the homomorphism i_*, this generator is mapped onto the cell chain $1 \cdot \sigma^k \in P_k(X)$, and when ∂ is applied, transforms it into $1 \times \times \partial\sigma^k$. Now let us trace how this generator moves along the upper side of the square. The element 1 passes under j into $1 \in \mathbf{Z} = H_{k-1}(S^{k-1})$ (since j is the isomorphism). The subsequent mapping φ_* transforms the generator of $\mathbf{Z}$ into an element of the group

$$\mathbf{Z} \oplus \cdots \oplus \mathbf{Z} = H_{k-1}(X^{k-1}/X^{k-2}) = H_{k-1}(\bigvee S^{k-1}).$$

To each generator of $P_{k-1}(X) = H_{k-1}(\bigvee S^{k-1})$ corresponds a cell σ^{k-1}. Clearly, the coefficient of this cell in the image of the unit with respect to φ_* equals exactly the degree of the through mapping $S^{k-1} \to S^{k-1}$, i.e., to $[\sigma^k:\sigma^{k-1}]$. The theorem is proved.

Thus, we have obtained a simple rule for computation of groups of singular homology of the cell complex. It turns out that for this it suffices to consider the chain complex of cell chains uniquely recovered from the cell structure of X, and furthermore, to compute explicitly the boundary operators in this complex, for which it suffices to compute incidence coefficients of pairs of cells of neighbor dimensions. After this, it is necessary to compute the homology groups of the obtained complex. These groups happen to be groups of singular homology of X. This construction is so transparent (and in many cases easily computable) that it is sometimes taken for the definition of groups of cell homology of a complex.

Definition 4.1. Let X be a finite CW-complex; let $\{P_k(X)\}$ be groups of cell chains of X; and let $\partial_k: P_k(X) \to \to P_{k-1}(X)$ be homomorphisms defined by the formula $\partial_k \sigma^k = \Sigma[\sigma^k:\sigma^{k-1}]\sigma^{k-1}$. Then homology groups of this complex, i.e., groups $\operatorname{Ker} \partial_k / \operatorname{Im} \partial_{k+1}$, are called groups of cell homology of X.

As we have shown above, to compute these groups it is not necessary to know the singular homology of X since all objects mentioned in Definition 4.1 admit purely geometric description (cells, characteristic mappings, incidence coefficients, and such). Theorem 4.1 may be reformulated now as follows: for a finite cell, complex groups of its singular and cell homology are isomorphic.

Corollary. If a topological space X is presentable in two ways in the form of a finite CW-complex, then groups of cell homology of X do not depend on the cell decomposition of X, since they are isomorphic to groups of singular homology of X.

This simple corollary is extremely useful in concrete computations. If the problem of computation of the homology of a concrete space X arises, we must select its simplest possible presentation in the form of a CW-complex and then compute its cell homology.

4.6. Examples of Computations of Groups of Cell Homology

Example 1. The sphere S^n admits the simplest cell decomposition (i.e., the presentation in the form of a CW-complex): $\sigma^0 \cup \sigma^n$, where σ^0 is a point and σ^n is a complement to σ^0 in S^n. Clearly, $P_0 = P_n = \mathbf{Z}$ and ∂ is trivial (for $n > > 1$). Whence, $H_i(S^n) = \mathbf{Z}$ for $i = 0, n$, and $H_j(S^n) = 0$ for $j \neq 0, n$. Groups of cell homology coincide with groups of singular homology computed in Lemma 4.1, as expected.

Example 2. The real projective space $\mathbf{R}P^n$. Recall that one of its realizations is in the form of the set of sequences $x = (x_0, x_1, \ldots, x_n)$, where x_i are real numbers and at least one of the x_i's is nonzero considered up to a nonscalar multiple. The simplest cell decomposition of $\mathbf{R}P^n$ is as follows: for σ^k we must take all sequences x such that $x_k \neq 0$, $x_{k+1} = \ldots = = x_n = 0$. Then in each dimension k we obtain exactly one cell σ^k, i.e., $\mathbf{R}P^n = \sigma^0 \cup \sigma^1 \cup \ldots \cup \sigma^n$. Hence, $P_k(\mathbf{R}P^n) = \mathbf{Z}$, where

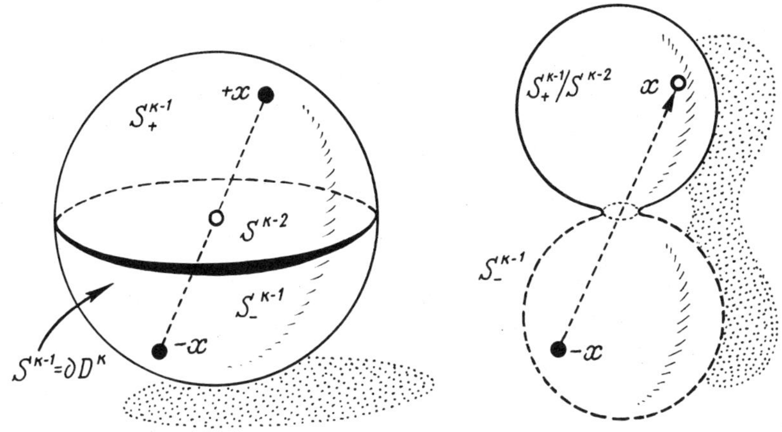

a b

Fig. 23

$0 \leq k \leq n$. It only remains to compute the boundary operator $\partial_k: \mathbf{Z} = P_k \to P_{k-1} = \mathbf{Z}$. Consider the cell decomposition in detail. In Fig. 23 the closure of the cell σ^k, i.e., $\mathbf{R}P^k$, is depicted in the form of a k-dimensional ball whose boundary – the sphere S^{k-1} – is factorized with respect to the $\mathbf{Z}_2$-action, i.e., the generator of $\mathbf{Z}_2$ is represented by the transformation $x \to -x$, which is the inversion of the sphere with respect to the origin. In other words, $\mathbf{R}P^k$ is obtained from D^k after the identification of the antipodal points on its boundary. After this, the boundary of the ball D^k, i.e., S^{k-1}, is mapped in $X^{k-1}/X^{k-2} = \mathbf{R}P^{k-1}/\mathbf{R}P^{k-2} = S^{k-1}$ as follows. Let us present S^{k-1} in the form of the union of three nonintersecting subsets $S_+^{k-1} \cup S^{k-2} \cup S_-^{k-1}$, where S_+^{k-1} and S_-^{k-1} are open hemispheres (upper and lower) and S^{k-2} is an equator. The mapping $\partial D^k = \partial\sigma^k = S^{k-1} \to S^{k-1} = S_+^{k-1}/S^{k-2}$ is defined by the formulas

$$x \to x \quad \text{for} \quad x \in S_+^{k-1},$$
$$x \to * \quad \text{for} \quad x \in S^{k-2},$$
$$-x \to x \quad \text{for} \quad -x \in S_-^{k-1}.$$

Thus we have obtained the continuous mapping $h: S^{k-1} \to S^{k-1} = S_+^{k-1}/S^{k-2}$, which is a diffeomorphism on each of the subsets S_+^{k-1} and S_-^{k-1}. It remains to find the degree of h. Clearly, the preimage of each point $x \in S_+^{k-1}$ consists of two points: of x itself and of its antipodal point on S^{k-1}.

Hence, the degree is either two or zero (depending on whether the mapping $x \to -x$ preserves the orientation of S^{k-1} or not). Hence, we must find the degree of the auxiliary mapping α: $S^{k-1} \to S^{k-1}$, where $\alpha(x) = -x$, i.e., α is the symmetry with respect to the origin.

Lemma 4.6. The degree of the mapping $\alpha(x) = -x$ of S^{k-1} onto itself equals $(-1)^k$.

This immediately implies that the degree of h equals 2 for even k and 0 for odd k. Thus, $[\sigma^k:\sigma^{k-1}] = 0$ for odd k and 2 for even k. This means that boundary operators in the chain complex $\{P_k(X), \partial_k\}$ are of the form $\partial\sigma^{2n} = 2\sigma^{2n-1}$, $\partial\sigma^{2n-1} = 0$. Therefore, we have proved the following statement.

Proposition 4.1. Groups of singular (and cell) homology of $\mathbf{R}P^n$ are of the form

H_0	H_1	H_2	H_3	H_4	$\cdots$	H_{n-1}	H_n	
$\mathbf{Z}$	$\mathbf{Z}_2$	0	$\mathbf{Z}_2$	0	$\cdots$	$\mathbf{Z}_2$	0	for even n
$\mathbf{Z}$	$\mathbf{Z}_2$	0	$\mathbf{Z}_2$	0	$\cdots$	0	$\mathbf{Z}_2$	for odd n

In conclusion, let us prove Lemma 4.6. Consider the sphere S^{k-1} and fix an arbitrary tangent orthoframe $e(x) = (e_1, \ldots, e_{k-2}, e_{k-1})$ at x. Under the mapping $\alpha: x \to -x$ this frame is transformed into $e(-x) = (-e_1, \ldots, -e_{k-1})$ (we assume that the sphere is naturally embedded in the Euclidean space). We must compare orientations induced in the sphere by these two frames. For this, let us connect points x and $-x$ by a meridian γ such that at x its velocity vector is e_{k-1} (see Fig. 24). Let us accomplish a smooth deformation (transport) of $e(x)$ from x to the point $-x$ moving along the path γ so that e_{k-1} is always tangent to γ. Then at $-x$ we obtain two frames $(-e_1, \ldots, -e_{k-2}, -e_{k-1})$ and $(-e_1, \ldots, -e_{k-2}, e_{k-1})$. Clearly, the mutual orientation is defined by the sign $(-1)^{k-2}$. The lemma is proved.

Example 3. The complex projective space $\mathbf{C}P^n$. Points of $\mathbf{C}P^n$ are sequences of $n + 1$ complex numbers $(z_0, \ldots, z_n)$, where at least one of the numbers is nonzero subject to a nonzero multiple. In analogy with Example 2, we may construct the following cell decomposition of $\mathbf{C}P^n$.

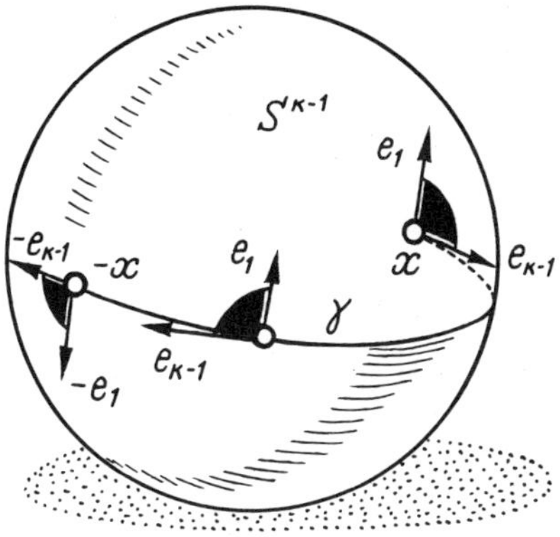

Fig. 24

For σ^{2k} (where 2k stands for the real dimension), where $0 \leqq k \leqq n$, we consider the set of points defined by conditions $z_k \neq 0$, $z_{k+1} = \ldots = z_n = 0$. We advise the reader to construct explicitly the characteristic mapping of σ^{2k}. Thus, groups of cell chains are of the form $P_i(CP^n) = 0$ for $i = 2s + 1$ or $i > 2n$ and $P_i(CP^n) = \mathbf{Z}$ for $i = 2s$. Since all boundary operators are, evidently, zero (by dimensional considerations), then $H_i(CP^n) = P_i(CP^n)$.

Example 4. A two-dimensional compact connected closed orientable manifold. Any such manifold M_g^2 of genus g is homeomorphic to a two-dimensional sphere with g handles (see Sec. 1, Example 2) and admits the following simplest cell decomposition $\sigma^0 \cup \left(\bigcup_{i=1}^{2g} \sigma_i^1\right) \cup \sigma^2$. This implies that homology groups of M_g^2 are of the form $H_0 = \mathbf{Z}$, $H_1 = \mathbf{Z} \oplus \ldots \oplus \mathbf{Z}$ (2g times), $H_2 = \mathbf{Z}$. The verification is left to the reader as a simple exercise.

Problem. Compute the homology of a nonorientable compact closed connected two-dimensional manifold.

The above-defined groups of singular (cell) homology do not exhaust the collection of "homological invariants" that enable one to discern CW-complexes. In particular, a homology theory may be constructed with the use of the so-called singular (cell) chains with coefficients in an arbitrary Abelian group A. This means that, following the lines of Definitions 2.1 and 2.2, we may consider chains as linear combinations with coefficients from A. Clearly, all the subsequent constructions and definitions are generalized automatically, yielding the definition of groups $H_k(X, A)$ and $H_k(X, Y, A)$, called homology groups with coefficients in the group A. As for A = **Z**, results concerning cell homology are valid; in particular, these latter groups may be defined as homology groups of the chain complex constructed from groups of cell chains $\{\sum_i a_i \sigma_i^k,\ a_i \in A\} = P_k(X, A)$ (with coefficients in A) and homomorphisms ∂_k, where

$$\partial_k\left(\sum a_i\sigma_i^k\right) = \sum_i a_i \partial_i \sigma_i^k = \sum_i a_i \sum_j [\sigma_i^k : \sigma_j^{k-1}]\sigma_j^{k-1}$$

$$= \sum_{i,j} a_i [\sigma_i^k : \sigma_j^{k-1}]\sigma_j^{k-1}.$$

Homology groups $H_k(X)$ which we have studied earlier may be denoted following our new notation by $H_k(X, \mathbf{Z})$. However, we shall often omit the definition of coefficient group except in cases when the final result depends essentially on A.

Problem. In Examples 1-4 (above), compute homology groups with coefficients in A = **R**, **Q**, $\mathbf{Z}_2$, $\mathbf{Z}_p$, where **R** is the field of real numbers, **Q** is the field of rational numbers, and p is prime, such that $p \neq 2$.

In this problem we have taken the most generally used groups (to say nothing of **Z**) and in what follows we will use them sometimes.

5. COHOMOLOGY

5.1. Singular Cochains and the Operator δ

Let X be a cell complex and $C_k(X)$ the group of k-dimensional singular chains of X. Let A be a chosen Abelian group.

Definition 5.1. A singular cochain of X with coefficients in A is a homomorphism of $C_k(X)$ in A. The natural operation of addition of cochains transforms the set of cochains into an Abelian group which will be denoted by $C^k(X, A)$ and called the group of singular cochains of X.

Consider the boundary operator $\partial: C_k(X) \to C_{k-1}(X)$. Let $h \in C^{k-1}(X, A)$ be an arbitrary cochain, i.e., a homomorphism $h: C_k(X) \to A$. Then the singular cochain $\delta h \in C^k(X, A)$ is uniquely defined by the formula $\delta h(\alpha) = h(\partial\alpha)$, i.e., $\delta h: C_k(X) \to A$. In other words, the cochain δh is defined from the following diagram:

$$\begin{array}{ccc} \alpha \in C_k(X) & \longrightarrow & C_{k-1}(X) \\ & {\scriptstyle \delta h}\searrow \quad \swarrow{\scriptstyle h} & \\ & A & \end{array}$$

The operator $\delta: C^{k-1}(X, A) \to C^k(X, A)$ will sometimes be denoted by δ_{k-1}.

<u>Definition 5.2</u>. The operator $\delta_{k-1}: C^{k-1} \to C^k$ is called the coboundary operator. It is the operator adjoint to the boundary operator ∂ (see above).

5.2. Cohomology Groups

Since $\partial^2 \equiv 0$, it is evident that $\delta^2 \equiv 0$. Hence, we obtain the sequence of groups and connecting homomorphisms of the form $C^0(X; A) \xrightarrow{\delta_0} C^1(X; A) \xrightarrow{\delta_1} C^2(X; A) \to \ldots$, where $\delta_k \delta_{k-1} \equiv 0$ for any k. It is natural to call this sequence (in analogy with the homology case) a cochain complex. Following the reasoning of the preceding section, let us consider groups Ker δ_k and Im δ_{k-1} and construct the group $H^k(X; A) = \text{Ker}\,\delta_k / \text{Im}\,\delta_{k-1}$.

<u>Definition 5.3</u>. Groups $H^k(X; A)$, where $k \geq 0$, are called groups of singular cohomology of the space X with coefficients in Abelian group A. Elements of the group $B^k = \text{Im}\,\delta_{k-1}$ are called coboundaries and elements of the group $Z^k = \text{Ker}\,\delta_k$ are called cocycles.

When X is arcwise connected, its groups of singular cohomology are easy to compute.

<u>Lemma 5.1</u>. If X is arcwise connected, then $H^0(X; A) = A$.

Proof. Note that there are no zero-dimensional coboundaries since there are no chains of negative dimensions. Furthermore, a chain $\alpha \in C^0(X; A)$ is a cycle if it maps the whole group $C_0(X)$ into one element of A. In fact, Definition 5.1 implies that α, being a homomorphism of $C_0(X)$, is uniquely defined by its values on elementary zero-dimensional chains, i.e., actually, on points of X. Hence, α is presentable in the form of a function on X with values in A. Suppose there are points $x, y \in X$ such that $\alpha(x) \neq \alpha(y)$. Then consider a one-dimensional singular simplex Δ^1 that joins these points. We have

$$\delta_\alpha(\Delta^1) = \alpha\partial\Delta^1 = \alpha(x - y) = \alpha(x) - \alpha(y) \neq 0.$$

This contradicts the fact that α is a cocycle. Thus, all zero-dimensional cochains are constant functions on X and $H^0(X; A) = A$.

If X consists of several components of the arcwise connection, then $H^0(X; A) = A \oplus \ldots \oplus A$ (N times), where N is the number of components.

As in the homology case, groups of relative cohomology are naturally defined. Suppose Y is a closed subcomplex in the complex X; then $C_k(Y) \subset C_k(X)$. Denote by $C^k(X, Y)$ the group of all homomorphisms $\alpha: C_k(X) \to A$ that vanish on $C_k(Y)$. Clearly, $\delta C^k(X, Y) \subset C^{k+1}(X, Y)$; hence the groups $H^k(X, Y; A) = \mathrm{Ker}\, \delta/\mathrm{Im}\, \delta$ arise which are called groups of relative cohomology. In complete analogy with the homology case, the exact sequence of the pair (X, Y) arises. We skip details of the construction of this sequence and leave them to the reader as a useful exercise. Let us give only the final form of this exact sequence (prove its exactness!):

$$\cdots \to H^k(X, Y; A) \to H^k(X; A) \to H^k(Y; A) \xrightarrow{\delta} H^{k+1}(X, Y; A) \to \cdots .$$

If the triple of complexes $X \supset Y \supset Z$ is given, then the exact cohomology sequence of a triple arises:

$$\cdots \to H^k(X, Y; A) \to H^k(X, Z; A) \to H^k(Y, Z; A) \xrightarrow{\delta} H^{k+1}(X, Y; A) \ldots .$$

As in the homology case, we prove that groups of singular cohomology are homotopic invariants. Following the reasoning of Sec. 4, we may define groups of singular cohomology of a complex. For this we introduce groups of cell cochains $P^k(X; A)$ defined as groups $H^k(X^k, X^{k-1}; A)$, where X^k is a k-dimensional skeleton of the CW-complex X. The exact cohomology sequence of the triple (X^{k+1}, X^k, X^{k-1}) gives rise to a coboundary operator $\delta: P^k(X; A) \to P^{k+1}(X; A)$. After this we define groups of cell homology of the complex X as groups Ker δ/Im δ for a cochain complex $\{P^k(X; A), \delta\}$. The theorem similar to Theorem 4.1 holds.

Theorem 5.1. For a finite CW-complex X, groups of singular cohomology $H^k(X; A)$ and groups of cell cohomology of X are isomorphic and are finitely generated Abelian groups if the coefficient group A is also.

The proof follows the reasoning of Sec. 4 and is left to the reader.

5.3. Cohomology Groups with Coefficients in a Field

In applications, cohomology groups with coefficients in a field F play an important role. The cases $F = \mathbf{R}, \mathbf{Q}, Z_p$ (where p is prime) will be the most common for us.

Let X be a finite CW-complex. Consider groups $H^k(X; F)$ and $H_k(X; F)$ [we do not distinguish between isomorphic groups of cell and singular (co)homology due to Theorems 4.1 and 5.1]. The group of cell chains $P_k(X; F)$ is a finite-dimensional linear space over F, i.e., $P_k(X; F) = F \oplus \ldots \oplus F$. Clearly, subgroups of cycles and boundaries are linear subspaces of this space; hence Z_k/B_k is a finite-dimensional linear space over F. The group of cell chains $P^k(X; F)$, i.e., the group of F-linear mappings of the group $P_k(X; F)$ in F, may be considered as the space of linear functionals on $P_k(X; F)$ with values in F. In other words, groups $P_k(X; F)$ and $P^k(X; F)$ are conjugate finite-dimensional linear spaces over F. The boundary and coboundary operators ∂ and δ are conjugate linear operators. This means that Ker ∂ is conjugate to $P^k(X; F)$/Im δ and Im ∂ is conjugate to $P^{k-1}(X; F)$/Ker δ. Thus, the space $H_k(X; F)$ = Ker ∂/Im ∂ is conjugate to the kernel of the projection, hence to the space Ker δ/Im δ = $= H^k(X; F)$. Thus we have proved the following simple statement.

Statement 5.1. Groups $H_k(X; F)$ and $H^k(X; F)$ for a finite CW-complex X are dual linear spaces; in particular, they have the same dimension over F.

Statement 5.2. For a finite CW-complex X we have

$$H_k(X; \mathbf{Q}) = H_k(X; \mathbf{Z}) \otimes \mathbf{Q}.$$

This means that if $H_k(X; \mathbf{Z}) = (\bigoplus_1^n \mathbf{Z}) \oplus S$, where S is a finite group [such a presentation of the Abelian group $H_k(X; \mathbf{Z})$ is uniquely defined], then $H_k(X; \mathbf{Q}) = \bigoplus_1^n \mathbf{Q}$. Similarly,

$$H_k(X; \mathbf{R}) = H_k(X; \mathbf{Z}) \otimes \mathbf{R} = \bigoplus_1^n \mathbf{R} = \mathbf{R}^n.$$

Proof. Let us prove the statement for $F = \mathbf{Q}$, leaving the case $F = \mathbf{R}$ to the reader. The evident embedding $\mathbf{Z} \to \mathbf{Q}$ induces the embedding $P_k(X; \mathbf{Z}) \to P_k(X; \mathbf{Q})$. The following commutative diagram arises:

$$\begin{array}{ccc} P_k(X; \mathbf{Z}) & \xrightarrow{\partial} & P_{k-1}(X; \mathbf{Z}) \\ \downarrow & & \downarrow \\ P_k(X; \mathbf{Q}) & \longrightarrow & P_{k-1}(X; \mathbf{Q}) \end{array}$$

This implies that an element of $P_k(X; \mathbf{Z})$ is a cycle in $P_k(X; \mathbf{Z})$ if it is a cycle in $P_k(X; \mathbf{Q})$ (with respect to the embedding). This means that $Z_k(X; \mathbf{Z}) = Z_k(X; \mathbf{Q}) \cap P_k(X; \mathbf{Z})$. Furthermore, for any $a \in P_k(X; \mathbf{Q})$ there is an integer N such that $Na \in P_k(X; \mathbf{Z})$ (since chains are finite linear combinations). This implies that if $a \in P_k(X; \mathbf{Q})$ is a boundary, then Na is a boundary for a certain N in $P_k(X; \mathbf{Z})$. Thus, the homomorphism $H_k(X; \mathbf{Z}) \to H_k(X; \mathbf{Q})$ induced by the embedding $\mathbf{Z} \to \mathbf{Q}$ has a finite kernel and, at the same time, any element of $H_k(X; \mathbf{Q})$ being multiplied by a certain integer number belongs to the image of $H_k(X; \mathbf{Z})$. This means that the mapping

$$H_k(X; \mathbf{Z}) \otimes \mathbf{Q} \to H_k(X; \mathbf{Q}) \otimes \mathbf{Q} = H_k(X; \mathbf{Q})$$

is an isomorphism. The statement is proved.

The corresponding cohomology statement is proved by the exact same arguments.

Statement 5.3. For a finite complex X we have $H^k(X; \mathbf{Q}) = H^k(X; \mathbf{Z}) \oplus \mathbf{Q}$ and $H^k(X; \mathbf{R}) = H^k(X; \mathbf{Z}) \oplus \mathbf{R}$. Note that since $\dim H^k(X; \mathbf{R}) = \dim H_k(X; \mathbf{R})$, then for a finite complex X, we always have $\operatorname{rank} H^k(X; \mathbf{Z}) = \operatorname{rank} H_k(X; \mathbf{Z})$.

As we will see in what follows, in many cases (co)homology groups with real (or rational) coefficients are the easiest to compute. The computation of (co)homology groups with integer coefficients is usually a little more complicated. In practice, groups $H_n{}^k(X; \mathbf{R})$ play an important role. It is convenient to introduce groups $H^*(X; \mathbf{R}) = \bigoplus_{0 \leq k \leq n} H^k(X; \mathbf{R})$, where $n = \dim X$. Unlike $H_*(X; \mathbf{R})$, the group $H^*(X; \mathbf{R})$ may be endowed with an important extra property — multiplication. Since practically all main examples of complexes that we will deal with subsequently are smooth manifolds, let us recall the definition of multiplication in this special case. In [1, Chapter 6], the definition of de Rham cohomology is given in terms of exterior differential forms.

Recall that the exterior differential form w on the smooth manifold M is closed if $dw = 0$ and it is exact if it is presentable in the form $w = dw'$, where w' is a form. The quotient space of the space of a closed form of degree k by the subspace of exact forms is called the k-dimensional (de Rham) cohomology group of the manifold M. There is an important theorem whose proof will be omitted though, since we will not need it in what follows.

Theorem 5.2. If M is a smooth compact closed manifold, then groups of its cell (singular) cohomology with real coefficients are isomorphic to the de Rham cohomology groups.

This implies that the group $H^*(X; \mathbf{R})$ may be identified with the de Rham cohomology group. The exterior derivation of forms induces an operation on cohomology classes that transforms $H^*(M; \mathbf{R})$ into an algebra with unit. This multiplication has the following properties (resulting from the corresponding properties of the exterior forms): 1) associativity; 2) distributivity; 3) the functorial property, i.e., $f^*(w_1 \wedge w_2) = f^*w_1 \wedge f^*w_2$, where $f: M \to N$ is a smooth mapping; and 4) anticommutativity, i.e., $w^{(k)} \wedge w^{(p)} = (-1)^{kp} w^{(p)} \wedge w^{(k)}$, where k and p stand for degrees of homogeneous mul-

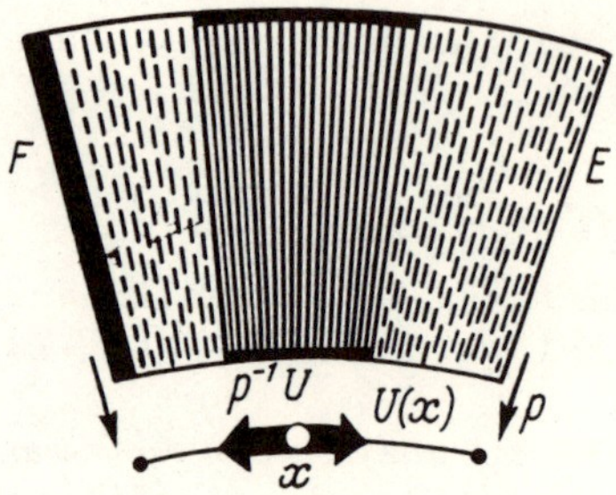

Fig. 25

tiples. This ring structure in $H^*(M; \mathbf{R})$ is extremely useful in many concrete computations, as will be illustrated in what follows. In fact, $H^*(X; \mathbf{R})$ admits a multiplication with properties 1-4 for any complex X (not only for manifolds). Moreover, if the coefficient group A is a ring with unit, then $H^*(X; A)$ is an algebra with unit whose multiplication satisfies 1-4 for any complex X.

6. BUNDLES

6.1. Definition of a Locally Trivial Bundle

In many applied problems after they are reformulated in geometric terms, a topological space E arises endowed with a special structure whose main property may be roughly formulated as follows: the space E is a "continuous" union of its subspaces (fibers), homeomorphic to the same space F, and locally, this decomposition is quite simple; namely, any subset of points of E composed of "neighbor fibers" is homeomorphic to the direct product of the fiber F by a topological space (see Fig. 25). This fibration of the space E into fibers F is often extremely useful in the study of the geometry of the space E. Now let us give the precise mathematical definition.

Definition 6.1. Suppose we are given the topological spaces E, B, F and a continuous mapping p of the space E on the space B. Let us say that the quadruple (E, B, F, p) is a locally trivial bundle if for any point $b \in B$ there is an open neighborhood $U \subset B$ and a homeomorphism $\varphi U: p^{-1}U \to U \times F$ such that the diagram

$$\begin{array}{ccc} p^{-1}U & \xrightarrow{\varphi_U} & U \times F \\ & p\searrow \quad \swarrow\pi & \\ & U & \end{array}$$

where $\pi: U \times F \to F$ is the natural projection of $U \times F$ onto F, commutes, i.e., if $\pi \varphi_U = p$. Denote by $p^{-1}U$ the pre-image of U with respect to p. In this situation, E is usually called the (total) space of the bundle, F is called the fiber, B, the base, and p, the projection.

In the literature, the mapping p itself is sometimes called a bundle and the notation $p: E \xrightarrow{F} B$ is used.

Definition 6.2. The fibration $p: E \xrightarrow{F} B$ is trivial if E is homeomorphic to $B \times F$ so that the diagram

$$\begin{array}{ccc} E & \xrightarrow{\alpha} & B \times F \\ & p\searrow \quad \swarrow\pi & \\ & B & \end{array}$$

commutes, i.e., $p = \pi\alpha$, where α is a homeomorphism and π is the natural projection.

The latter condition means that α transforms each fiber F into one point of the base so that different fibers are transformed into different points. Thus, we require that the homeomorphism $\alpha: E \to B \times F$ be consistent with the fibration of E into the union of bundles. The origin of the term

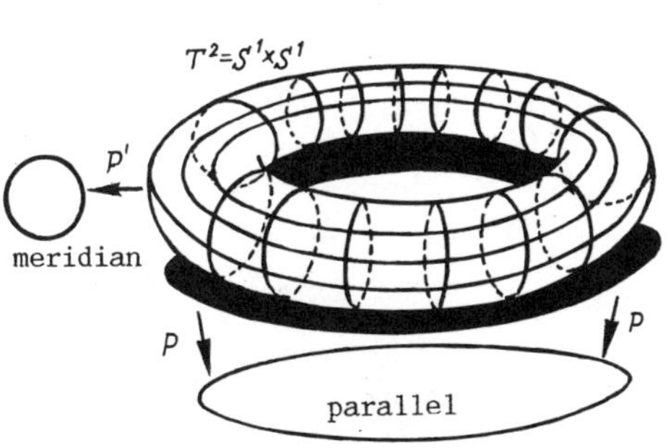

Fig. 26

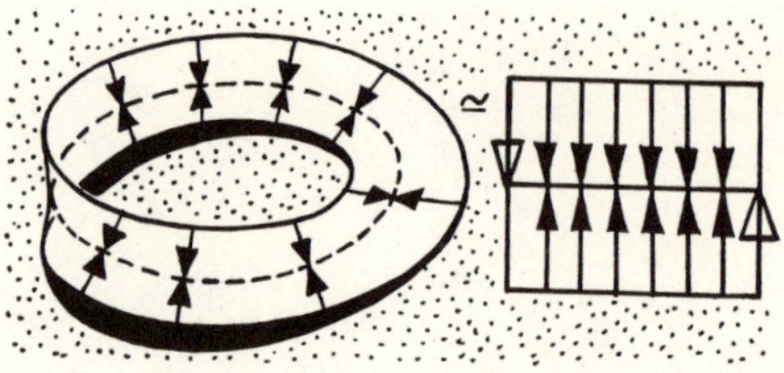

Fig. 27

"local triviality" is clear from Definition 6.1, namely, the bundle $p^{-1}U \to U$ is trivial if U is a sufficiently small neighborhood. Let us give examples of locally trivial bundles.

6.2. Examples of Bundles

Example 1. The direct product $E = B \times F$ is a bundle with respect to both the projections $p: B \times F \to B$ and $p': B \times F \to F$ (see Fig. 26).

Example 2. The Moebius band E, where F is the segment I and B is the circle S^1. Then $p: E \to B$ is the projection of the Moebius band onto its median (see Fig. 27). Since E is a nonorientable manifold (see, e.g., [1]), this bundle is not trivial because otherwise we would have obtained a homeomorphism of the Moebius band with the orientable manifold, i.e., the cylinder $S^1 \times I$ (recall that orientability is preserved with respect to homeomorphisms).

Example 3. Let $f: M^k \to N^p$ be a smooth mapping of a connected compact smooth manifold M onto the smooth compact manifold N, so that $k \geq p$, and f is a regular mapping in the sense that its differential $df|_x$ has the maximum rank equal to p at each point $x \in M$. Then, as is known from a standard course of differential geometry and topology (see, e.g., [1]), each fiber $F_y = f^{-1}(y)$, where $y \in N$, is a smooth $(k-p)$-dimensional submanifold in M; hence, the implicit function theorem being taken into account, we obtain that $f: M \to N$ defines a locally trivial bundle.

We are interested in this example because most bundles that we will encounter have the same structure. Among these bundles, an important class is naturally distinguished which is related to homogeneous bundles. Let $\mathfrak{G}$ be a compact Lie group with $\mathfrak{H} \subset \mathfrak{G}$ a closed subgroup (see [1], Chapter 4,

Sec. 3]). Consider the set $\mathfrak{G}/\mathfrak{H}$ of cosets $g\mathfrak{H}$ with respect to $\mathfrak{H}$. This set is naturally endowed with the induced topology (neighbor cosets may be defined as shifts of $\mathfrak{H}$ by neighbor elements $g, g' \in \mathfrak{G}$). Thus we obtain a continuous mapping $p: \mathfrak{G} \to \mathfrak{G}/\mathfrak{H}$, where $p(g) = (g\mathfrak{H}) \in \mathfrak{G}/\mathfrak{H}$. The topological space $\mathfrak{G}/\mathfrak{H}$ may be naturally endowed with the structure of a smooth manifold. We will not prove this statement here, since in all subsequent concrete examples this statement will be evident.

6.3. Geometry of the Hopf Bundle

Example 4. The Hopf bundle. In practice, one often comes across these bundles which are a special case of Example 3. Consider the sphere S^3 with its natural realization in the complex space $\mathbf{C}^2 = \mathbf{C}^2(z_1, z_2)$ as the set of points (z_1, z_2) such that $|z_1|^2 + |z_2|^2 = 1$. Consider the standard two-dimensional sphere S^2 as the plane $\mathbf{R}^2$ completed by an infinite point, and let us identify S^2 with the complex projective line $\mathbf{C}P^1$. Let us construct a continuous mapping $p: S^3 \to S^2$ defined by the formula $p(z_1, z_2) = z_1/z_2$. It is easy to see that p is a smooth mapping and that it is regular at all points. Hence, p defines a locally trivial bundle with the base S^2, total space S^3, and fiber S^1, since $p(e^{i\varphi}z_1, e^{i\varphi}z_2) = p(z_1, z_2)$. The Hopf bundle may be described quite

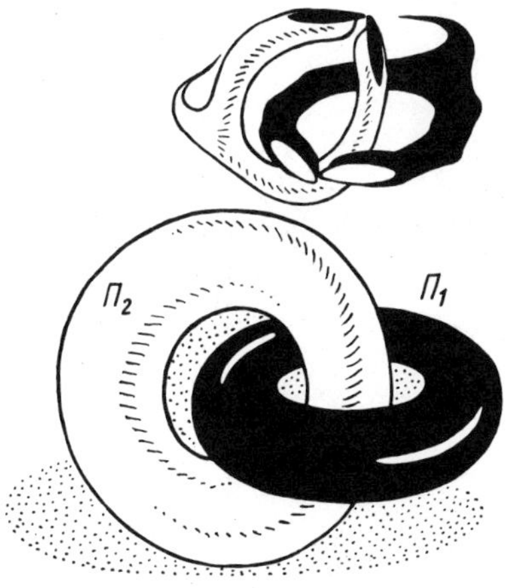

Fig. 28

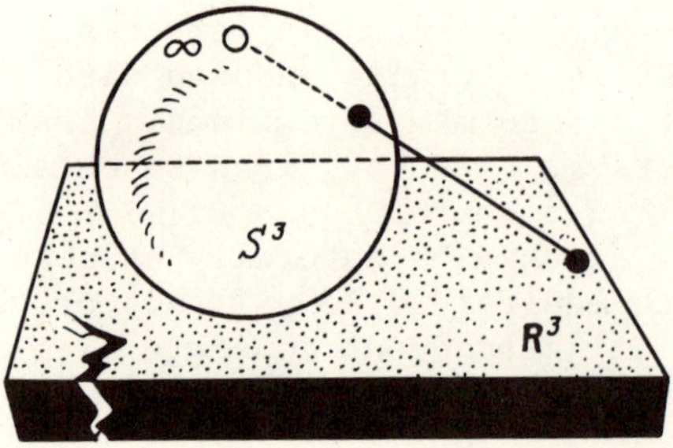

Fig. 29

clearly if we realize S^3 as the union of two solid tori. In fact, consider in $C^2(z_1, z_2)$ the cone K, defined by the equation $|z_1| = |z_2|$. Its intersection with S^3 is evidently a two-dimensional torus $T^2 = S^1 \times S^1$, since

$$K \cap S^3 = \left\{(z_1, z_2),\ z_1 = \frac{e^{i\varphi}}{\sqrt{2}},\ z_2 = \frac{e^{i\psi}}{\sqrt{2}}\right\}.$$

Hence the sphere splits into the union of two closed subsets Π_1 and Π_2, defined by the inequalities $\Pi_1 = \{|z_1| \leq |z_2|\}$, $\Pi_2 = \{|z_1| \geq |z_2|\}$ and intersecting via a two-dimensional

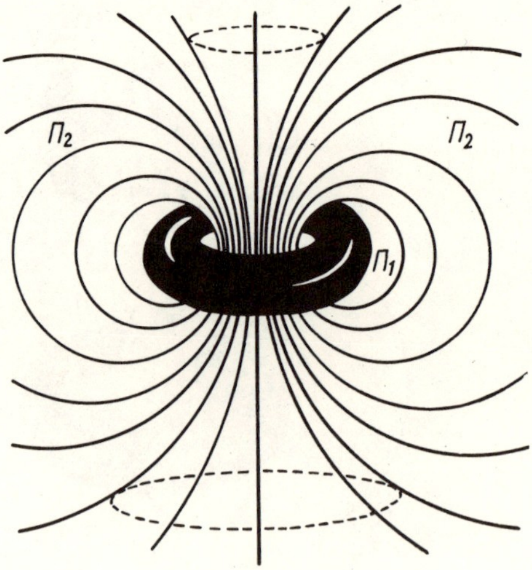

Fig. 30

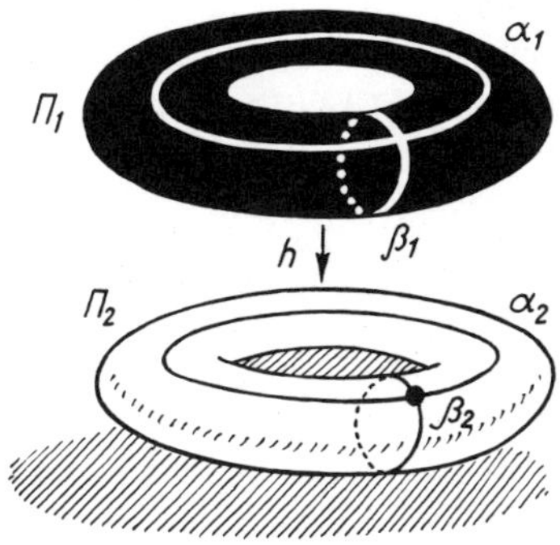

Fig. 31

torus T_2. Each of Π_1, Π_2 is homeomorphic to a solid torus $S^1 \times D^2$ (see Fig. 28). This picture is conditional in the sense that it shows the solid tori not in S^3 but in $\mathbf{R}^3$. To make the picture exact, it is necessary to present S^3 as the space $\mathbf{R}^3$, completed by an infinite point (see Fig. 29). In this presentation, a solid torus Π_2 may be considered naturally embedded in $\mathbf{R}^3$ (see Fig. 30) and then its complement in $S^3 = \mathbf{R}^3 \cup \infty$ is homeomorphic to another solid torus Π_2. If we select in each of the solid tori standard parallels and meridians and denote them by α_1, β_1 and α_2, β_2, respectively (see Fig. 31), then the glueing of the solid tori by their boundary T^2 (resulting in S^3) is defined by the diffeomorphism $h: S^1 \times S^1 \to S^1 \times S^1$ which transforms parallels into me-

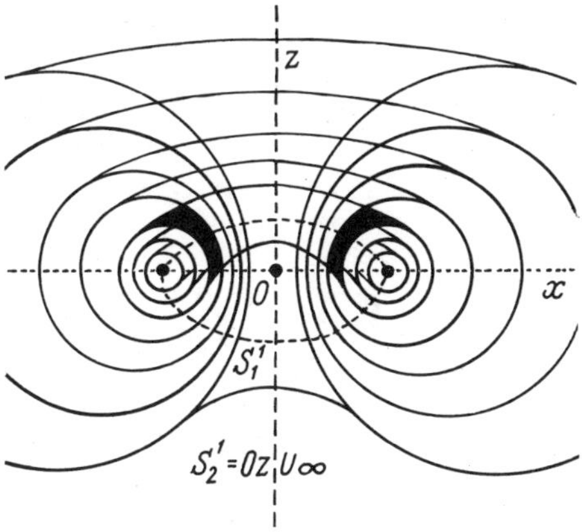

Fig. 32

ridians and vice versa, i.e., the induced mapping of one-dimensional homology groups $H_1 = \mathbf{Z} \oplus \mathbf{Z}$ of T^2 is defined by the following integer matrix:

$$h_* = \begin{pmatrix} 0 & 1 \\ -1 & 0 \end{pmatrix}, \qquad h(\alpha_1) = -\beta_2, \qquad h(\beta_1) = \alpha_2.$$

This glueing operation is depicted in Fig. 30.

We now describe the Hopf fibration of S^3. For this, it suffices to produce explicitly a smooth fibration of S^3 onto fibers – circles. Let us stratify S^3 presented in the form $\mathbf{R}^3 \cup \infty$ into a family of "concentric" tori as depicted in Fig. 32. To make the picture clear, we have dissected $\mathbf{R}^3$ by a two-dimensional plane (z, x) to show the needed fibration of S^3. Two extreme situations occur. The first one is as follows: "concentric" tori retract onto the standard circle that belongs to the plane (y, x). Denote this circle by S_1^1, depicted in Fig. 32 by a dashed curve. The other extreme situation is as follows: expanding tori tend to the vertical axis O_z which is in fact their "limit," since, as we remember, the situation occurs in the three-dimensional sphere which is presented as $\mathbf{R}^3$ with an infinite point (i.e., all the "infinity" is glued into a point). Therefore, the axis O_z depicts the circle S_2^1 in S^3. Clearly, S_1^1 is an axis of Π_1 and S_2^1 is an

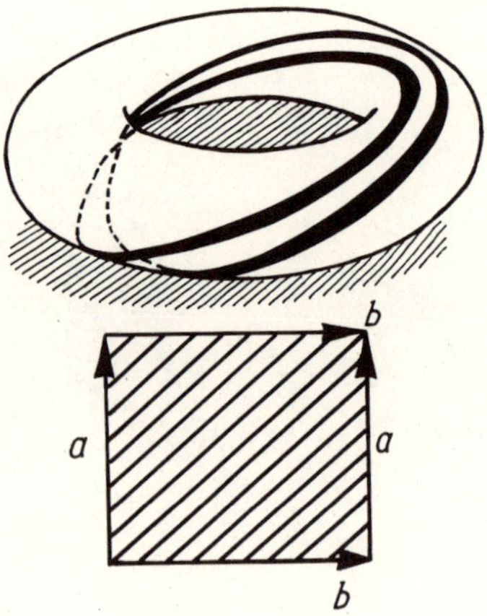

Fig. 33

axis of Π_2. To describe fibers of the Hopf fibration, fix a two-dimensional torus that makes a fibration of $\mathbb{R}^3 \cup \infty$ (see Fig. 32). On this torus define a smooth trajectory, i.e., a circle that bypasses the torus once around its parallel and once around its meridian, i.e., the trajectory of the type (1, 1) as we will say in what follows. The closed trajectory that bypasses a torus p times around a parallel and q times around a meridian is called a trajectory of type (p, q). Later, turning this trajectory of type (1, 1) by an orthogonal transformation that preserves O_z, we fibrate T^2 into the union of nonintersecting circles of type (1, 1) (see Fig. 33). The same fibration is depicted in Fig. 33 as a torus, i.e., as a square with identifications on its boundary corresponding to the word $aba^{-1}b^{-1}$. Let us construct the same bundle on each torus, the union of which constitutes the three-dimensional sphere. Since these tori do not intersect, then neither do the constructed fibrating circles that belong to different tori. In both limiting cases, when tori tend to S_1^1 and S_2^1, circles of the type (1, 1) that fibrate these tori also tend to circles S_1^1 and S_2^1, respectively. Thus, we have explicitly constructed the fibration of S^3 onto circles that define the Hopf bundle (verify!). Observe an important property of this bundle: any two of its fibers are linked, i.e., a two-dimensional film (e.g., a disk) with one of the fibers as a boundary necessarily intersects another fiber. This intersection consists of one point. To prove this, it suffices to scrutinize Fig. 33, where two fibers that belong to a torus are depicted (see Fig. 34). The same obvious considerations imply that any two circles of type (1, 1) that belong to different concentric tori are linked. Two "limit"

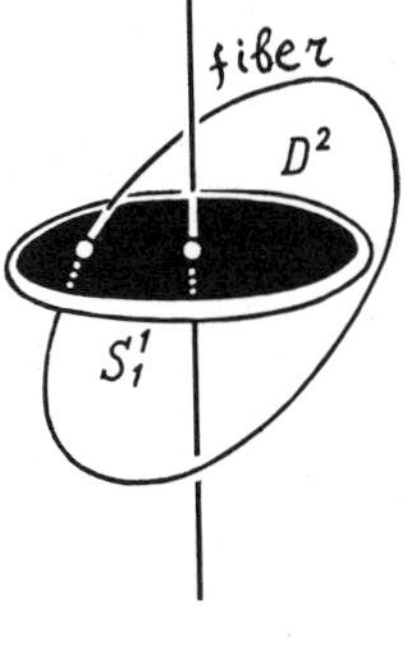

Fig. 34

Fig. 35

circles-fibers S_1^1 and S_2^1 are evidently linked (see Fig. 32). The constructed description of the Hopf bundle makes it easy to understand its topological structure; namely, we see into which points of the two-dimensional sphere (base) S^2 fibers-circles are mapped. In fact, consider the standard disk D^2 in the plane (x, y) with the boundary S_1^1. Then it is evident that any circle-fiber different from S_1^1 necessarily intersects this disk at exactly one point (see Fig. 35). When the fiber tends to S_1^1, the point of the intersection of this fiber with D^2 tends to the boundary of the disk, and when finally the fiber coincides with S_1^1, we must identify this circle $S_1^1 = \partial D^2$ with a point, making D^2 a two-dimensional sphere, each point of the sphere corresponding to a single circle-fiber of our bundle. In this situation, the fiber S_2^1 corresponds to the center of D^2 (see Fig. 35). Thus, the Hopf bundle is described as follows. We must fibrate $S^3 = \mathbf{R}^3 \cup \infty$ into circles as depicted in Fig. 33 and then assign to each of them the point of intersection of the circle with D^2 (or assign the whole boundary of the disk if $S^1 = S_1^1$). Note that the Hopf bundle is not trivial, since otherwise (i.e., if $S^3 = S^1 \times S^2$) we would have $H_1(S^3; \mathbf{Z}) = \mathbf{Z}$ (prove!); this is impossible by Lemma 4.1, which claims $H_1(S^3; \mathbf{Z}) = 0$.

We have dwelt on the Hopf bundle because this example shows various important effects of general bundles. Consider one more typical example of a nontrivial bundle encountered in applications.

6.4. Geometry of the Bundle of Unit Tangent Vectors to a Sphere

Example 5. Let E be the space of all unit tangent vectors to an even-dimensional sphere S^{2n}, let the base B be S^{2n}, and assume that the projection $p:E \to B$ assigns to each tangent vector its initial point, i.e., the point of tangency to S^{2n}. Clearly, it is a smooth locally trivial bundle with the fiber S^{2n-1} (see Fig. 36).

Lemma 6.1. The bundle $p:E \to S^{2n}$ is nontrivial.

Proof. Suppose the contrary, namely, $E = S^{2n} \times S^{2n-1}$. The commutativity of the diagram from Definition 5.1 means that there is a nonzero section of this bundle, i.e., there is a continuous mapping $h:S^{2n} \to E$ such that $ph = 1_{S^{2n}}$ (the identity mapping of the base B). But this means that at each point $b \in S^{2n}$ there is uniquely defined a nonzero tangent

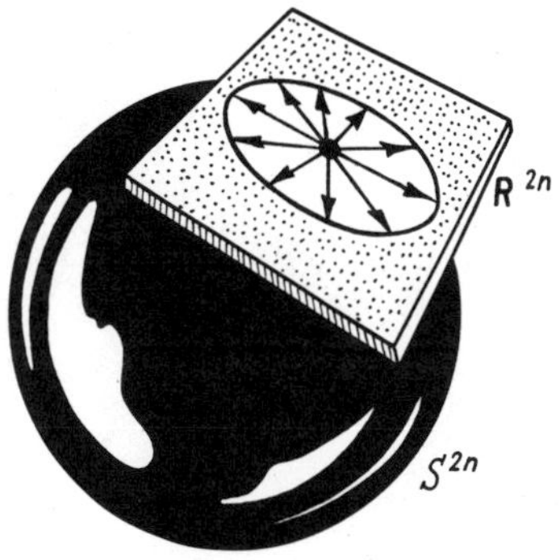

Fig. 36

vector h(b) (of unit length), which is a continuous function at the point b. Thus, it remains to show that on S^{2n} there is no continuous tangent vector field that does not vanish at all points of the sphere. This statement follows from a general theorem on the index of a vector field on a sphere which has other interesting corollaries. Therefore, we will detail the proof of this theorem.

Recall the definition of the index of a vector field. Let v(x) be a smooth vector field on the smooth manifold M^n with only a finite number of singular points [i.e., points x such that v(x) = 0]. Then all these points are isolated, i.e., for each of them there is a sufficiently small open neighborhood that does not contain other zeros of the field. In M^n, consider the sphere S_ε^{n-1} of a sufficiently small radius ε with the center at the singular point x_0. We may assume that $v(x) \neq 0$ on the whole sphere S_ε^{n-1}. Proceeding on to the local coordinates $x_1, \ldots, x_n$ in the neighborhood of x_0, we may assume that S_ε^{n-1} belongs to a domain in $\mathbb{R}^n$; hence, there is defined a smooth mapping $f: S_\varepsilon^{n-1} \to S^{n-1}$ defined by the formula $f(x) = v(x)/|v(x)|$, where $|v(x)|$ is the Euclidean length of v(x).

<u>Definition 6.3</u>. The index $\operatorname{ind}_{x_0} v$ of the singular point x_0 of the vector field v is the integer number deg f, i.e., the degree of the mapping $f: S_\varepsilon^{n-1} \to S^{n-1}$. The sum of indices of singular points of a field over all its singular points is called the index of the vector field and is denoted by ind v.

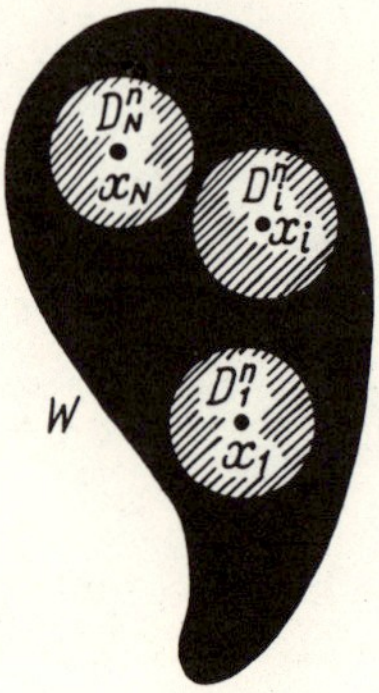

Fig. 37

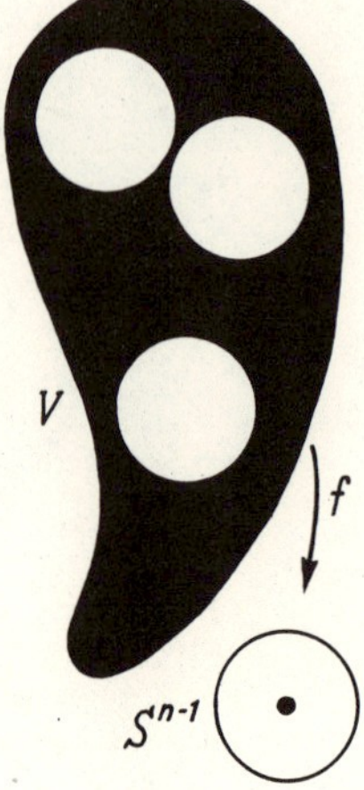

Fig. 38

By Theorem 1 of Sec. 3, Chapter 6 in [1], the index of a singular point and the index of a field are well defined; in particular, they do not depend on the choice of local coordinates in a neighborhood of a singular point of a vector field.

Theorem 6.1. Let $W^n \subset \mathbf{R}^n$ be a bounded domain in $\mathbf{R}^n$ such that the boundary ∂W of W is a smooth compact closed submanifold in $\mathbf{R}^n$ of dimension $n - 1$. Let v be a vector field on W such that $v(x) \neq 0$ on ∂W and with only a finite number of singular points in W. Then the index of this vector field equals deg F, where $F: \partial W \to S^{n-1}$ is the smooth mapping defined by the formula $F(x) = v(x)/|v(x)|$, where $x \in \partial W$.

Corollary 6.1. If vector fields v_1 and v_2 on W satisfy $\deg F(v_1) = \deg F(v_2)$ on ∂W, then indices of these fields coincide. In particular, if v_2 is obtained from v_1 by a smooth homotopy g_t such that $g_t v_1(x) \neq 0$ on ∂W for $0 \leq t \leq 1$, then indices of v_1 and v_2 coincide.

Proof of Theorem 6.1. Let $x_1, \ldots, x_N$ be singular points of v and $N < \infty$. Suppose S_i^{n-1} is a sphere with its center at x_i and of sufficiently small radius such that $D_i^n \cap D_j^n$ for $i \neq j$, where D_i^n is an n-dimensional ball with S_i^{n-1} as the boundary (see Fig. 37). Consider the domain V that is obtained from W by rejection of all balls D_i^n, where $1 \leq i \leq N$ (see Fig. 38). Let us construct the smooth mapping $f: V \to$

$\to S^{n-1}$ putting $f(x) = v(x)/|v(x)|$, where S^{n-1} is the standard sphere of unit radius in R^n. This mapping is well defined since v is nonzero at all points of V. On S^{n-1}, consider the standard Riemannian volume form w, invariant with respect to all orthogonal transformations of R^n. Then $\int_{S^{n-1}} w$ is equal to the volume of the sphere S^{n-1} which we will denote by a_{n-1}. The form w, being an exterior differential form on a sphere, induces the exterior differential form f^*w on V. Hence, we may consider the integral $\int_V d(f^*(w))$, where d is the exterior differential (see [1], Chapter 6, Sec. 1]). By Stokes' formula ([1], Chapter 6, Sec. 2), we have $\int_V df^*w = \int_{\partial V} f^*w$, where ∂V is the boundary of V (the sign depends on the choice of orientation). But, ∂V splits into the disjoint union of (n − 1)-dimensional submanifolds $\partial V = \partial W \cup (-\cup S_i^{n-1})$ [the sign "−" stands for the orientation opposite to the induced one] (see Fig. 37). Thus we get the equality

$$\int_V df^*w = \int_{\partial W} f^*w = -\sum_{1\leq i\leq N} \int_{S^{n-1}} f^*w.$$

Properties of d (Chapter 6, Sec. 1) imply that $df^* = f^*d$. Thus, $\int_V df^*w = \int_V f^*(dw)$. Now, recall that w is the form of the maximal dimension on S^{n-1}; hence $dw = 0$, i.e., $\int_V df^*w = 0$. It follows that $\int_{\partial W} f^*w = \sum_{1\leqq i\leqq N} \int_{S_i^{n-1}} f^*w$. By Theorem 2 of [1] (pp. 394-395), we have

$$\int_{\partial W} f^*w = \deg f|_{\partial W} \cdot \int_{S^{n-1}} w = a_{n-1} \cdot \deg f|_{\partial W}.$$

Similarly, we get $\int_{S_i^{n-1}} f^*w = a_{n-1} \cdot \deg f|\partial W = a_{n-1}$. Therefore, our equality reduces to the form

$$a_{n-1} \cdot \deg f|_{\partial W} = a_{n-1} \sum_{1\leq i\leq N} \deg f|_{S_i^{n-1}}.$$

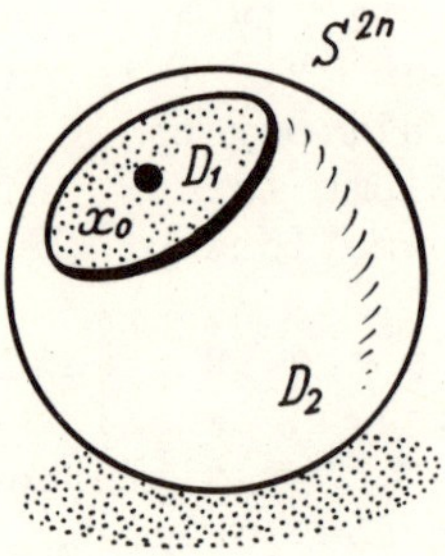

Fig. 39

Definitions of the index of a vector field and of the mapping $F:\partial W \to S^{n-1}$ finally imply $\deg F = \sum_{1 \leq i \leq N} \mathrm{ind}_{x_i} v = \mathrm{ind}\, v$. The theorem is proved.

Proof of Corollary 6.1 follows immediately from the theorem on homotopic invariance of the degree of a smooth mapping ([1], Chapter 6, Sec. 3).

Theorem 6.2. Any continuous tangent vector field v on S^{2n} has at least one singular point, i.e., a point x_0 such that $v(x_0) = 0$.

Proof. Suppose the contrary, i.e., that there is a continuous vector field v on S^{2n} that does not vanish at any point. By Theorem 1 of [2] (p. 478), there is a however-close approximation of v by a smooth tangent vector field that will be denoted, for the sake of brevity, by the same symbol v. Since the initial field was nonzero everywhere on the sphere, then so is the approximating smooth field. Theorem 6.2 is now a corollary of a general statement on the index of a vector field on a sphere.

Theorem 6.3. Let v be an arbitrary smooth tangent vector field on S^k with perhaps only isolated singular points. Then the index of v is $1 + (-1)^k$, i.e., it equals zero on an odd-dimensional sphere and equals 2 on an even-dimensional sphere.

This immediately implies Theorem 6.2, since assuming the contrary, we have constructed on S^{2n} a smooth nonsingular tangent field, i.e., having a zero index, which is impossible by Theorem 6.3. This contradiction means that the initial vector field on S^{2n} has at least one singular point.

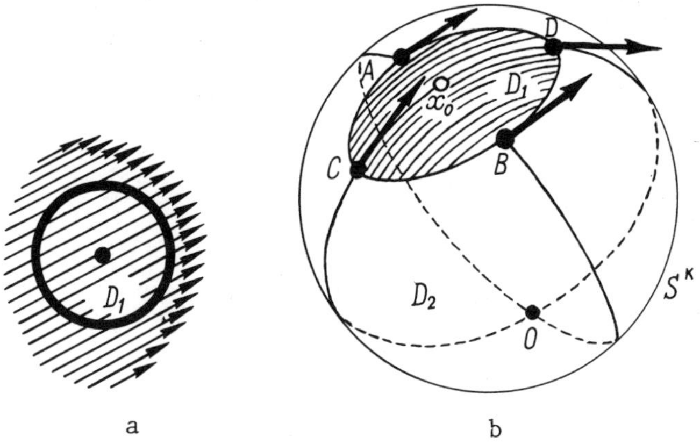

Fig. 40

Proof of Theorem 6.3. Let $x_0 \in S^k$ be an arbitrary point and D_1 a disk in S^k of sufficiently small radius having x_0 as the center (see Fig. 39). Let us present S^k as the union of two closed disks $S^k = D_1 \cup D_2$, where D_2 is the closed complement to D_1 in S^k (see Fig. 39). Let $x_0 \in S^k$ be an arbitrary nonsingular point of v, i.e., $v(x_0) \neq 0$. Since v is smooth, we may assume that v is nonzero on the whole D_1 and, moreover, is however-close to the "parallel field," i.e., integral trajectories of v are however-close to the bunch of parallel lines (see Fig. 40, where the case of a two-dimensional sphere is depicted). Since v does not have singular points on D_1, then all singular points of v (if they exist) belong to D_2. This disk may be transformed with respect to the "unfolding diffeomorphism" (e.g., by the standard stereographic projection with center at x_0) into the standard disk in $\mathbf{R}^k$ with center at 0. In Fig. 40, this operation is equivalent to the following: we must reject D_1 from the sphere and "unfold" the remaining disk and put it on the plane tangent to S^k at 0. We must trace images of vectors of the field v on the boundary of D_2 after this operation, i.e., after we introduce on D_2 the standard Euclidean coordinates with center in D_2. Clearly, from the point of view of these Euclidean coordinates in D_2, the field on its boundary is of the form depicted in Fig. 41. For the sake of descriptiveness, the case of the two-dimensional sphere is considered again. Even now it is intuitively clear that, for instance, in the two-dimensional case, the existence of such a curl of a field

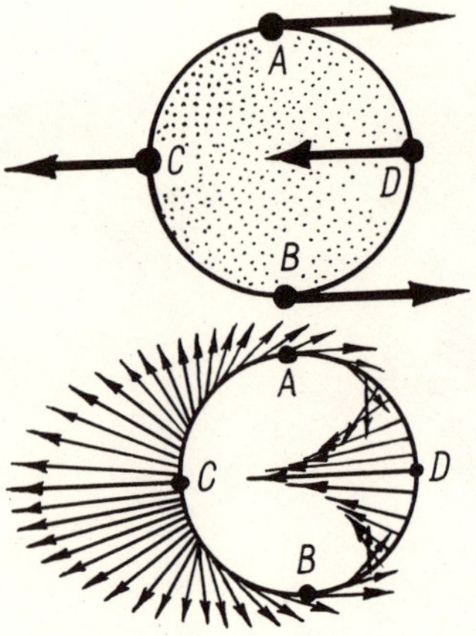

Fig. 41

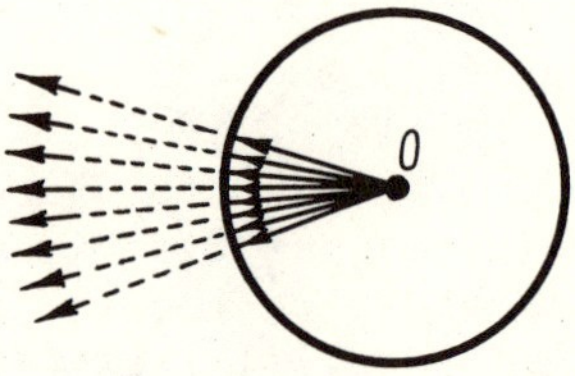

Fig. 42

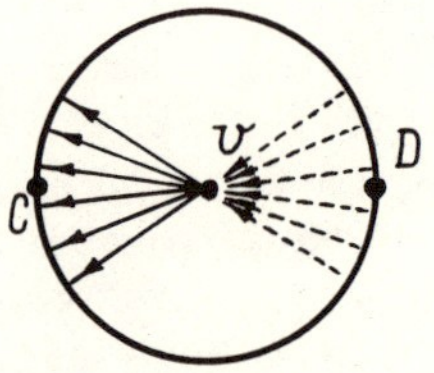

Fig. 43

on the boundary of D_2 means that in the interior of the disk there is at least one singular point. Now we will prove this fact, making use of Theorem 6.1. In fact, the disk $D_2 \subset \mathbf{R}^k$ may be identified with the domain W of Theorem 6.1. Then ∂W is S^{k-1} and the smooth field v on ∂W is as depicted in Fig. 41 (for the two-dimensional case). By Theorem 6.1, to compute ind v on $D_2 \subset \mathbf{R}^k$, it suffices to compute the degree of the mapping $F: S^{k-1} \to S^{k-1}$ defined by the formula $F(x) = v(x)/ /|v(x)|$. It also suffices to consider an arbitrary regular point x and compute the signs of Jacobians of the mapping F in all pullbacks of x (regular points). Since the degree of the mapping does not depend on the choice of a regular point, then we may take for convenience a point C of S^{k-1} as such a point. In this situation, we identify two copies of S^{k-1}

and assume that F transforms S^{k-1} into itself (see Fig. 41). From the construction of v, it follows that pre-images of C are only two points on S^{k-1}, i.e., C itself and the antipodal point D. The mapping F at C is as depicted in Fig. 42, i.e., its differential at the tangent plane to the sphere at C is defined by the diagonal matrix with positive entries on the main diagonal. In particular, the determinant of this transformation is positive, i.e., the sign of the Jacobian at C is a plus. At the point D, the situation is opposite. In fact, each vector v close to D is mapped onto the vector v' with O as the source obtained from v by the parallel transformation onto O (see Fig. 43). This means that dF at D may be expressed on the plane tangent to the sphere as the diagonal matrix with negative entries on the diagonal. In other words, we deal with the mapping of S^{k-1} which transforms each point to its antipodal point with respect to the center. Above, in Lemma 4.6, we have computed the degree of such a mapping $\alpha(x) = -x$ of S^{k-1} onto itself. This degree equals $(-1)^k$. Thus, the sign of the Jacobian at D equals $(-1)^k$. Finally, the degree of F equals $1 + (-1)^k$, proving Theorem 5.3.

In particular, on a two-dimensional sphere, any continuous tangent field has at least one singular point. In popular literature, this theorem is sometimes called "the theorem on a hedgehog": it is impossible to comb a hedgehog that is rolled into a ball so that all its needles are tangent to it.

Thus, returning to bundles, we see that Lemma 5.1 is completely proved, i.e., the bundle $p:E \to S^{2n}$ is not trivial. The similar bundle $p:E \to S^1$ is trivial, being homeomorphic to the direct product $S^1 \times S^0$ (where S^0 is a pair of points, the zero-dimensional sphere).

Example 6. The bundle $p:E \xrightarrow{F} B$ is a covering if its fiber F is discontinuous, i.e., consists of a discrete collection of points; see [2].

If, for example, F is a finite set of points, then each point of the base has only a finite and identical number of pre-images with respect to p.

7. SOME METHODS OF COMPUTATION OF (CO)HOMOLOGY (SPECTRAL SEQUENCES)

7.1. Filtration of a Complex

To compute (co)homology groups, there are different methods, but an effective device based on the study of spectral sequences is the outstanding one. This method is based on a certain filtration and is essentially as follows. If X is a complex, then we may present it as the union of subsets $\{X_p\}$, where $X_p \subset X_{p+1}$ (for any p), and it is desirable that each subset X_{p+1} not differ much from the previous X_p, e.g., from the (co)homology point of view. If it is possible to construct such a chain $\{X_p\}$, one can proceed step by step and trace how (co)homologies differ by obtaining some information on (co)homology of the whole X. It turns out that this is a natural idea: to split the long process of computation of (co)homology of X into the sequence of elementary operations; it is possible to convert the corresponding algorithm into an algebraic—topological device whose structure can be explicitly described in many cases.

Let X be a topological space with the collection of subspaces $\{X_i\}$ such that $\phi = X_{-1} \subset X_0 \subset X_1 \subset \ldots \subset X_{N-1} \subset X_N = X$. For example, if X is a CW-complex, then for X_i we may take the i-dimensional skeleton X^i of X. Consider, for the sake of definiteness, the case of homology, since all the constructions described below are easy to extend to the case of cohomology.

Consider the group $C_k(X)$ of k-dimensional singular chains of X and assume that X_i are closed subspaces in X. Evidently,

$$0 \subset C_k(X_0) \subset C_k(X_1) \subset \cdots \subset C_k(X_{N-1}) \subset C_k(X_N) = C_k(X).$$

If X is a complex, then it is sometimes convenient to take as X_i subcomplexes of X. In these cases, instead of singular chains $C_k(X)$, we may consider cell chains. But since the constructions of these sections are meaningful only for complexes, we will only use singular chains and homology.

Definition 7.1. Let us say that the element $a \in C_k(X)$ is of degree i if $a \in C_k(X_i)$ and $a \notin C_k(X_{i-1})$. If the sequence of subspaces (see above) is defined, we will say that in $C_k(X)$ the filtration is defined.

Clearly, the group $C_k(X_i)$ contains only elements of filtration not greater than i. We will consider only finite filtrations, i.e., such that $N < \infty$. Sometimes the number k is called the total degree of $a \in C_k(X)$ and i is called its filtered degree. Starting from the sequence $\{C_k(X_i)\}$, let us construct a new sequence consisting of groups $C_k(X_i)/$ $/C_k(X_{i-1})$ that will be denoted by $E_0^{i,k-1}$. Set $k - i = j$. The results of Sec. 4 imply that $E_0^{i,j} = C_k(X_i, X_{i-1})$ [the group of relative chains]. Above we have considered several times the boundary operator ∂_k defined on groups of relative chains $\partial_k: C_k(X_i, X_{i-1}) \to C_{k-1}(X_i, X_{i-1})$. Denote this operator by $d_0^{i,j}$. Furthermore, we will see that these new notations are convenient and enable one to operate freely with the algebraic structures that we are going to describe. Thus, we obtain homomorphisms $d_0^{i,j}: E_0^{i,j} \to E_0^{i,j-1}$. The properties of the boundary operator immediately imply that $d_0^{i,j} d_0^{i,j+1} =$ $= 0$. Therefore, the algebraic chain complex

$$\cdots \longrightarrow E_0^{i,j+1} \xrightarrow{d_0^{i,j+1}} E_0^{i,j} \xrightarrow{d_0^{i,j}} E_0^{i,j-1} \longrightarrow$$

arises. Consider the homology of this complex. The previous definitions imply that $\operatorname{Ker} d_0^{i,j}/\operatorname{Im} d_0^{i,j+1} = H_k(X_i,$ $X_{i-1})$. Denote this group by $E_1^{i,j}$.

7.2. Recovering the Spectral Sequence from the Filtration

We have defined above the collection of groups $E_0^{i,j}$ and $E_1^{i,j}$. Groups $E_1^{i,j}$ are homology groups of the algebraic complex constructed from $E_0^{i,j}$. Let us continue this process by now passing on to the homology groups of the complex constructed from $E_1^{i,j}$. Denote these homology groups by $E_2^{i,j}$, etc. Therefore, for any pair of indices i, j, where $i + j = k$, we will construct the chain $E_0^{i,j}$, $E_1^{i,j}$, $\ldots$, $E_r^{i,j}$, $\ldots$. This will be the spectral sequence that we wish to construct.

Definition 7.2. Let $Z_r^{i,j}$ be a subgroup (of r-cycles) in $E_0^{i,j} = C_k(X_i)/C_k(X_{i-1})$, consisting of elements α with the following property: $\alpha \in Z_r^{i,j}$ iff α, considered as the class with respect to $C_k(X_{i-1})$, contains a representative $a \in C_k(X_i)$

whose boundary ∂a (where ∂ is a homological operator of the boundary) has a filtration r less than that of a, i.e., $\partial a \in C_{k-1}(X_{i-r})$.

To comprehend this definition, consider the following special cases:

a) Let $r = 0$; then, evidently, $Z_0^{i,j} = E_0^{i,j}$.

b) Let $r = 1$; then $\alpha \in Z_1^{i,j}$ means that there exists an $a \in \alpha$ such that $\partial a \in C_{k-1}(X_{i-1})$, i.e., a is a cycle in $C_k(X_i, X_{i-1}) = C_k(X_i)/C_k(X_{i-1})$. In other words, $Z_1^{i,j} = Z_k(X_i, X_{i-1})$ [the subgroup of relative cycles].

Definition 7.2 implies that $Z_{r+1}^{i,j} \subset Z_r^{i,j}$ for any r. Hence, in $E_0^{i,j}$ there is the chain of embedded subgroups $Z_\infty^{i,j} \subset \ldots \subset Z_{r+1}^{i,j} \subset Z_r^{i,j} \subset \ldots \subset Z_0^{i,j} = E_0^{i,j}$.

<u>Definition 7.3</u>. Let $B_r^{i,j}$ be a subgroup (of r-boundaries) in $E_0^{i,j} = C_k(X_i)/C_k(X_{i-1})$ consisting of elements α with the following property: $\alpha \in B_r^{i,j}$ if this element, considered as the class with respect to $C_k(X_{i-1})$, contains a representative $a \in C_k(X_i)$ such that $a = \partial b$, where $b \in C_{k+1}(X_{i+r-1})$.

Consider some special cases.

a) Let $r = 0$; then $\alpha \in B_0^{i,j}$ means that $\alpha \in E_0^{i,j} = C_k(X_i)/C_k(X_{i-1})$ contains a representative $a \in C_k(X_i)$ such that $a = \partial b$, where $b \in C_{k+1}(X_{i-1})$, i.e., $a \in C_k(X_{i-1})$. But this means that $\alpha = 0$. Thus, $B_0^{i,j} = 0$.

b) Let $r = 1$; then $\alpha \in B_1^{i,j}$ means that $\alpha \in E_0^{i,j} = C_k(X_i, X_{i-1})$ contains a representative $a \in C_k(X_i)$ such that $a = \partial b$, where $b \in C_{k+1}(X_i)$, i.e., $B_1^{i,j}$ is the subgroup of relative boundaries in $C_k(X_i, X_{i-1})$. Finally, $B_1^{i,j} = B_k(X_i, X_{i-1})$.

Definition 7.3 implies that $B_r^{i,j} \subset B_{r+1}^{i,j}$. Hence, in $E_0^{i,j}$ the chain of embedded subgroups arises:

$$0 = B_0^{i,j} \subset B_1^{i,j} \subset \cdots \subset B_r^{i,j} \subset B_{r+1}^{i,j} \subset \cdots \subset B_\infty^{i,j}.$$

<u>Remark</u>. Although we use the notation $Z_\infty^{i,j}$ and $B_\infty^{i,j}$, since the filtration is finite (see above), groups $Z_r^{i,j}$ and

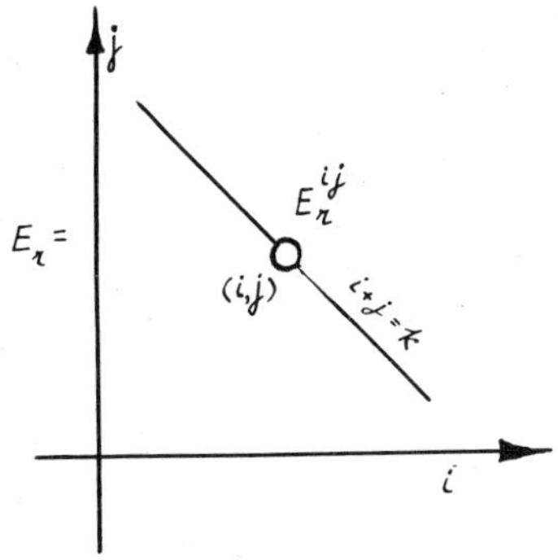

Fig. 44

$B_r^{i,j}$ do not vary for sufficiently large r and these "stable" groups will be denoted by $Z_\infty^{i,j}$ and $B_\infty^{i,j}$.

Clearly, $B_\infty^{i,j} \subset Z_\infty^{i,j}$. Thus, we obtain the following long chain of inclusions:

$$0 = B_0^{i,j} \subset B_1^{i,j} \subset \cdots \subset B_\infty^{i,j} \subset Z_\infty^{i,j} \subset \cdots$$
$$\subset Z_1^{i,j} \subset Z_0^{i,j} = E_0^{i,j} = C_k(X_i, X_{i-1}).$$

In particular, $B_r^{i,j} \subset Z_r^{i,j}$ for any r.

Definition 7.4. Put $E_r^{i,j} = Z_r^{i,j}/B_r^{i,j}$ and $E_r = \bigoplus_{i,j} E_r^{i,j}$.

Consider some special cases.

a) Let r = 0; then

$$E_0^{i,j} = Z_0^{i,j}/B_0^{i,j} = Z_0^{i,j} = C_k(X_i, X_{i-1})$$

(in the previous notation).

b) Let r = 1; then

$$E_1^{i,j} = Z_1^{i,j}/B_1^{i,j} = Z_k(X_i, X_{i-1})/B_k(X_i, X_{i-1}) = H_k(X_i, X_{i-1}).$$

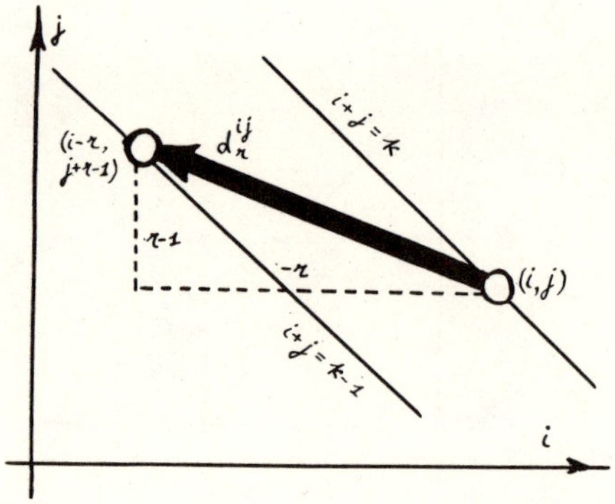

Fig. 45

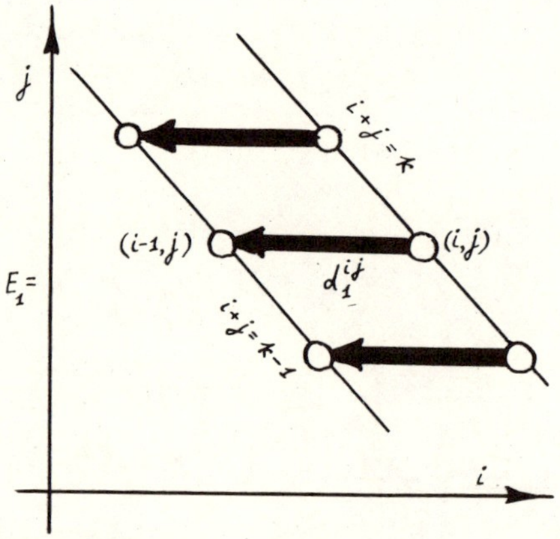

Fig. 46

Groups $E_r^{i,j}$ "decrease" with the growth of r. In fact, when r grows, then (see Definition 7.4) the "numerator" $Z_r^{i,j}$ decreases and the "denominator" $B_r^{i,j}$ increases. Since the filtration is finite, there exists a number r_0 such that for any pairs of indices where i + j = k there are isomorphisms $E_{r_0}^{i,j} = E_{r_0+1}^{i,j} = \ldots = E_\infty^{i,j}$, i.e., for $r > r_0$, groups $E_r^{i,j}$ stop diminishing and become stable. This "stable" group has been denoted by $E_\infty^{i,j}$. For clarity, we may assume that for fixed r groups, $E_r^{i,j}$ are numbered by integer points in the two-dimensional plane, i.e., the pairs (i, j) are considered as coordinates (see Fig. 44). In this situation, all groups $E_r^{i,j}$, such that i + j = k (if the total degree k is

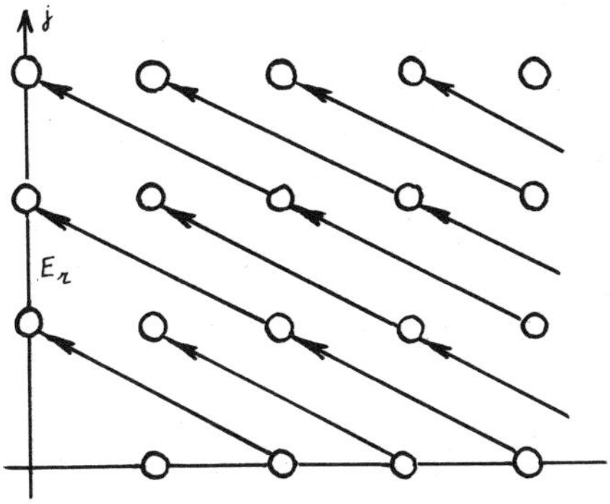

Fig. 47

fixed), are described by integer points in the plane defined by the equation $i + j = k$. Let us construct differentials of the special kind that connect different groups $E_r^{i,j}$.

Definition 7.5. The differential $d_r^{i,j}:E_r^{i,j} \to E_r^{i-r,j+r-1}$ is defined as follows. Let $\alpha \in E_r^{i,j} = Z_r^{i,j}/B_r^{i,j}$ and α' be its representative in the conjugate class with respect to $B_r^{i,j}$, i.e., $\alpha' \in Z_r^{i,j} \subset C_k(X_i)/C_k(X_{i-1})$. Furthermore, let $a \in C_k(X_i)$ be a representative of α' such that $b = \partial a$ is the filtration, which is no greater than $i - r$. Then the conjugate class of b', of the chain b, in $C_{k-1}(X_{i-r})/C_{k-1}(X_{i-r-1})$ belongs to $Z_r^{i-r,j+r-1}$ and defines in $E_r^{i-r,j+r-1}$ the element depending only on α. Denote this element by $d_r^{i,j}(\alpha)$.

As an obligatory exercise, the reader must verify that

a) $d_r^{i,j}(\alpha)$ is well defined,

b) $d_r^{i,j}$ is a homomorphism,

c) $d_r^{i-r,j+r-1}d_r^{i,j} \equiv 0$.

It is useful to consider the action of homomorphisms $d_r^{i,j}$ in our model in the two-dimensional plane. The differential $d_r^{i,j}$ maps $E_r^{i,j}$ situated on the line $i + j = k$ onto the group $E_r^{i-r,j+r-1}$ situated on the neighboring line $i + j = k - 1$ (see Fig. 45). In this process, the point (i, j) is shifted along the i-axis by r steps backward and is lifted along the j-axis by $r - 1$ steps upward.

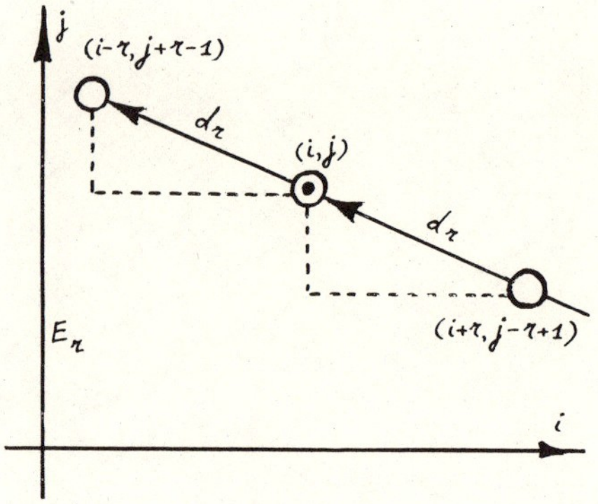

Fig. 48

For r = 1, the isomorphism $d_1{}^{i,j}:E_1{}^{i,j} \to E_1{}^{i-1,j}$ coincides with $\partial:H_k(X_i, X_{i-1}) \to H_{k-1}(X_{i-1}, X_{i-2})$, which is contained in the exact sequence of the triple (X_i, X_{i-1}, X_{i-2}). This follows from Definitions 7.2, 7.3, and 7.5 (verify!); see Fig. 46.

Finally, fix r and consider the collection of all groups $E_r{}^{i,j}$ (for different i, j). Define the group E_r setting $E_r = \bigoplus_{i,j} E_r{}^{i,j}$. This means that we take the direct sum of all groups situated at integer points of the plane (see Fig. 44). The homomorphisms $d_r{}^{i,j}$ defined above coil themselves naturally into one homomorphism (differential) $d_r:E_r \to E_r$ such that $d^2 \equiv 0$. Therefore, the group E_r is depicted by the table shown in Fig. 47 and the differential d_r is described by the collection of homomorphisms acting on elements of this planar table.

Definition 7.6. The sequence of groups and differentials E_r, d_r, r = 1, 2, ..., ∞, is called the spectral sequence of the filtration.

7.3. Main Algebraic Properties of Spectral Sequences

Theorem 7.1. The homology group of the algebraic complex (E_r, d_r) is isomorphic to E_{r+1}.

This means that $\operatorname{Ker} d_r/\operatorname{Im} d_r = E_{r+1}$, i.e., $E_{r+1}{}^{i,j} = \operatorname{Ker} d_r{}^{i,j}/\operatorname{Im} d_r{}^{i+r,j-r+1}$ (see Fig. 48).

Proof. Let $\alpha \in E_r{}^{i,j}$ and $d_r{}^{i,j}(\alpha) = 0$. Then the conjugate class $b' \in Z_r{}^{i-r,j+r-1}$ (see Definition 7.5) is the image of α under the action of $d_r{}^{i,j}$ and belongs to $B_r{}^{i-r,j+r-1}$, i.e., there is a representative $q \in b'$ such that $q = \partial t$, where $t \in C_k(X_{i-1})$. Let $a \in C_k(X_i)$ be the representative of the element $\alpha' \in Z_r{}^{i,j}$. Then $a - t \in C_k(X_i)$. We have made use of the fact that $C_k(X_{i-1}) \subset C_k(X_i)$. Since $t \in C_k(X_{i-1})$, then the extraction of this element does not affect the conjugate class in $C_k(X_i)/C_k(X_{i-1})$, i.e., $a - t$ is also a representative of $\alpha' \in Z_r{}^{i,j}$. Since the definition of $d_r{}^{i,j}$ does not depend on the choice of representative, then we might take from the very beginning $a - t$ instead of a. But then $\partial(a - t) \in C_{k-1}(X_{i-r-1})$. Since α' is a representative of $a - t$, then $\alpha' \in Z_{r+1}{}^{i,j}$. The conjugate class in $E_{r+1}{}^{i,j}$ that contains this representative will be denoted by $\bar{\alpha}$. Assigning to α the element $\bar{\alpha}$, we obtain the homomorphism of groups φ:$\operatorname{Ker} d_r{}^{i,j} \to E_{r+1}{}^{i,j}$. Starting directly from Definitions 7.5 and 7.3, we verify that: a) φ is well defined, i.e., $\bar{\alpha}$ depends only on α (and does not depend on the choice of its representatives); b) φ is an epimorphism; and, c) the kernel of φ is $\operatorname{Im} d_r{}^{i+r,j-r+1}$ (verify!).

Thus we get $\operatorname{Ker} d_r{}^{i,j}/\operatorname{Im} d_r{}^{i+r,j-r+1} = E_{r+1}{}^{i,j}$. The theorem is proved.

The group E_∞, constructed of stable group $E_\infty{}^{i,j}$, is of special interest. Recall that $E_\infty{}^{i,j} = Z_\infty{}^{i,j}/B_\infty{}^{i,j}$. Consider the natural embedding $X_i \subset X$ and the induced homomorphism $H_k(X_i) \to H_k(X)$; let $H_k(X)$ be the image of this homomorphism in $H_k(X)$. The following filtration arises:

$$0 = {}_{-1}H_k(X) \subset {}_0H_k(X) \subset \cdots \subset {}_NH_k(X) = H_k(X).$$

Proposition 7.1. $E_\infty{}^{i,j} = {}_iH_k(X)/{}_{i-1}H_k(X)$, where $i + j = k$.

Proof. Definiton 7.2 implies $Z_\infty{}^{i,j} = Z_k(X_i)/Z_k(X_{i-1})$. Definition 7.3 yields $B_\infty{}^{i,j} = B_k(X) \cap C_k(X_i)/B_k(X) \cap C_k(X_{i-1})$. Furthermore,

$${}_iH_k(X) = Z_k(X_i)/B_k(X) \cap C_k(X_i),$$

$${}_{i-1}H_k(X) = Z_k(X_{i-1})/B_k(X) \cap C_k(X_{i-1}).$$

Recall that if A and B are subgroups in an Abelian group C, then $(A + B)/B = A/A \cap B$, where the sum and the intersection is considered in C. Then $Z_\infty^{i,j}/B_\infty^{i,j} = {}_iH_k(X)/{}_{i-1}H_k(X)$. The proof is completed.

Summarizing all these statements, we obtain the following theorem.

Theorem 7.2. Let the complex (topological space) X be filtered by subspaces X_p, i.e., $\phi = X_{-1} \subset X_0 \subset \ldots \subset X_{N-1} \subset \subset X_N = X$. Then there are groups $E_r^{i,j}$, defined for $r \geq 0$ and all i, j (so that $E_r^{i,j} = 0$ for $i < 0$ and $j > N$), and homomorphisms (differentials) $d_r^{i,j}: E_r^{i,j} \to E_r^{i-r,j+r-1}$ (so that $d_r^{i-r,j+r-1} d_r^{i,j} = 0$) such that

a) $E_r^{i,j} = \operatorname{Ker} d_r^{i,j} / \operatorname{Im} d_r^{i+r,j-r+1}$,

i.e., E_{r+1} is the homology group for E_r with respect to d_r;

b) $E_0^{i,j} = C_{i+j}(X_i, X_{i-1})$,

i.e., $E_0^{i,j}$ coincides with the group of relative chains;

c) $$E_\infty^{i,j} = \frac{\operatorname{Im}\left(H_{i+j}(X_i) \to H_{i+j}(X)\right)}{\operatorname{Im}\left(H_{i+j}(X_{i-1}) \to H_{i+j}(X)\right)} = \frac{{}_iH_{i+j}(X)}{{}_{i-1}H_{i+j}(X)},$$

i.e., E_∞ is recovered from the filtration.

The last statement (c) needs several comments. As we have proved above, in $H_k(X)$ (where k is any fixed number), there exists the subgroup ${}_0H_k(X) = E_\infty^{0,k}$. Furthermore, in the quotient group $H_k(X)/E_\infty^{0,k}$ there exists the subgroup $E_\infty^{0,k}$, etc. Finally, the quotient group

$$(\cdots((H_k(X)/E_\infty^{0,k})/E_\infty^{1,k-1})/E_\infty^{2,k-2})\cdots)/E_\infty^{N-1,k-N+1}$$

is isomorphic to $E_\infty^{N,k-N}$. Thus, computing the direct sum of the groups $E_\infty^{i,j}$ situated on the diagonal $i + j = k$, i.e., computing $\bigoplus_{i+j=k} E_\infty^{i,j}$, we get enough information on $H_k(X)$.

Sometimes one says that the group $\bigoplus_{i+j=k} E_\infty^{i,j}$ is adjoint to $H_k(X)$ with respect to filtration by subgroups ${}_iH_k(X) \subset H_k(X)$. Denote this adjoint group by $RH_k(X)$. Let us enumerate several of the simplest properties of adjoint groups, leaving the proof to the reader. Let A be an Abelian group and let $0 \subset A_0 \subset A_1 \subset \ldots \subset A_N = A$ be its filtration. Let $RA = \bigoplus_i A_i/A_{i-1}$ be the adjoint group.

1) If RA is finitely generated, then so is A and their ranks coincide.

2) If RA is finite, then so is A and their orders coincide.

3) If among groups A_i/A_{i-1} only one is nonzero, then $RA = A$.

4) If all groups A_i/A_{i-1} are free Abelian, then $RA = A$.

5) If all groups A_i/A_{i-1} are vector spaces over the field F, then $RA = A$.

In what follows, property 5 is of special interest. It means that by computing homology with coefficients in a field, we obtain the group $H_k(X)$ immediately after computing all groups $E_\infty^{i,j}$, since $H_k(X) = \bigoplus_{i+j=k} E_\infty^{i,j}$. In other words, we must sum all groups on the diagonal $i + j = k$ in the table E_∞.

7.4. The Cohomological Spectral Sequence

It is clear that all the constructions made above and all considerations can be repeated for cohomology. We will not repeat previous sections and formulate here only the final theorem, which is similar to Theorem 7.2.

Theorem 7.3. Suppose the topological space X is filtered by subspaces X_p, i.e., $\phi = X_{-1} \subset X_0 \subset \ldots \subset X_{N-1} \subset X_N = X$. Then there are groups $E_r^{i,j}$ defined for $r \geq 0$ and all i, j (so that $E_r^{i,j} = 0$ for $i < 0$ and $j > N$) and homomorphisms (differentials) $d_r^{i,j}: E_r^{i,j} \to E_r^{i+r,j-r+1}$ (so that $d_r^{i+r,j-r+1} d_r^{i,j} = 0$) with the following properties:

a) $E_r^{i,j} = \operatorname{Ker} d_r^{i,j} / \operatorname{Im} d_r^{i-r,j+r-1}$;

b) $E_0^{i,j} = C^{i+j}(X_i, X_{i-1})$;

c) $E_\infty^{i,j} = \dfrac{\mathrm{Ker}(H^{i+j}(X) \to H^{i+j}(X_{i-1}))}{\mathrm{Ker}(H^{i+j}(X) \to H^{i+j}(X_i))} = \dfrac{{}_{i-1}H^{i+j}(X)}{{}_iH^{i+j}(X)}$,

i.e., the group $\bigoplus_{i+j=k} E_\infty^{i,j}$ (the sum of groups on the diagonal $i + j = k$ in the table E_∞) is adjoint to the group $H^k(X)$ with respect to its filtration by subgroups

$$0 = {}_NH^k(X) \subset \cdots \subset {}_0H^k(X) \subset {}_{-1}H^k(X) = H^k(X),$$

where ${}_pH^k(X)$ is the kernel of the mapping $H^k(X) \to H^k(X_p)$ induced by the embedding $X_p \subset X$.

The cohomological spectral sequence admits a graphical description similar to that introduced in Sec. 3. The only difference is in the direction of the differentials d_r (see Fig. 49).

7.5. The Spectral Sequence of a Bundle

Thus, with each filtration of a CW-complex (or, more generally of a topological space), the spectral sequence is na-

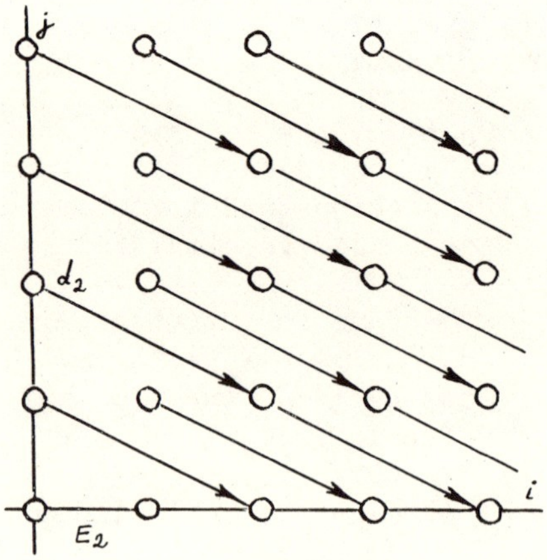

Fig. 49

turally connected, i.e., the chain of groups E_r and differentials D_r. This chain begins with the term (group) E_0, for the computation of which it suffices in principle to know the geometrical interaction of neighboring (with respect to the numeration) subspaces X_p and X_{p+1} of the filtration of X. This chain of groups E_r terminates with E_∞ which, as we have established in Sec. 7.3, is in close connection with the (co)homology of the whole X; more precisely, E_∞ is adjoint to the (co)homology of X. All of this opens several possibilities for computing the (co)homology of X, provided we can find in X a sufficiently simple filtration. The choice of filtration is at our disposal and we may modify it so as to simplify the spectral sequence. Of all the possibilities for construction of filtrations, there is one that plays an extremely important role in many concrete computations. This special filtration is possible when the space X is the total space of the bundle E over some base B with the fiber F. It turns out that, in this case, the spectral sequence obtains several extra properties that enable us to shorten considerably the computation of the (co)homology. Moreover, if E is presentable in the form of a bundle $E \xrightarrow{F} B$, then, in many cases, the base B and the fiber F are of simpler organization than the whole E, leading us to the following problem: how to compute the (co)homology of E when those of B and F are known.

Thus, let $p: E \xrightarrow{F} B$ be a bundle, where E, B, and F are CW-complexes and B is filtered in the simplest way, namely, $\phi = B^{-1} \subset B^0 \subset B^1 \subset \ldots \subset B^{n-1} \subset B^n = B$, where B^i is an i-dimensional skeleton of B and n = dim B. Clearly then, the no less natural filtration of E appears, namely, $\phi = E^{-1} \subset E^0 \subset \subset E^1 \subset \ldots \subset E^{n-1} \subset E^n = E$, where $E^i = p^{-1}(B^i)$, i.e., the preimage of B^i with respect to p (see Fig. 50). Consider the spectral sequence E associated with this filtration. It turns out that the term E_2 of this spectral sequence may be expressed quite simply in terms of the (co)homology of the base and the fiber. This situation defines, to a great extent, the properties of the whole spectral sequence.

The term E_0, due to Definition 7.4, consists of groups $E_0^{i+j} = P_{i+j}(E^i, E^{i-1})$, where P_{i+j} is the group of singular chains. Since we consider here CW-complexes, we will use cell chains and homology (for the sake of definiteness, we consider the homology case). Thus, E_0 is computed from the data on the topological interaction of pre-images of skeletons of the base. Now let us proceed to E_1.

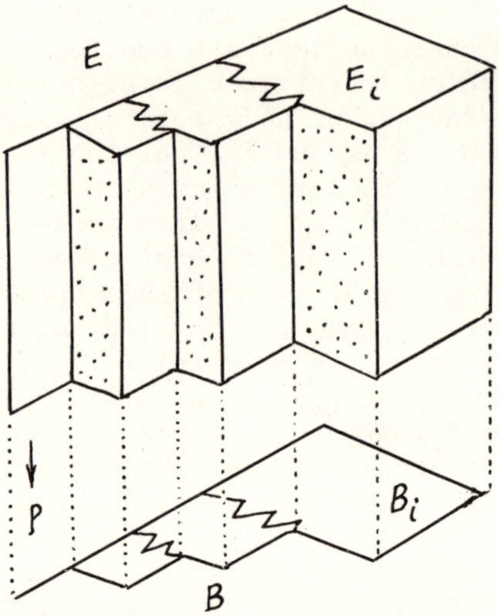

Fig. 50

As was proved above, $E_1^{i,j} = H_{i+j}(E^i, E^{i-1}) = H_{i+j}(E^i/ /E^{i-1})$. The purpose of this section is to give several methods of computation of the (co)homology of the total space of a bundle based on spectral sequences. Therefore, we will skip here the formal proof of theorems that enable us to express E_2 in terms of the (co)homology of the base and the fiber, since the concepts underlying this proof will not be used subsequently. (See this proof, for example, in [3]). The formulation of the theorem will be given for the case when the base B of the bundle is a simply connected complex, i.e., when $\pi_1(B) = 0$. For definitions and properties of the fundamental and higher homotopic groups, see, for example, [2, 4].

Theorem 7.4. Let $p: E \xrightarrow{F} B$ be a locally trivial bundle; E, B, and F be CW-complexes; and B be simply connected, i.e., $\pi_1(B) = 0$. Consider the spectral sequence of this bundle, i.e., the sequence corresponding to the filtration of the space E by pre-images of skeletons of the base described above. Then $E_2^{i,j} = H_i(B; H_j(F))$, i.e., the term E_2 is completely expressed in terms of the homology groups of B, with coefficients in homology groups of the fiber F. If we consider homology with coefficients in the field K, then $E_2^{i,j} = H_i(B) \otimes H_j(F)$ (tensor product). If homology with coefficients in $\mathbf{Z}$ (integer homology) are considered, then the for-

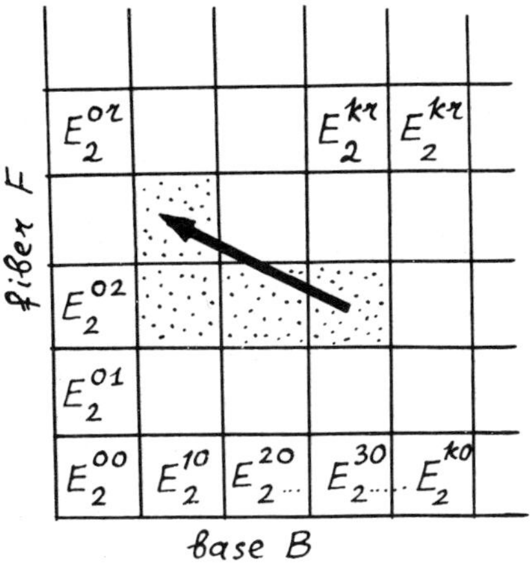

Fig. 51

mula $E_2^{i,j} = H_i(B) \otimes H_j(F)$ holds when either $H_*(B)$ or $H_*(F)$ are free Abelian groups (i.e., are torsion-free; there are no subgroups of finite order).

The table for E_2 is sometimes conveniently described in the form depicted in Fig. 51. In the lower row groups $E_2^{i,0} = H_i(B; H_0(F))$ exist. If F is connected, then $H_0(F) = A$ (the coefficient group) and $E_2^{i,0} = H_i(B; A)$, i.e., in the lower row homology of the base exists. In the left column, groups $E_2^{0,i} = H_0(B; H_i(F))$ exist, i.e., the homology of the fiber, if the base is simply connected: $E_2^{0,i} = H_i(F)$. Thus, when the coefficient group is a field or when the integer homology of the base or fiber are torsion-free, the term E_2 is easily computed in terms of the homology of the base and a fiber. However, to continue on to E_∞, some data on differentials in the spectral sequence is needed. The computation of these homomorphisms is one of the most important problems in this theory, since the properties of differentials depend sometimes on quite refined topological properties of the bundle. These differentials can be computed completely only in rare cases. However, the fact that the spectral sequence of a bundle exists is good enough, and sometimes it suffices to compute E_∞. We will give several examples below.

The differential d_2 acts on E_2, as is shown in Fig. 51 (knight's move). In the passage from E_2 to E_r, arrows that show the action of d_r become longer and tend to the limit defined by the line $i + j = \text{const}$ (see Fig. 52).

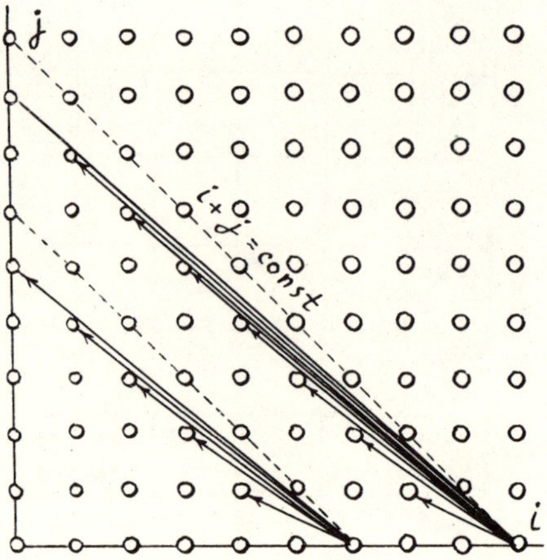

Fig. 52

The cohomological spectral sequence of a bundle is constructed by the same scheme and for the same filtration of the space E by pre-images of skeletons of the base also. For a bundle with a simply connected base, we have $E_2{}^{i,j} = H^i(B; H^j(F))$. The graphical diagram of E_r is constructed similarly for cohomology as for homology, the difference being in the direction of the arrows that depict differentials.

7.6. Multiplication in the Cohomological Spectral Sequence

There is one important distinction between homological and cohomological spectral sequences. Its source is the presence of the ring structure in the cohomology of complexes, whereas in homology there is no such universal ring structure. However, in some special cases, for example when the complex is a Lie group, there is a natural multiplicative structure in the homology as well. Consider the case of homology, and let the group of coefficients be a ring with unit, e.g., a field or **Z**. These cases will be the main ones. Then it turns out that a multiplicative structure arises in the spectral sequence, too. The following theorem, whose proof may be found, for example in [3], holds.

<u>Theorem 7.5</u>. For any $r \geqq 2$, the group $E_r = \bigoplus_{i,j} E_r{}^{i,j}$ may be endowed with a natural multiplicative structure, i.e.,

there is a mapping $E_r^{i,j} \otimes E_r^{\alpha,\beta} \to E_r^{i+\alpha,j+\beta}$ which defines the associative and distributive multiplication consistent with the action of differentials. This means that $d(a, b) = da\cdot b + (-1)^{i+j}a\cdot db$ for any $a \in E_r^{i,j}$, $b \in E_r^{\alpha,\beta}$ ($\cdot$ stands for the product of elements). The multiplication in E_r induces the multiplication in E_{r+1} and coincides with the operation $\cdot$ defined on that group.

Let us briefly explain. It turns out that the multiplication whose existence is stated in Theorem 7.5 is consistent with the natural multiplication arising in the terms E_2 and E_∞ due to Theorem 7.4. First, consider E_2. Since $E_2^{i,j} = H^i(B; H^j(F))$ (see above), then $E_2 = \bigoplus_{i,j} E_2^{i,j}$ may be expressed as $\bigoplus_i H^i(B; \bigoplus_j H^j(F))$. Since $\bigoplus_j H^j(F)$ is a ring, then the group E_2 is naturally transformed into a ring with multiplication generated by the multiplications in the cohomologies of the base and the fiber. This multiplication coincides with that described in Theorem 7.5 (on E_2), i.e., the ring $E_2 = \bigoplus_{i,j} E_2^{i,j}$ is isomorphic to the ring $\bigoplus_i H^i(B; \bigoplus_j H^j(F))$. Furthermore, in E_2, consider the subring $\bigoplus_i E_2^{i,0}$ (the lowest row of the table E_2). It turns out that this subring is isomorphic to the ring of the cohomology of the base, $H^*(B)$. The left column $\bigoplus_j E_2^{0,j}$ in E_2 is also a subring isomorphic to the cohomology ring of the fiber, $H^*(F)$.

If the ring of coefficients is a field, then E_2 is the tensor product $E_2 = H^*(B) \otimes H^*(F)$ and the ring structure on E_2 coincides with the ring structure of this product. A similar statement on the isomorphism of two rings holds: when integer cohomology of the base or the fiber are torsion-free, then the isomorphism $H^*(B; Z) \otimes H^*(F; Z) = E_2$ is that of rings. In these cases, the multiplication in the ring E_2 is defined as follows: $(a \otimes b)(x \otimes y) = (-1)^{j\alpha}(ax \otimes by)$, where $b \in E_2^{0,j}$, $x \in E_2^{0,\alpha}$. In what follows, we will consider the cases when $E_2 = H^*(B) \otimes H^*(F)$. Thus, the general scheme of computation of the (co)homology of E is reduced to the following. First we compute E_2 [if (co)homology of the base and the fiber are known]; then we compute the action of d_2 from algebraic-geometrical considerations. After that, we compute E_3, which consists of homology groups of the term E_2 with respect to d_2. Later, we define the action of d_3 on E_3,

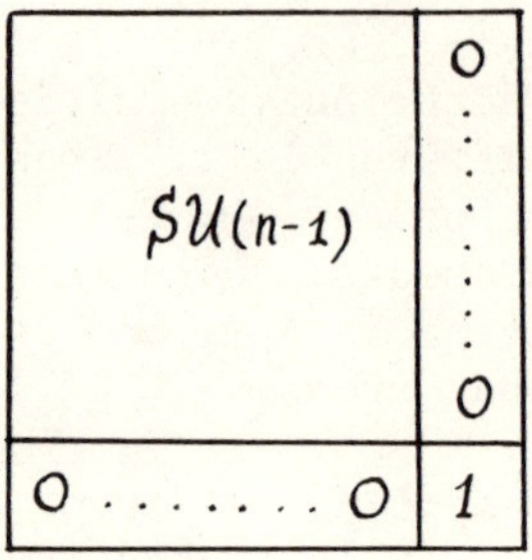

Fig. 53

compute E_4, and so on. Continuing this process, we can, in principle, compute the term E_∞ which is, as we know, adjoint to the (co)homology of E.

7.7. Some Examples of Computations using Spectral Sequences

As we have noted, the very fact that the spectral sequence of a bundle exists is good enough and in several cases it suffices to compute the (co)homology of the space E.

Example 1. Let us compute the (co)homology of the special unitary group SU_n, which is a compact smooth manifold. The group SU_n consists of unitary transformations $g:\mathbf{C}^n \to \mathbf{C}^n$, such that det g = 1. To apply the machinery developed above, it is necessary to express SU_n in the form of the total space of a bundle. There is such a bundle, actually. In SU_n let us distinguish the subgroup SU_{n-1} embedded naturally, i.e., realized by matrices depicted in Fig. 53. Let us consider the standard action of SU_n in $\mathbf{C}^n$ and pick the vector e_1 of an orthonormal basis in $\mathbf{C}^n$; let us consider its orbit with respect to this action, i.e., consider the set of points of the form $g(e_1)$, where $g \in SU_n$. Clearly, the vector e_1 runs over the standard sphere S^{2n-1}, which is an orbit of SU_n-action in $\mathbf{C}^n$. Consider the smooth mapping $p:SU_n \to S^{2n-1}$, defined by the formula $p(g) = g(e_1)$. In other words, we have assigned to each transformation g the image of e_1 with respect to the action of g. We obtain a locally trivial bundle whose base B is S^{2n-1}, the total space E is SU_n, and the fiber F is SU_{n-1}. In particular, the sphere S^{2n-1} is a homogeneous space in the sense of Example 3 from Sec. 6. Thus,

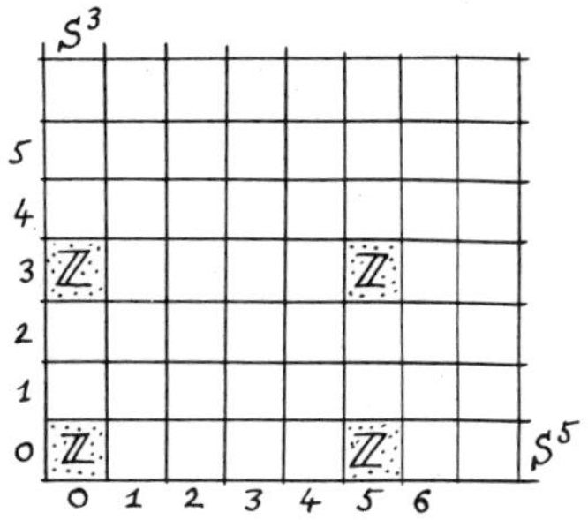

Fig. 54

$p: SU_n \xrightarrow{SU_{n-1}} S^{2n-1}$. For n = 2, the fiber consists of one point, i.e., $SU_2 = S^3$. Hence, the group SU_2 may be presented as the group of matrices $g = (-x/y, y/x)$, where $|x|^2 + |y|^2 = 1$. Clearly, the equation det g = 1 describes in $\mathbf{C}^2 = \mathbf{C}^2(x, y)$ a three-dimensional sphere. Since the homology of a sphere is known, the first step of induction is complete (see Sec. 4). For n = 3, we obtain the bundle $SU_3 \xrightarrow{SU_2} S^5$, i.e., $SU_3 \xrightarrow{S^3} S^5$. Since homology of the fiber and the base is known, we can write the term E_2 of the spectral sequence of this bundle. We have $E_2^{i,j} = H_i(S^5; \mathbf{Z}) \otimes H_j(S^3; \mathbf{Z})$, since the groups are torsion-free (see Fig. 54). This table readily implies that all differentials d_r are trivial from dimensional considerations, $d_2 = d_3 = \ldots = 0$. Thus, $E_2 = E_3 = \ldots = E_\infty$ and, in the term E_∞, on each line $i + j = k$, there is no more than one nontrivial term. Hence (see Sec. 7.3, property 3), the adjoint group coincides with the initial one. Hence, the homology of SU_3 is as follows: $H_0 = H_3 = H_5 = H_8 = \mathbf{Z}$, other H_i being zero. Clearly, $H_k(SU_3; \mathbf{Z}) = H_k(S^3 \times S^5; \mathbf{Z})$.

Now consider the case n = 4, i.e., $SU_4 \xrightarrow{SU_3} S^7$. The term E_2 is of the form depicted in Fig. 55. From dimensional considerations, we get $E_2 = E_3 = \ldots = E_\infty$, i.e., the spectral sequence is trivial once again; hence SU_4 has the following homology: $H_0 = H_3 = H_5 = H_7 = H_8 = H_{10} = H_{12} = H_{15} = \mathbf{Z}$, other H_i being zero. This means that $H_*(SU_4; \mathbf{Z}) = H_*(S^3 \times S^5 \times S^7; \mathbf{Z})$.

Consider the case n = 5. Here the simple considerations that enabled us to prove the triviality of all differentials

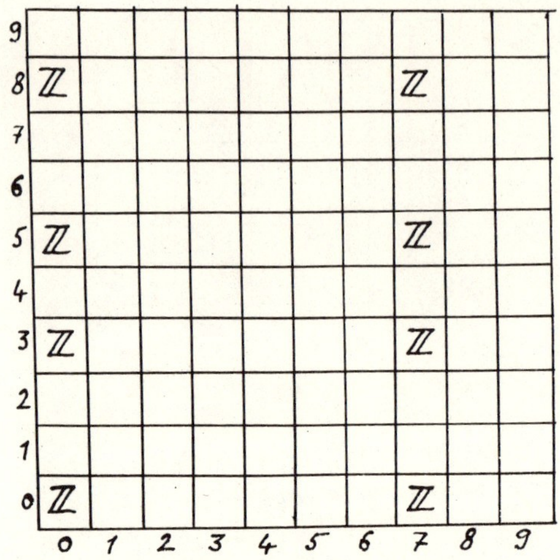
9
8
7
6
5
4
3
2
1
0
0 1 2 3 4 5 6 7 8 9

Fig. 55

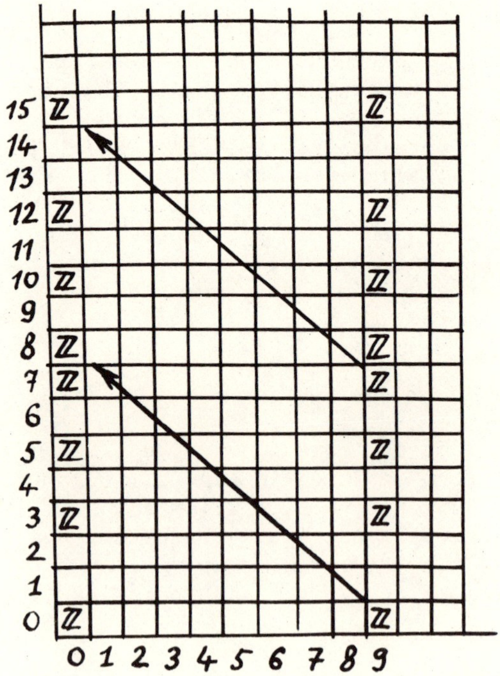
15
14
13
12
11
10
9
8
7
6
5
4
3
2
1
0
0 1 2 3 4 5 6 7 8 9

Fig. 56

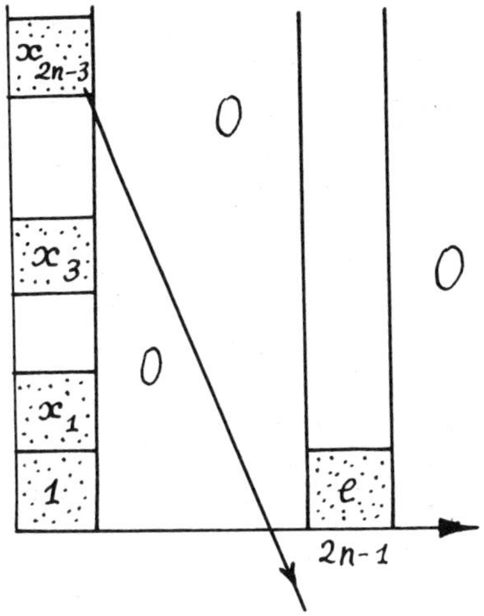

Fig. 57

are not valid. In fact, consider the bundle $SU_5 \xrightarrow{SU_4} S^9$. Then the term E_2 is depicted by the table in Fig. 56. From dimensional considerations, it follows that $d_2 = d_3 = \ldots = d_8 = 0$, but d_9 can be nonzero, since its action $d_9{}^{9,0}$: $E_9{}^{9,0} \to E_9{}^{0,8}$ might be nontrivial (both groups are isomorphic to **Z**). In terms of the homological spectral sequence, it is not possible to establish triviality of d_9 (though actually $d_9 = 0$). Hence, we must apply the data to the cohomological spectral sequence. So let us proceed to the cohomological sequence and make use of the fact that the already-computed homology of SU_2, SU_3, and SU_4 are torsion-free. This means (see above) that $H^k(SU_n; \mathbf{Z}) = H_k(SU_n; \mathbf{Z})$ for n = 2, 3, 4, and all k. Now we will prove the following theorem.

<u>Statement 7.1</u>. $H_*(SU_n; \mathbf{Z}) = H_*(S^3 \times S^5 \times \ldots \times S^{2n-1}; \mathbf{Z})$; $H^*(SU_n; \mathbf{Z}) = H^*(S^3 \times S^5 \times \ldots \times S^{2n-1}; \mathbf{Z})$. The ring $H^*(SU_n; \mathbf{Z})$ is isomorphic to the exterior algebra $\Lambda(x_3, x_5, \ldots, x_{2n-1})$ in indeterminates $x_3, x_5, \ldots, x_{2n-1}$, where $\deg x_k = k$. Furthermore, $H^*(U_n; \mathbf{Z}) = \Lambda(x_1, x_3, x_5, \ldots, x_{2n-1})$.

<u>Proof</u>. Let us apply the induction to n. For n = 2, the statement is evident, since $SU_2 = S^3$, $H^*(S^3) = \Lambda(x_3)$. Suppose our theorem is proved for all $S \leqq n - 1$, i.e., $H^*(SU_{n-1}; \mathbf{Z}) = \Lambda(x_3, x_5, \ldots, x_{2n-3})$. Let us take the next step and consider the bundle $SU_n \xrightarrow{SU_{n-1}} S^{2n-1}$. The term E_2

of the cohomological sequence of this bundle is depicted in Fig. 57. In the left column, all generators of $\Lambda(x_3, x_5, \ldots, x_{2n-3})$ are shown. The other elements (products of generators) are not depicted. As we can see from the table, all differentials $d_2, d_3, \ldots, d_8$ are trivial (also see above). Earlier we dwelt on the case n = 5 of the cohomological differential $d_9{:}E_9^{0,8} \to E_9^{9,0}$, where $E_9^{0,8} = E_9^{9,0} = \mathbf{Z}$. In the left column, the ring of cohomology of the fiber exists, i.e., the exterior algebra $\Lambda(x_3, x_5, x_7)$. In dimension 8 the group whose generator is of the form $x_3 \cdot x_5$ exists, i.e., factors with respect to multiplicative generators x_3 and x_5. Hence, $d_9(x_3 \cdot x_5) = d_9 x_3 \cdot x_5 + x_3 \cdot d_9 x_5 = 0$. Thus, the "dubious" differential d_9 (for n = 5) becomes trivial. Other differentials in higher dimensions are evidently trivial. In the general case for arbitrary n, the unique "dubious" differential is $d_{2n-1}{:}E_{2n-1}^{0,2n-2} \to E_{2n-1}^{2n-1,0}$. Since in dimension $2n - 2$ in the ring $H^*(SU_{n-1})$ there are no multiplicative generators, then all elements of dimension $2n - 2$ are divisible by generators of lower dimensions. Applying d_{2n-1} to these products, and since in lower dimensions differentials d_{2n-1} are trivial, then so is the mapping $d_{2n-1}{:}E_{2n-1}^{0,2n-2} \to E_{2n-1}^{2n-1,0}$, as required. All other differentials are trivial from dimensional considerations. Thus, $E_2 = \ldots = E_\infty$. It follows that $H^*(S^{2n-1}; \mathbf{Z}) \otimes H^*(SU_{n-1}; \mathbf{Z}) = E_\infty$, where the equality signifies the isomorphism of rings (see Sec. 7.6). Since E_∞ is adjoint to $H^*(SU_n; \mathbf{Z})$, then $RH^*(SU_n; \mathbf{Z}) = H^*(S^{2n-1}) \otimes H^*(SU_{n-1})$. Note that on each line $i + j = k$ in E_∞, there are no more than two nontrivial groups and these groups are free (as Abelian groups). Suppose these groups are situated in places (i_1, j_1) and (i_2, j_2), where $i_1 + j_1 = i_2 + j_2 = k$. Then $E_\infty^{i_1, j_1} = {}^{i_1}H^k(E; \mathbf{Z})$, $E_\infty^{i_2, j_2} = {}^{i_2}H^k(E; \mathbf{Z})$. Since all these groups are free (Abelian), then $H^*(SU_n; \mathbf{Z})$ is obtained from E_∞ by summation of the groups situated on diagonals of the form $i + j = k$ (see the properties of adjointness). This implies the additive isomorphism

$$H^*(SU_n; \mathbf{Z}) = \Lambda(x_3, x_5, \ldots, x_{2n-1}) = H^*(S^{2n-1} \times SU_{n-1}).$$

But this isomorphism is also a ring one, since on the diagonal $i + j = k$ we pass to the quotient with respect to the group in the second column ${}^{i_1}H^k(E; \mathbf{Z})$ which is obtained from the first one (if $i_1 > i_2$) by tensor multiplication by x_{2n-1}, i.e., it consists only of products. The case of U_n is an

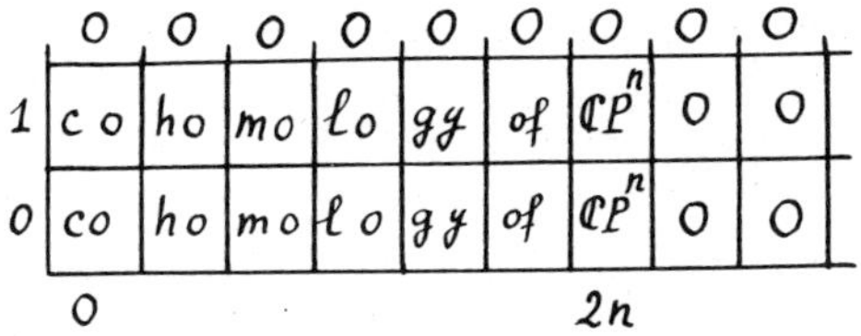

Fig. 58

analogous one. Moreover, note that U_n is homeomorphic (as a topological space) to the direct product $S^1 \times SU_n$. The theorem is proved.

In several cases, the spectral sequence enables one to compute the (co)homology of the base when those of the fiber and the total space are known. Let us give one such example.

Example 2. Let us compute the cohomology ring of the complex projective space CP^n. For this, consider the bundle $S^{2n-1} \xrightarrow{S^1} CP^n$. The bundle is arranged as follows: we must realize S^{2n+1} as the standard sphere in C^{n+1} and assign to each point $(z_0, \ldots, z_n) \in S^{2n+1}$ the point $(\lambda z_0, \ldots, \lambda z_n) \in \in \mathbf{C}P^n$, where $\lambda = e^{i\varphi}$. Since CP^n is simply connected (see [2]), then the term E_2 of the cohomological spectral sequence is of the form depicted in Fig. 58. By dimensional considerations, $E_3 = E_4 = \ldots = E_\infty = H^*(S^{2n+1})$, whence $E_3{}^{i,j} = 0$ for all (i, j) except $(0, 0)$ and $(2n, 1)$. In this situation, $E_3{}^{2n,1} = Z$; hence $E_2{}^{1,0} = 0$, $E_2{}^{2n,1} = Z$, and the differential $d_2{}^{i,1}: E_2{}^{i,1} \to E_2{}^{i+2,0}$ is an isomorphism for $i = 0, 1, \ldots, 2n - 1$. Since $E_2{}^{i,0} = E_2{}^{i,1}$, then $E_2{}^{i,0} = E_2{}^{i,1} = \mathbf{Z}$ for $i = 0, 2, \ldots, 2n$. Other $E_2{}^{i,J}$ are trivial. Hence, $H^k(CP^n; \mathbf{Z}) = \mathbf{Z}$ for $k = 0, 2, \ldots, 2n$, the other groups being zero. Thus, the term E_2 is of the form depicted in Fig. 59. Here $e_i \in H^{2i}(\mathbf{C}P^n; Z)$ and $e \in H^1(S^1; \mathbf{Z})$ are additive generators. Clearly, they may be expressed so that $d_2{}^{2i,1}(e_i, e) = e_{i+1}$ (since $d_2{}^{2i,1}$ is an isomorphism; see above). Then $e_{i+1} = d_2{}^{2i,1}(e_i, e) = e_i \cdot d_2{}^{0,1}(e) = e_i e_1$. This implies that $e_i = e_1{}^i$ for $i = 1, 2, \ldots, n$. Thus we have proved the following statement.

Statement 7.2. There is a ring isomorphism $H^*(CP^n, \mathbf{Z}) = \mathbf{Z}[e_1]/(e_1{}^{n+1})$, where $\dim e_1 = 2$.

$e_0 e$	0	$e_1 e$	0	$e_2 e$	0	....	0	$e_n e$	0
e_0	0	e_1	0	e_2	0	....	0	e_n	0
0	1	2	3	4	5		$2n-1$	$2n$	

Fig. 59

Remark. Denote by $\mathbf{Z}[e_1]$ the polynomial ring in one indeterminate e_1 of degree 2 with coefficients of $\mathbf{Z}$, and by $(e_1{}^{n+1})$ the ideal generated by $e_1{}^{n+1}$. The quotient ring $\mathbf{Z}[e_1]/(e_1{}^{n+1})$ is sometimes called the ring of truncated polynomials.

Example 3. By similar considerations we can prove the following theorem.

Statement 7.3. The ring of the cohomology of the real projective space $\mathbf{R}P^n$ is as follows:

$$H^*(\mathbf{R}P^{2k};\mathbf{Z}) = \mathbf{Z}_2[e]/(e^{k+1}),$$

where deg e = 1;

$$H^*(\mathbf{R}P^{2k+1};\mathbf{Z}) = \left(\mathbf{Z}_2[e]/(e^k)\right) \otimes \Lambda(b),$$

where deg e = 1, deg b = 2k + 1;

$$H^*(\mathbf{R}P^n;\mathbf{Z}_2) = \mathbf{Z}_2[e]/(e^{n+1}),$$

where deg e = 1.

The proof is left to the reader as a useful exercise.

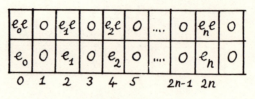

Chapter 2

CRITICAL POINTS OF SMOOTH FUNCTIONS ON MANIFOLDS

8. CRITICAL POINTS AND GEOMETRY OF LEVEL SURFACES

8.1. Definition of Critical Points

In the first part of this work we have proved the classification theorem of two-dimensional connected closed manifolds. We have established that they constitute two infinite series: spheres with handles and spheres with Moebius bands (see [1], Chapter 4, Sec. 5). An interesting problem is that of classification of three-dimensional closed manifolds. Unlike the two-dimensional case, a similar classification is still lacking. Nevertheless, the set of all three-dimensional manifolds admits a quite simple description useful in various applications. It turns out that it is possible to produce those "elementary bricks" from which any such manifold is constructed. In what follows we will prove this theorem, but for this, a geometrical apparatus will be needed, which is important in itself, since it has applications that transcend the limits of description of three-dimensional manifolds. Therefore, we will concentrate now on studying the relationships between the geometry of manifolds. Let f be a smooth real-valued function on the smooth manifold M of dimension n. Consider the vector field grad f.

<u>Definition 8.1</u>. The point $x_0 \in M$ is called critical or stationary for the function f if grad $f(x_0) = 0$. The value $f(x_0)$ is called the critical value of f.

At the critical point x_0, the matrix of second partial derivatives of f is defined, i.e., $d^2f = \left(\frac{\partial^2 f\,(x_0)}{\partial x_i \partial x_j}\right) = (f_{x_i x_j})$, called the Hessian of f.

Definition 8.2. The critical point x_0 of f is called nondegenerate if the matrix d^2f is nonsingular at x_0.

This definition does not depend on the choice of a local coordinate system. In fact, if there is given a regular change of coordinates $x_i = x_i(y_1, \ldots, y_n)$ for $i \leq i \leq n$, then the matrix d^2f is transformed as follows:

$$f_{x_i x_j}(x_0) = \frac{\partial y_p}{\partial x_i}\frac{\partial y_q}{\partial x_j} f_{y_p y_q}(x_0)\ ,$$

since $f_{x_i}(x_0) = 0$ (x_0 is a critical point). Hence, the regularity of $(f_{x_i x_j})$ implies the regularity of $(f_{y_i y_j})$. The matrix d^2f is evidently symmetric. Hence, d^2f defines the symmetric bilinear form on the plane $T_{x_0}M$ tangent to M at x_0. This form will be denoted by the same symbol d^2f.

Definition 8.3. The index $\operatorname{ind}_f x_0$ of the critical point x_0 is the maximum dimension of a linear subspace in $T_{x_0}M$, where d^2f is negative definite. The degree of degeneracy of the critical point x_0 is the dimension of the null subspace in $T_{x_0}M$, i.e., consisting of all vectors a, such that $d^2f(a, b) = 0$ for any $b \in T_{x_0}M$.

In other words, the degree of degeneracy coincides with the number of zero eigenvalues of d^2f and the index is the number of negative eigenvalues. Clearly, the point x_0 is nondegenerate if the degree of degeneracy is zero (see Fig. 60). All future discussion will deal mostly with nondegenerate critical points.

Definition 8.4. The smooth function f on M is called the Morse function if all its critical points are nondegenerate.

In [2], p. 488, the following statement is proved.

Proposition 8.1. On any smooth compact manifold, a Morse function exists. Morse functions are dense in the space of all smooth functions on a manifold. Each Morse

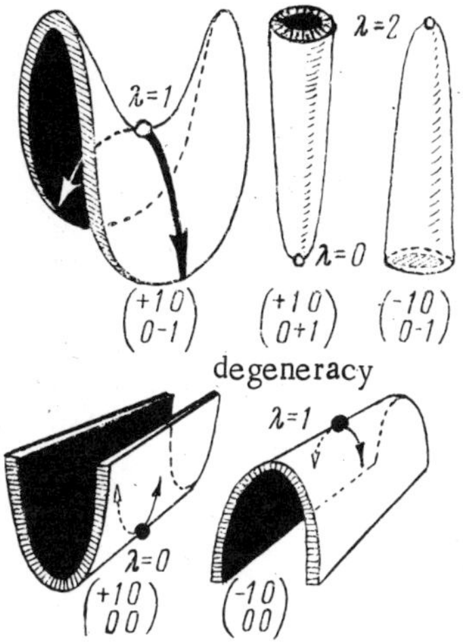

Fig. 60

function on a compact manifold has only a finite number of critical points; in particular, all of them are distinct. In the set of all Morse functions, there is a dense subset consisting of functions such that to each critical value of such a function corresponds a unique critical point on a manifold (values of the function in different critical points are distinct).

8.2. The Canonical Presentation of a Function in a Neighborhood of a Nondegenerate Critical Point

In this section, we will prove one technical but rather useful theorem which enables us to study functions in neighborhoods of a nondegenerate critical point.

Proposition 8.2. Let f be a smooth function on M and x_0 a nondegenerate critical point. Then, in an open neighborhood of x_0, there are local regular coordinates $y_1, \dots, y_n$ such that, in these coordinates, the function f is expressed in the form $f(y) = -y_1^2 - \dots - y_\lambda^2 + y_{\lambda+1}^2 + \dots + y_n^2$, where λ is the index of the critical point.

Remark. The meaning of this statement is that, unlike a Taylor series expansion in a neighborhood of x_0, in the expression of f established by the theorem, there are no terms of order higher than 2. It turns out that all terms of higher order may be removed by an appropriate coordinate transform.

Proof. Since we are interested in a small neighborhood of x_0, we may assume from the start that the function $f(x) = f(x_1, \ldots, x_n)$ is defined on the disk (ball) $D_\varepsilon^n(0)$ of radius ε and $f(0) = 0$, where 0 is the critical point of g. Then we claim that there are smooth functions $g_1, \ldots, g_n$ such that $f(x) = x_1 g_1 + \ldots + x_n g_n$ and $g_i(0) = \partial f(0)/\partial x_i$. In fact, the following obvious identity holds: $f(x) = \int_0^1 \frac{d}{dt} f(tx)\, dt = f(1\cdot x) - f(0\cdot x) = f(x)$. After a parameter transformation, we get

$$f(x) = \sum_i \int_0^1 \frac{\partial f(tx)}{\partial x_i} x_i dt = \sum_i x_i \int_0^1 \frac{\partial f(tx)}{\partial x_i} dt = \sum_i x_i g_i(x),$$

where $g_i(x) = \int_0^1 \frac{\partial f(tx)}{\partial x_i} dt$. Evidently, $g_i(0) = 0$, since grad $f(0) = 0$. Furthermore, applying this statement to functions g_i, we get that there are smooth functions h_{ij} such that $g_i = \sum_j x_j h_{ij}$. Therefore, we have presented f in the form $f = \sum_{i,j} x_i x_j h_{ij}$, where we may assume that $h_{ij} = h_{ji}$. Furthermore, we claim that $h(0) = (h_{ij}(0)) = (f_{x_i x_j}(0))$. From the equalities it follows that

$$g_i(x) = \int_0^1 \frac{\partial f(tx)}{\partial x_i} dt = \sum_j x_j \int_0^1 \frac{\partial g_i(tx)}{\partial x_j} dt =$$

$$= \sum_j x_j \int_0^1 \frac{\partial}{\partial x_j} \left(\int_0^1 \frac{\partial f(t\tau x)}{\partial x_i} d\tau \right) dt =$$

$$= \sum_j x_j \int_0^1 \int_0^1 \frac{\partial^2 f(t\tau x)}{\partial x_i \partial x_j} d\tau dt = \sum_j x_j h_{ij}.$$

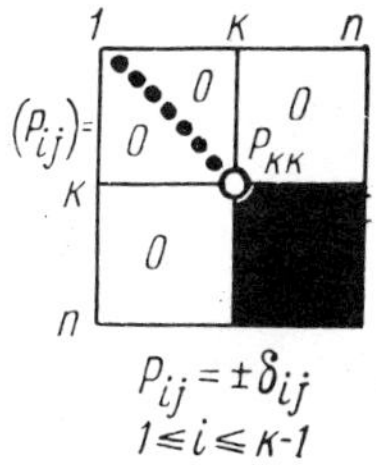

Fig. 61

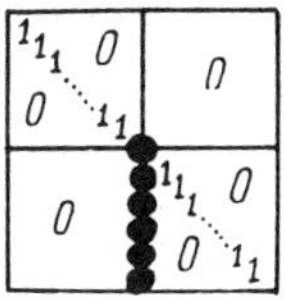

Fig. 62

Hence, $h_{ij}(0) = f_{x_ix_j}(0)$. Furthermore, we will proceed by induction. Suppose we have found coordinates $y_1, \ldots, y_n$ such that f is of the form

$$f(y) = \pm y_1^2 \pm \ldots \pm y_{k-1}^2 + \sum_{i,j \geqslant k} y_i y_j P_{ij},$$

where $P_{ij}(y)$ constitute a symmetric matrix invertible at 0. The first step of induction (for k = 1) is already done, since for k = 1 it suffices to take (h_{ij}) for (P_{ij}). Thus, let us proceed with the next step. Let us rewrite f in the form

$$f = \pm y_1^2 \pm \ldots \pm y_{k-1}^2 + P_{kk} y_k^2 + \sum_{i,j \geqslant k} y_i y_j P_{ij},$$

where (i, j) ≠ (k, k). The square (n × n)-matrix P_{ij} is depicted in Fig. 61. Since this matrix is symmetric and invertible, there exists a linear transformation of coordinates $y_k, \ldots, y_n$ such that the matrix is reduced to a diagonal one at the origin. We may assume from the start that coordinates $y_k, \ldots, y_n$ were chosen just so, implying $P_{kk}(0) \neq$ $\neq 0$. Consider the function $q(y) = \sqrt{|P_{kk}(y)|}$ and make the coordinate transformation (y) → (z) by the formulas

$$\begin{cases} z_i = y_i \quad \text{for} \quad 1 \leqslant i \leqslant k-1; \; k+1 \leqslant i \leqslant n; \\ z_k = q(y) \cdot \left(y_k + \sum_{i>k} y_i \dfrac{P_{ik}}{P_{kk}} \right). \end{cases}$$

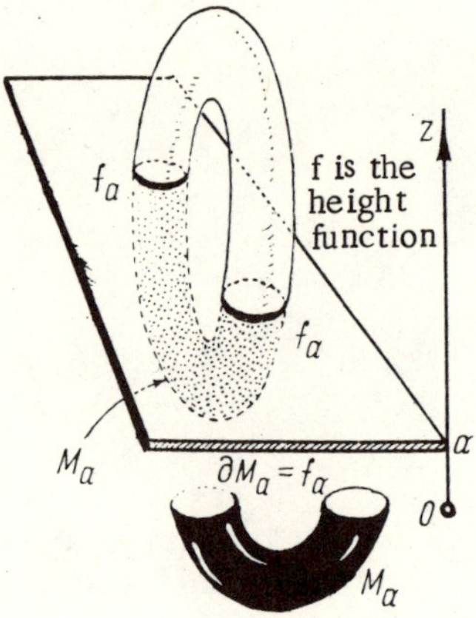

Fig. 63

Let us find the Jacobian of the transformation at 0 (see Fig. 62). Clearly,

$$\left.\frac{\partial z_k}{\partial y_k}\right|_O = q(O) = \sqrt{|P_{kk}(O)|} \neq 0, \text{ i.e., } \det\left(\frac{\partial z}{\partial y}\right) = \frac{\partial z_k}{\partial y_k} \neq 0.$$

By the explicit function theorem, functions $z_1, \ldots, z_n$ may be chosen as local regular coordinates in a sufficiently small neighborhood of 0. Hence, we get

$$f(z) = \sum_{i \leqslant k-1} \pm z_i^2 + P_{kk}\frac{z_k^2}{q^2} - 2P_{kk}\frac{z_k}{q}\sum_{i>k} y_i \frac{P_{ik}}{P_{kk}} +$$
$$+ P_{kk}\left(\sum_{i>k} y_i \frac{P_{ik}}{P_{kk}}\right)^2 + 2\left(\frac{z_k}{q} - \sum_{i>k} y_i \frac{P_{ik}}{P_{kk}}\right)\sum_{i>k} y_i P_{ik} +$$
$$+ \sum_{i,j \geqslant k+1} y_i y_j P_{ij} = \pm z_1^2 \pm \ldots \pm z_k^2 + \sum_{i,j \geqslant k+1} z_i z_j \widetilde{P}_{ij}.$$

We have completed the second step of induction, proving the theorem.

Thus, it is always possible to choose coordinates so that the function is described as a quadratic function reduced to the diagonal form in a neighborhood (not only at the critical point).

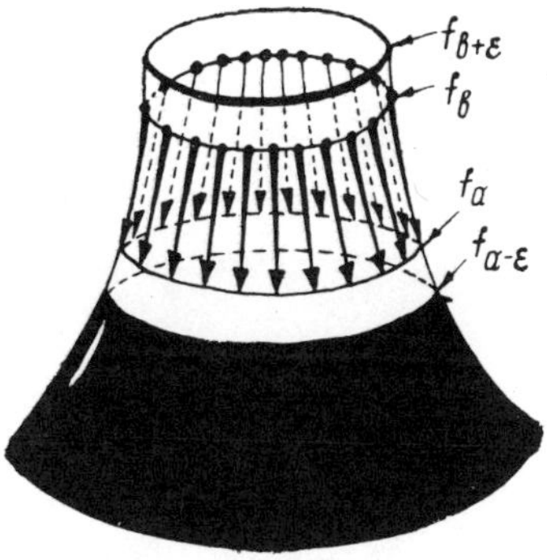

Fig. 64

8.3. The Topological Structure of Level Surfaces of a Function in Neighborhoods of Critical Points

Let f be a smooth function on M. Denote by $f_a = f^{-1}(a)$ the level surface of f corresponding to the value a; set $f_a = \{x \in M, f(x) = a\}$ and $M_a = \{x \in M, f(x) \leqq a\}$, i.e., M_a consists of all points of M, where the value of f does not exceed a. The boundary of M_a is f_a. When a is a regular value of f, i.e., a is not critical, then, by the implicit function theorem, the surface f_a is a smooth submanifold in M of dimension $n - 1$ and M_a is a smooth n-dimensional manifold with boundary f_a (see Fig. 63).

Lemma 8.1. Let f be a smooth function on a compact closed manifold M, and assume that the segment $[a, b]$ (where $a < b$) does not contain critical values of f, i.e., the set $f^{-1}(a, b)$ belonging to M has no critical points of f. Then, manifolds f_a and f_b are diffeomorphic and, in addition, so are manifolds M_a and M_b (these last two manifolds have boundaries).

Proof. Since M is compact, there exists a sufficiently small number $\varepsilon > 0$ such that $[a - \varepsilon, b + \varepsilon]$ also does not contain critical values of f. As is known from a course in geometry and topology, any compact manifold may be embedded into a finite-dimensional Euclidean space; therefore, on M, a Riemannian metric may be defined (e.g., the metric induced by the Euclidean one). Let us fix a positive-definite metric and consider the vector field $v(x) = -\text{grad } f(x)$ on M. On the manifold $f^{-1}[a - \varepsilon, b + \varepsilon]$, this field has no singularities (zeros) and v is orthogonal to the level hypersurfaces

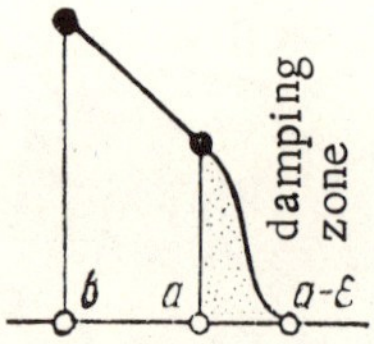

Fig. 65

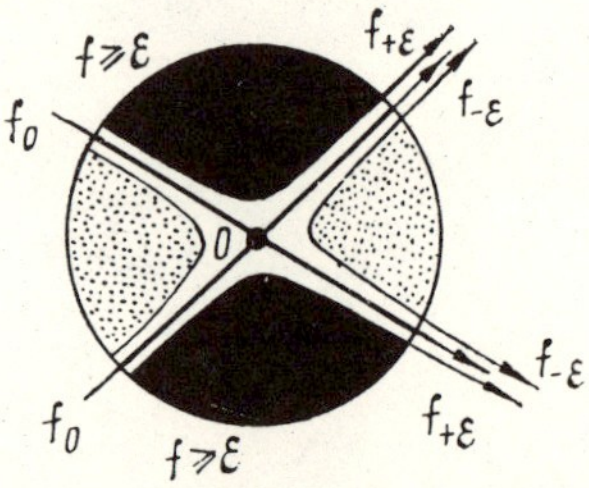

Fig. 66

$f^{-1}(t)$, where $a \leqq t \leqq b$. Consider integral trajectories of v that begin on the submanifold f_b and terminate on f_a (see Fig. 64). Since M is compact, it is possible to define a smooth deformation of the surface f_b along these trajectories into the surface f_a. This implies that f_a and f_b are diffeomorphic, since the inverse deformation is obviously constructed. In the same way, we verify that M_a and M_b are diffeomorphic, since $f^{-1}[a, b]$ is diffeomorphic to $f_a \times I$, where I is a segment. To extend this diffeomorphism, defined on the fiber $a \leqq f \leqq b$, it suffices to retreat by ε "downard" from f_a and to slow the movement of points along integral trajectories on the segment from f_a to $f_{a-\varepsilon}$. The descending fiber f_b will gradually slow its gliding until it stops at the moment $a - \varepsilon$. This break (damping) may be fulfilled with the help of a smooth time transform defined by a smooth graph depicted in Fig. 65. The lemma is proved.

Let us study the level surfaces around critical points of f. Let x_0 be a nondegenerate critical point $f(x_0) = 0$. By Proposition 8.2, in a sufficiently small neighborhood U of x_0, coordinates $x_1, \ldots, x_n$ may be introduced so that $f = -x_1^2 - \ldots -x_\lambda^2 + x_{\lambda+1}^2 + \ldots x_n^2$. We will assume that the center O of U coincides with x_0 and $f(O) = 0$. Consider three

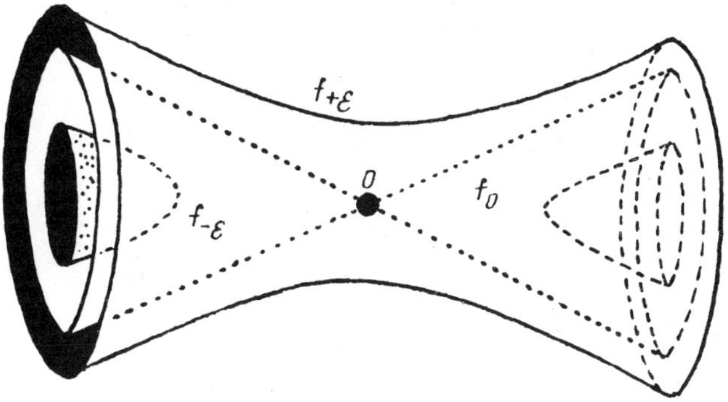

Fig. 67

hypersurfaces $f_{-\varepsilon}$, f_0, $f_{+\varepsilon}$, where $\varepsilon > 0$ is a sufficiently small number. Surfaces are defined by quadratic equations $-x_1^2 - \ldots -x_\lambda^2 + x_{\lambda+1}^2 + x_n^2 = \{-\varepsilon, 0, +\varepsilon\}$ in coordinates on U. Here λ is the index of the critical point. The surface f_0 is a cone with vertex at 0 and the surfaces $f \pm \varepsilon$ are hyperboloids. In Fig. 66, the case $n = 2$, $\lambda = 1$ is depicted and in Fig. 67, the case $n = 3$, $\lambda = 1$ is depicted.

Lemma 8.2. Let there exist only one critical point of index λ in the fiber $f^{-1}[-\varepsilon, +\varepsilon] = M_{+\varepsilon} \setminus M_{-\varepsilon}$. Then $M_{+\varepsilon}$ is homotopically equivalent to the finite CW-complex obtained from $M_{-\varepsilon}$ by attaching a cell σ^λ of dimension λ to the boundary $f_{-\varepsilon} = \partial M_{-\varepsilon}$.

Sometimes one says that $M_{+\varepsilon}$ has the homotopic type of the CW-complex $M_{-\varepsilon} \cup \sigma^\lambda$.

Proof. We have constructed a continuous deformation φ_t: $M_{+\varepsilon} \to M_{+\varepsilon}$, where $\varphi_0 = 1_{M_{+\varepsilon}}$ and $\varphi_1 : M_{+\varepsilon} \to M_{-\varepsilon} \cup \sigma^\lambda$, such that φ_t is identical on $M_{-\varepsilon}$. The definition of homotopically equivalent spaces (see Sec. 1) implies that the existence of such a deformation proves the lemma. Proceeding by induction in values of f, let us begin with critical points of the index $\lambda = 0$ (points of minimum). For minimum points, the statement of the lemma is evident. Consider again the vector field $v = -\operatorname{grad} f$ and construct the deformation connected with integral trajectories of this field. Outside of U, let us make points of the fiber $M_{+\varepsilon} \setminus M_{-\varepsilon}$ glide along trajectories of v (see Fig. 68). In the neighborhood U, the deformation is arranged otherwise and is also depicted in Fig. 68. The seg-

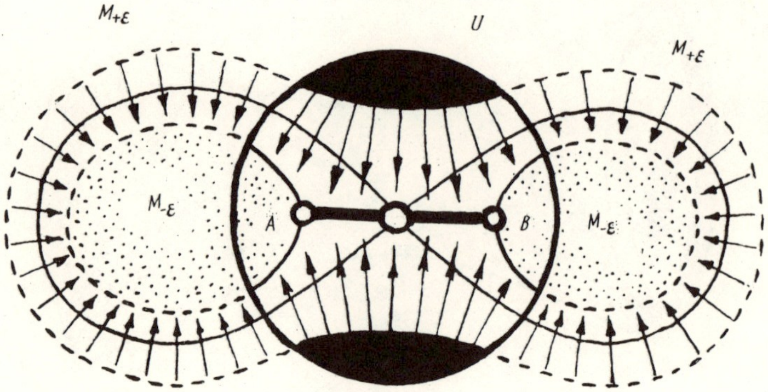

Fig. 68

ment AB describes the disk $D^{\lambda}(x_1, \ldots, x_{\lambda})$ with the boundary (i.e., the sphere $S^{\lambda-1}$) smoothly embedded in the boundary of $M_{-\varepsilon}$. In Fig. 68, the boundary of D^{λ} is depicted as the pair A, B. The result of the deformation is shown in Fig. 68. The lemma is proved.

8.4. A Presentation of a Manifold in the Form of a CW-Complex Connected with the Morse Function

It turns out that the Morse function on M defines the natural presentation of M as a CW-complex. Different Morse functions define, generally speaking, different presentations.

Theorem 8.1. Let f be a Morse function on the smooth compact closed connected manifold M. Then M is homotopically equivalent to a finite CW-complex such that to each critical point of index λ corresponds one cell of dimension λ. In other words, in this CW-complex there are as many cells as there are critical points of the function and the dimension

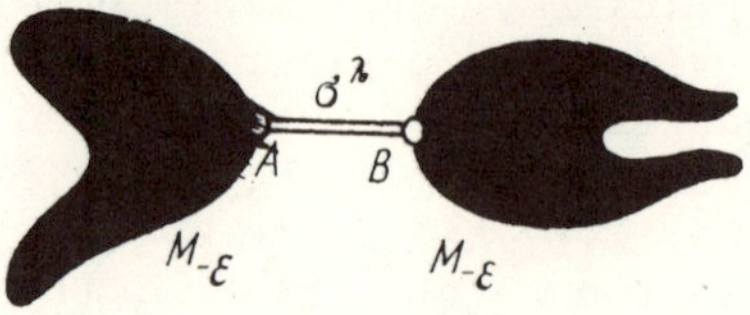

Fig. 69

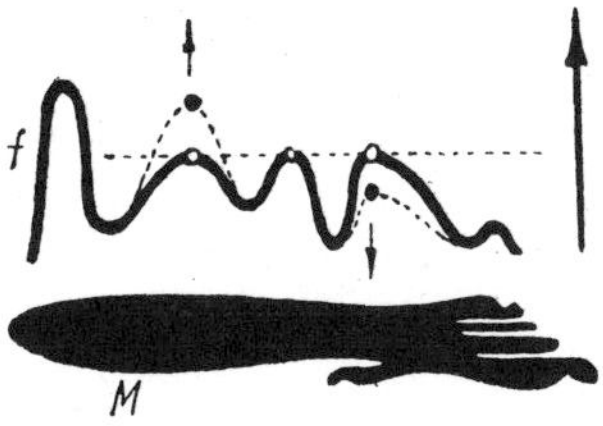

Fig. 70

of each cell equals the index of the corresponding critical point. Since on any compact smooth connected manifold a Morse function exists (Proposition 8.1), any such manifold admits the presentation described in Theorem 8.1.

Proof. Since M is compact, then the number of critical points of f is finite. If on each critical level of the function there is only one critical point, this statement follows immediately from Lemma 8.2. If on one critical level there are several critical points, then, as is clear from the proof of Lemma 8.2, it is possible to assume that several cells are attached to $M - \varepsilon$ simultaneously, since these points are isolated. The other method is that we may consider a small perturbation of the function in the neighborhood of critical points belonging to one critical level (see [2], p. 490). From Proposition 8.1, it is possible to deduce (cf. the proof in Sec. 11) that this perturbation transforms f into a new function $\tilde{f}$ with the same number of critical points and with the same indices, but situated on different critical levels (see Fig. 70). The theorem is proved.

It is important to note that Theorem 8.1 enables one to recover the homotopic type of a manifold from the function quite ambiguously. We may compute only generally the number of cells and their dimensions, but can say nothing about how they are attached to each other, which describes the degree of ambiguity with which the CW-structure of a manifold is restored.

8.5. Attachment of Handles and the Decomposition of a Compact Manifold into the Sum of Handles

We consider the attachment of cell σ^λ to the boundary of $M_{-\varepsilon}$ in more detail. As we have seen, this operation changes

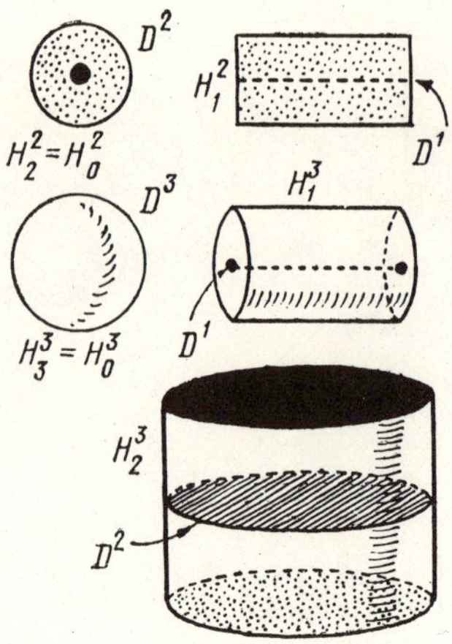

Fig. 71

the homotopic type of $M_{-\varepsilon}$ after we cross the critical level. In the preceding section we purposely roughened the process of reconstruction of level surfaces in order to distinguish the "homotopic part" of this reconstruction. Now we will study the change of $M_{-\varepsilon}$ from a more refined, differential point of view. For this, we will introduce the operation of attachment of handles.

Definition 8.4. The handle of dimension n and index λ is the direct product of two disks $H_\lambda^n = D^\lambda \times D^{n-\lambda}$. The disk D^λ of dimension λ is sometimes called the axis of the handle.

The handle H_λ^n is a smooth manifold with the boundary

$$\partial H_\lambda^n = \partial\,(D^\lambda \times D^{n-\lambda}) = (\partial D^\lambda \times D^{n-\lambda}) \cup (D^\lambda \times \partial D^{n-\lambda}) =$$
$$= (S^{\lambda-1} \times D^{n-\lambda}) \cup (D^\lambda \times S^{n-\lambda-1})$$

(see Fig. 71). Let $KP \subset V^q$ be a smoothly embedded compact manifold in the Riemannian manifold V^q. At each point $x \in K$ consider the normal disk of radius ε and center at x (consisting of segments of length ε orthogonal to V in K). The union of such disks is called the tubular neighborhood $N_\varepsilon K$. Then there exists a sufficiently small $\varepsilon > 0$ such that $N_\varepsilon K$ is a smooth q-dimensional submanifold in V^q with the boundary $\partial N_\varepsilon K$, which is a smooth (q − 1)-dimensional submanifold of V^q. In particular, $\partial N_\varepsilon K$ is stratified onto spheres S^{q-p-1} of radius ε with centers on KP (see [2], p. 498).

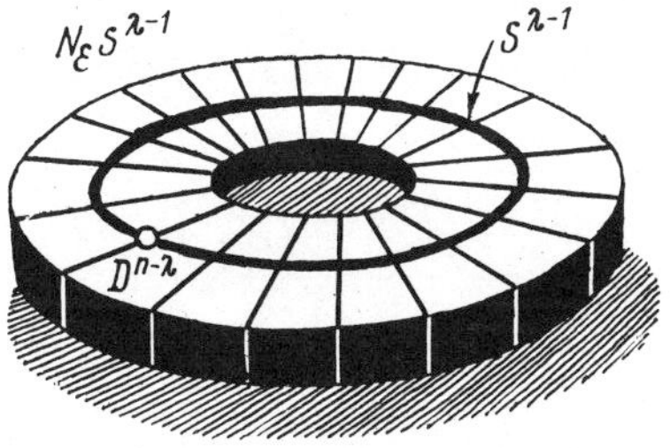

Fig. 72

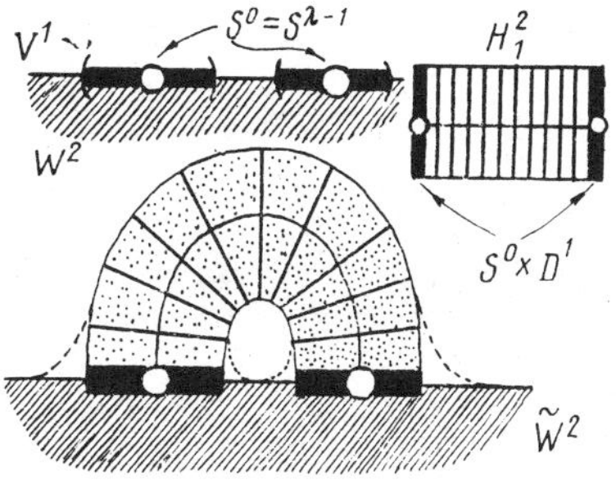

Fig. 73

Define the attachment of the cell H_λ^n to the manifold W^n with the boundary $V^{n-1} = \partial W^n$. Let $S^{\lambda-1} \subset V^{n-1}$ be a smoothly embedded sphere such that its sufficiently small tubular neighborhood $N_\varepsilon S^{\lambda-1}$, of radius $\varepsilon > 0$, is presentable as $S^{\lambda-1} \times D^{n-\lambda}$, where $D^{n-\lambda}$ is the normal disk of dimension $n - \lambda$ and radius ε (see Fig. 72). If such a sphere $S^{\lambda-1}$ belongs to the boundary V^{n-1} of W^n, then it is possible to construct a new manifold $\tilde{W}^n$ with the boundary $\tilde{V}^{n-1} = \partial\tilde{W}^n$ by considering the attachment of W^n to H_λ^n with respect to the mapping $g: S^{\lambda-1} \times D^{n-\lambda} \to N_\varepsilon S^{\lambda-1} \cong S^{\lambda-1} \times D^{n-\lambda}$, which is a diffeomorphism of the tubular neighborhood $N_\varepsilon S^{\lambda-1}$ and the part of the boundary $S^{\lambda-1} \times D^{n-\lambda} \subset \partial H_\lambda^n$. This operation is depicted in Fig. 73 for the handle H_1^2, in Fig. 74 for the handle H_1^3, and in Fig. 75 for the handle H_2^3. Smoothing "angles" that appear at points $x \in \partial N_\varepsilon S^{\lambda-1} = S^{\lambda-1} \times S^{n-\lambda-1}$, we obtain the new smooth manifold W^n, with the smooth boundary $\tilde{V}^{n-1}$. For example, in Fig. 73, this smoothing operation is shown by dots.

It turns out that this quite simple operation of attachment of handles is the elementary procedure whose iteration enables one to construct any compact manifold, starting with a finite number of points, the elementary bricks being handles.

Theorem 8.2. Any smooth compact connected closed manifold M^n is diffeomorphic to the union of a finite number of points $\{H_\lambda^n\}$ corresponding to critical points of a Morse function on M^n. In this situation, to each critical point of index λ there corresponds exactly one handle H_λ^n.

Proof. On M, consider an arbitrary Morse function each level surface of which contains exactly one critical point (see Proposition 8.1). Since Lemma 8.1 implies that M_a is

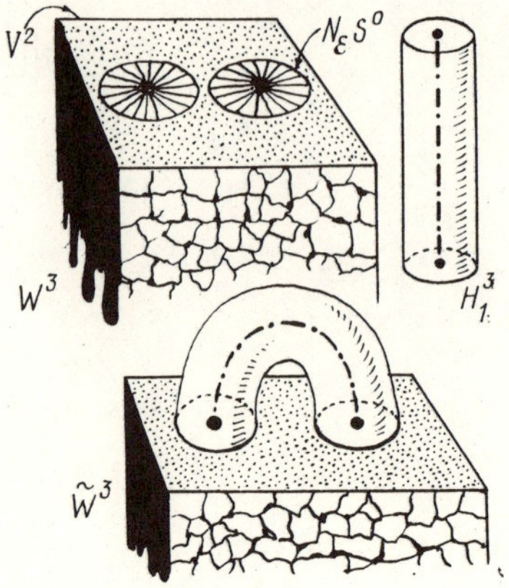

Fig. 74

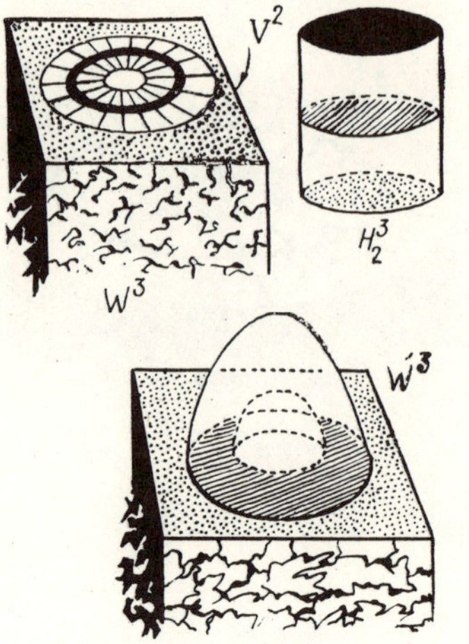

Fig. 75

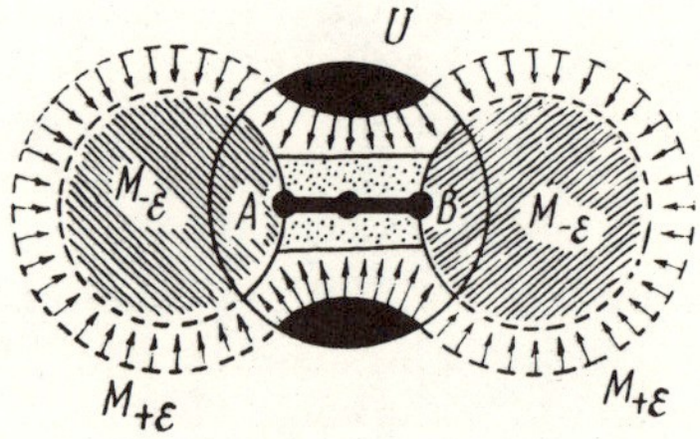

Fig. 76

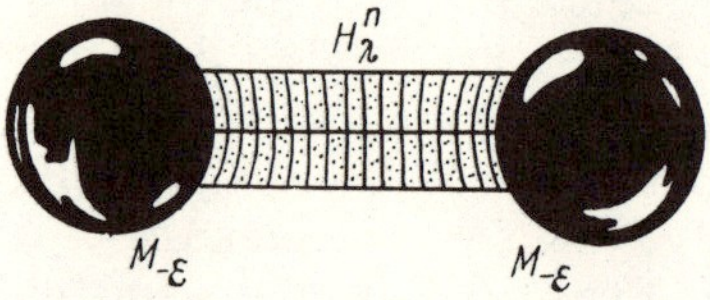

Fig. 77

diffeomorphic to M_b for $a < b$, when $[a, b]$ contains no critical values of the function, it suffices to study the change in $M_{-\varepsilon}$ during the crossing of a critical point of index λ. Let us make use of the already constructed operation of contraction of $M_{+\varepsilon}$ along integral trajectories of v (see proof of Theorem 8.1) but, unlike Sec. 4, we will not complete this contraction but stop a little short. The geometrical procedure is shown in Fig. 76. The result of the deformation is depicted in Fig. 77. Since deformation is a diffeomorphism, then $M_{+\varepsilon} = M_{-\varepsilon} \cup H_\lambda^n$, as required.

Note that the axis of H_λ^n is the λ-dimensional disk D^λ consisting of integral trajectories of the field $v = -\text{grad } f$, starting from the critical point x_0 of v. In a sense, the converse statement is also true, namely, if, on the contrary,

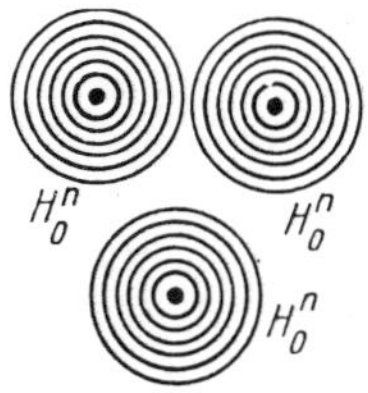

Fig. 78

the decomposition of the compact manifold M as the sum of handles is given, then a Morse function f on M may be constructed such that the associated decomposition of M as the sum of handles coincides with the initial one. The proof is given by induction of the number of handles and the value of their index. In fact, handles $\{H_0^n\}$ may be identified with disks $\{D^n\}$ whose centers are minimum points of the constructed function f (i.e., points of index zero); see Fig. 78. The function f is constructed by producing its smooth level surfaces in M (evidently, f will not be uniquely defined). Thus, as hypersurfaces f_a in disks $D^n = H_0^n$, we will take concentric spheres with centers at local minima of f. Now, suppose that the desired function is constructed on the smooth manifold defined by $f \leq a$ with the boundary $V^{n-1} = \{f = a\}$, and the next in number, handle H_λ^n, is attached to the boundary V^{n-1}. It is necessary to extend the function onto this handle. This extension is shown in Fig. 79 by producing level hypersurfaces. For this, it follows to slightly modify the ini-

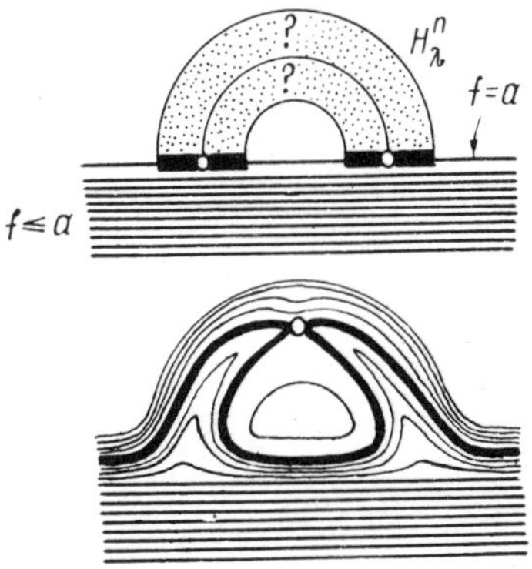

Fig. 79

tial function in a neighborhood of the boundary V^{n-1} and to extend the obtained protrusions of level surfaces inside of the handle. In this process, one critical point of index λ appears in the handle. This completes the proof, since the obtained function is again constant on the boundary of the new manifold, which enables us to continue the process onto the following handles.

9. POINTS OF BIFURCATION AND THEIR CONNECTION WITH HOMOLOGY

9.1. Definition of Points of Bifurcation

The fact that the definition of a Morse function on M enables one to present M in the form of a CW-complex whose cells correspond to critical points of the function indicates to us the possibility of a connection between the number of critical points, their indices, and the homology characteristics of the manifold, since the latter are defined by the decomposition of M into the union of cells. Such a connection actually exists, not only for smooth functions on manifolds but for continuous functions on arbitrary CW-complexes. In this section, we will consider continuous functions on the finite CW-complex X. We will also assume (for simplicity) that X is a space with a metric.

Definition 9.1. The point $x \in X$ is called regular for the function f on X if there is an open neighborhood U on this point such that U is homeomorphic to $f^{-1}(a) \times I$ [where I is the unit segment and $a = f(x)$], so that the following diagram commutes:

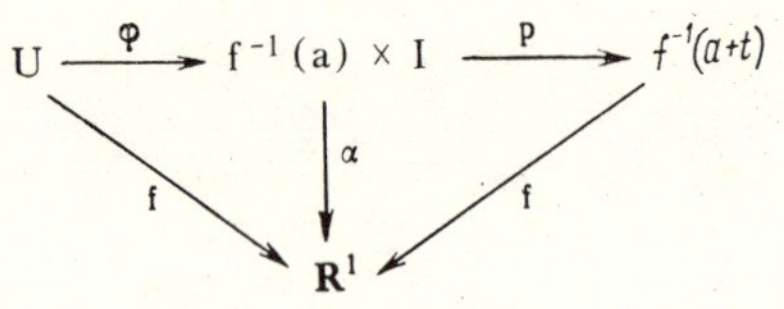

In this diagram, φ is a homeomorphism, p is a projection onto the multiple, and the mapping $f:f^{-1}(a) \to \mathbf{R}^1$ transforms $f^{-1}(a)$ into one point a; the mapping α whose existence is postulated must be continuous and $f = \alpha\varphi$.

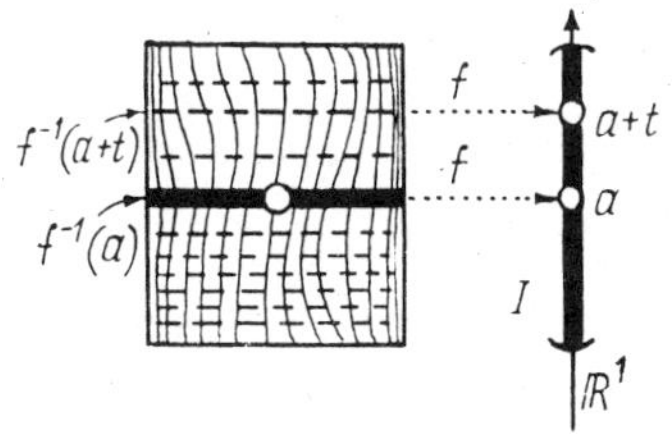

Fig. 80

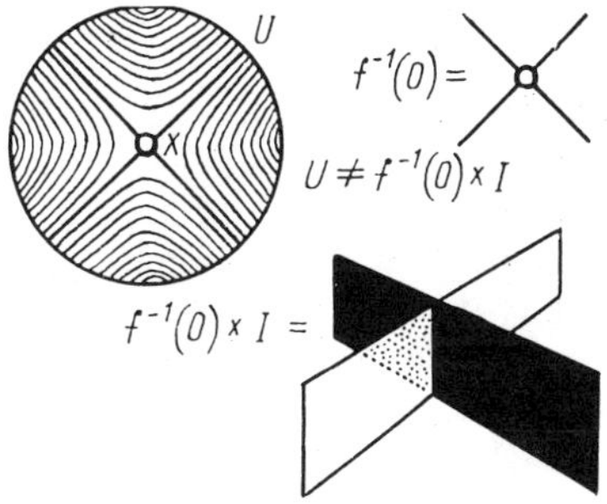

Fig. 81

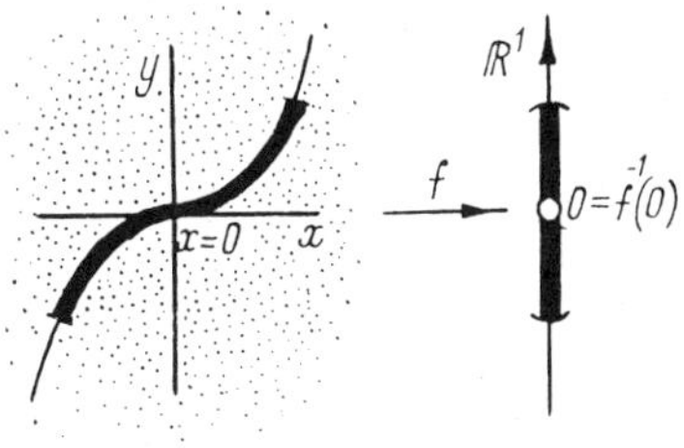

Fig. 82

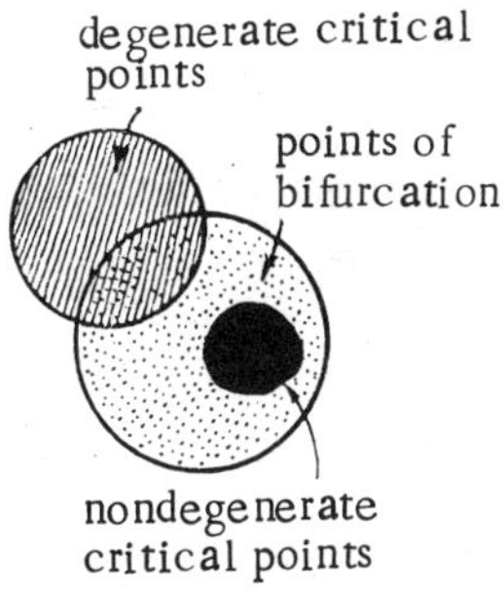

Fig. 83

The commutativity of the diagram means actually that φ is "fiberwise," i.e., surfaces $(f^{-1}(a), t)$ (in the presentation of U in the form of the direct product) must coincide with surfaces $f^{-1}(a + t)$ (see Fig. 80).

Definition 9.2. The point $x \in X$ is called the point of bifurcation of the function f if x is not regular for this function.

Let us consider the simplest examples. If $X = M$ is a smooth manifold and f is a smooth function on M, then any point that is regular for f, in the sense of the theory of smooth mappings (see [1]), is regular in the sense of Definition 9.1. If $x \in M$ is a nondegenerate critical point for the smooth function f, then evidently x is a point of bifurcation (see Fig. 81). At the same time, if $x \in M$ is a degenerate critical point of f, then it must not be a point of bifurcation. In fact, consider $M = \mathbf{R}^{-1}$ and take $f(x) = x^3$. Then

$x = 0$ is a degenerate critical point, but it is a regular point in the sense of Definition 9.1 (see Fig. 82).

Since all subsequent applications are related to functions on manifolds, we will assume for simplicity that $X = M$ is a smooth manifold and f is a smooth function on M. In Fig. 83, the relationship between degenerate critical points, points of bifurcation, and nondegenerate critical points is conditionally described. Suppose that f has on M only a finite number of points of bifurcation (in particular, all of them are isolated). Consider the homology groups $H_*(M; A)$, where A is a field of coefficients (i.e., $\mathbf{R}$ or Z_p). In what follows, the reference to A will be omitted and we will write just $H_*(M)$. Assume $\beta_k = \dim H_k(M)$ for the dimension of $H_k(M)$ (over the field), sometimes called the k-th Betti number.

Definition 9.3. The Poincaré polynomial of M is $P(M, t) = \sum_{k=0}^{n} \beta_k(M) t^k$, where $n = \dim M$.

Definition 9.4. The value c is bifurcational for f if the level surface $f^{-1}(c)$ contains at least one point of bifurcation.

If f is a Morse function, then bifurcational and critical values coincide. Since there is only a finite number of bifurcational points (see above), all of them are isolated. Let $\{x_\alpha\}$ be a set of bifurcational points on the level surface $f^{-1}(c_\alpha)$. Consider $M_{c_\alpha} = \{x \in M: f(x) \leq c_\alpha\}$. The relative homology groups $H_k(M_{c_\alpha}, M_{c_\alpha} \setminus \{x\}_\alpha)$ are, as we will now establish, important invariants of points of bifurcation of f. Here $H_k(M_{c_\alpha}, M_{c_\alpha} \setminus \{x\}_\alpha)$ is understood (since $\{x\}_\alpha$ are discrete) as the group $H_k(M_{c_\alpha}, M_{c_\alpha} \setminus U\{x\}_\alpha)$, where $U\{x\}_\alpha$ is the set of sufficiently small open neighborhoods of points $\{x\}_\alpha$ (see Fig. 84). Consider integer numbers $\beta_k(M_{c_\alpha}, M_{c_\alpha} \setminus \{x\}_\alpha)$, where $\beta_k(X, Y) = \dim H_k(X, Y)$ (over the field).

Definition 9.5. The Poincaré polynomial of f on M is

$$Q(M, f, t) = \sum_{\alpha=1}^{N} \sum_{\{x\}_\alpha} \sum_{k=0}^{n} \beta_k(M_{c_\alpha}, M_{c_\alpha} \setminus \{x\}_\alpha) t^k.$$

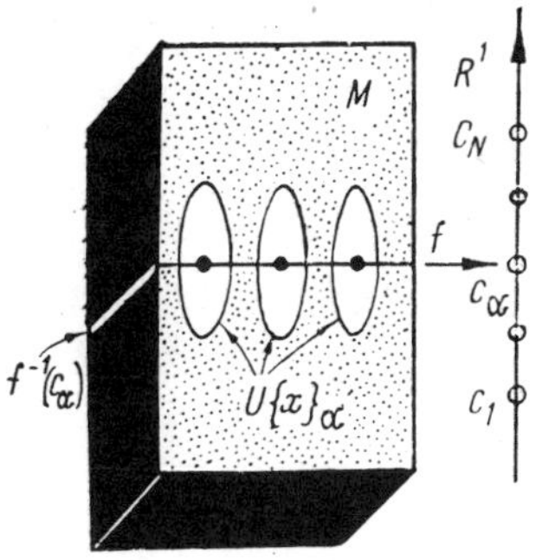

Fig. 84

9.2. A Theorem That Connects the Poincaré Polynomials of the Function and That of the Manifold

Theorem 9.1. Let Q(M, f, t) and P(M, t) be Poincaré polynomials of the function and the manifold. Then the difference Q − P is divisible by 1 + t and the ratio (Q − P)/(1 + + t) is a polynomial with negative integer coefficients.

Proof. Step 1. Consider numbers $a < b$ in the domain of values of f on M such that on the segment $[a, b]$ there are no bifurcational values of f. Then M_b is contracted onto M_a and $H_*(M_b, M_a) = 0$. In fact, if f is a Morse function, then the statement has been proved in Sec. 8. If f is not a Morse function, we must apply Definition 9.1 of regular points. Since all points of the "fiber" $f^{-1}[a, b]$ are regular, then $f^{-1}[a, b]$ may be covered by a finite number of open neighborhoods U, with the foliation by trajectories—pre-images of the segment with respect to φ in each of U. These trajectories may successfully play the role of integral trajectories of the field v = −grad f used in Sec. 8 for the construction of

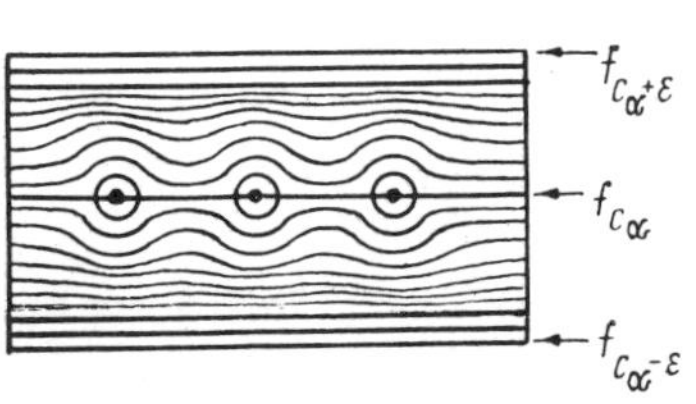

Fig. 85

contracting deformation. Definition 9.1 implies that these trajectories from different neighborhoods may be made consistent on intersections.

Step 2. For a sufficiently small $\varepsilon > 0$, there are isomorphisms $H_k(M_{c_\alpha}, M_{c_\alpha} \setminus \{x\}_\alpha) = H_k(M_{c_\alpha+\varepsilon}, M_{c_\alpha-\varepsilon})$. This follows from the preceding step and the definition of $H_k(M_{c_\alpha}, M_{c_\alpha} \setminus \{x\}_\alpha)$ (see Fig. 85).

Step 3. Consider the following three types of the Poincaré polynomials:

a) $P(M_a) = \sum_k \beta_k(M_a) t^k$,

b) $P(M_b, M_a) = \sum_k \beta_k(M_b, M_a) t^k$,

where $a < b$, i.e., $M_a \subset M_b$,

c) $P(\operatorname{Im} \partial) = \sum_k (\dim \operatorname{Im} \partial_{k+1}) t^k$,

where $\partial_{k+1}: H_{k+1}(M_b, M_a) \to H_k(M_a)$ is the boundary operator in the exact homological sequence of the pair and $\operatorname{Im} \partial_{k+1} \subset H_k(M_a)$.

Step 4. We claim that

$$P(M_b, M_a) - (P(M_b) - P(M_a)) = (1+t) P(\operatorname{Im} \partial).$$

To prove this, we will consider the exact homological sequence of the pair, namely,

$$\cdots \to H_{k+1}(M_b, M_a) \xrightarrow{\partial_{k+1}} H_k(M_a) \xrightarrow{i} H_k(M_b) \xrightarrow{j}$$
$$\to H_k(M_b, M_a) \xrightarrow{\partial_k} H_{k-1}(M_a) \to \cdots .$$

Since the sequence is exact, we get the following system of relations:

$$\beta_k(M_b, M_a) = \dim \operatorname{Im} j + \dim \operatorname{Im} \partial_k;$$
$$\dim \operatorname{Im} j = \beta_k(M_b) - \dim \operatorname{Im} i =$$

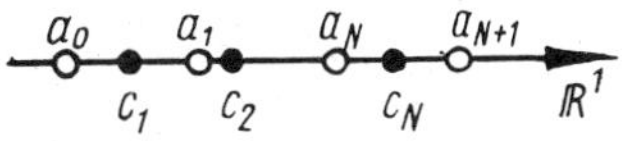

Fig. 86

$$=\beta_k(M_b)-(\beta_k(M_a)-\dim \operatorname{Im}\partial_{k+1})=$$
$$=\beta_k(M_b)-\beta_k(M_a)+\dim \operatorname{Im}\partial_{k+1};$$
$$\beta_k(M_b, M_a)-\dim \operatorname{Im} j=\beta_k(M_b, M_a)-(\beta_k(M_b)-\beta_k(M_a))-$$
$$-\dim \operatorname{Im}\partial_{k+1}=R_k-\dim \operatorname{Im}\partial_{k+1}=\dim \operatorname{Im}\partial_k;$$
$$R_k=\beta_k(M_b, M_a)-(\beta_k(M_b)-\beta_k(M_a)).$$

Therefore, $t^k R_k = t^k \dim \operatorname{Im} \partial_{k+1} + t(t^{k+1}\cdot \dim \operatorname{Im} \partial_k)$, i.e., $\sum_k t^k R_k = (1+t)P(\operatorname{Im}\partial)$, as required.

<u>Step 5</u>. Consider all bifurcational values $c_1, \ldots, c_N$ of f and numbers $a_0, a_1, \ldots, a_N, a_{N+1}$ such that $a_0 < c_1$; $a_i < c_i < a_{i+1}$; $c_N < a_{N+1}$, i.e., bifurcational values $\{c_i\}$ are separated by values $\{a_i\}$ which are no longer bifurcational (see Fig. 86). From Step 4 we may write for each i the following relation:

$$P(M_{a_{i+1}}, M_{a_i})-(P(M_{a_{i+1}})-P(M_{a_i}))=(1+t)P(\operatorname{Im}\partial)_i,$$

where $P(\operatorname{Im} \partial)_i$ is the polynomial with nonnegative integer coefficients. Summing these equalities with respect to all i from 0 to N + 1, we get

$$\sum_i P(M_{a_{i+1}}, M_{a_i})-P(M_{a_{N+1}})+P(M_{a_0})=(1+t)K(t),$$

where K(t) is a polynomial with nonnegative integer coefficients. Recall that

$$P(M_{a_{i+1}}, M_{a_i})=P(M_{c_i}, M_{c_i}\setminus\{x\}_i)$$

by Step 2. It remains to note that $P(M_{a_{N+1}}) = P(M)$, since a_{N+1} may be considered so large that $a_{N+1} > \max_{x\in M} f(x)$ and, therefore, $M_{a_{N+1}} \equiv M$. Furthermore, $P(M_{a_0}) = 0$, since a_0 may be chosen so that $a_0 < \min_{x \in M} f(x)$, i.e., $M_{a_0} = \phi$, and in the definition of the Poincaré polynomial the summation with respect to k began from k = 0. Hence, the equality obtained above is rewritten as $Q(M, f) - P(M) = (1 + t)K(t)$, which completes the proof of the theorem.

9.3. Several Corollaries

The proven theorem has several nice corollaries which will be partially enumerated; in particular, we will derive classical Morse inequalities. Let **R** be the field of coefficients and f be a smooth function on M; then $P(M, t) = \sum_k \beta_k t^k$. Let us express the polynomial $Q(M, f, t)$ in the "reduced" form, i.e., collect together terms with the same degree of t. We get $Q = \sum_k \mu_k t^k$. Numbers μ_k are called Morse numbers of the smooth function f on M. These numbers have a descriptive interpretation when f is a Morse function on a manifold.

<u>Proposition 9.1</u>. If f is a Morse function on M, then μ_k is equal to the number of critical points of index k.

<u>Proof</u>. Let x_0 be the unique critical point of index k on the surface $f^{-1}(c) = f_c$. Let us compute

Fig. 87

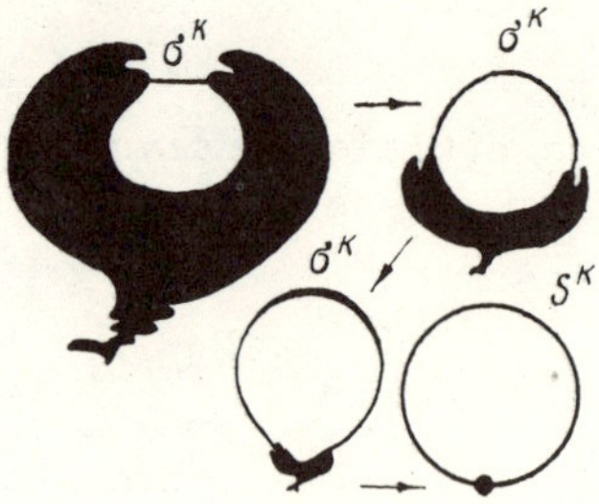

Fig. 88

$$\sum_i \dim H_i (M_c, M_c \setminus x_0) t^i.$$

The proof of Theorem 9.1 implies that $H_i(M_c, M_c \setminus x_0) = H_i(M_{c+\varepsilon}, M_{c-\varepsilon})$. But for a Morse function the topological structure of $M_{c+\varepsilon}$ and its relationship with the set $M_{c-\varepsilon}$ is well known. In fact, $M_{c+\varepsilon}$ is obtained from $M_{c-\varepsilon}$ by attachment of the handle H_k^n of index k, and from the homotopic point of view, $M_{c+\varepsilon}$ is $M_{c-\varepsilon} \cup \sigma^k$, where σ^k is the cell of dimension k (see Fig. 87). Therefore,

$$H_i(M_c, M_c \setminus x_0) = H_i(M_{c+\varepsilon}, M_{c-\varepsilon}) = H_i(M_{c-\varepsilon} \cup \sigma^k, M_{c-\varepsilon}) = \\ = H_i(M_{c-\varepsilon} \cup \sigma^k / M_{c-\varepsilon}) = H_i(\bar{\sigma}^k / \partial \sigma^k) = H_i(S^k)$$

for $i > 0$ (see Fig. 88). Hence, $\sum_i \dim H_i (M_c, M_c \setminus x_0) t^i = \sum_i \dim H_i (S^k) t^i = t^k$. Here we have made use of the fact that $H_0(X, Y) = 0$. Therefore, in the coefficient of t^k in Q, each critical point of index k contributes a unit; hence the coefficient μ_k of t^k indicates the number of such points. The statement is proved.

Proposition 9.2. For the Morse function f on M, the inequality $\mu_k \geqq \beta_k$ holds for any k.

In other words, Betti numbers β_k of M are estimated from below Morse numbers μ_k.

Proof. Theorem 9.1 and Proposition 9.1 imply

$$Q(M, f) - P(M) = \sum_k (\mu_k - \beta_k) t^k = (1 + t) K(t),$$

where the polynomial on the right-hand side has nonnegative coefficients. The proposition is proved.

The topological meaning of this statement is clear: there are no more closed cycles in dimension k than the total number of cells of dimension k. Proposition 9.2 follows from the fact that to each cell corresponds one critical point of a Morse function.

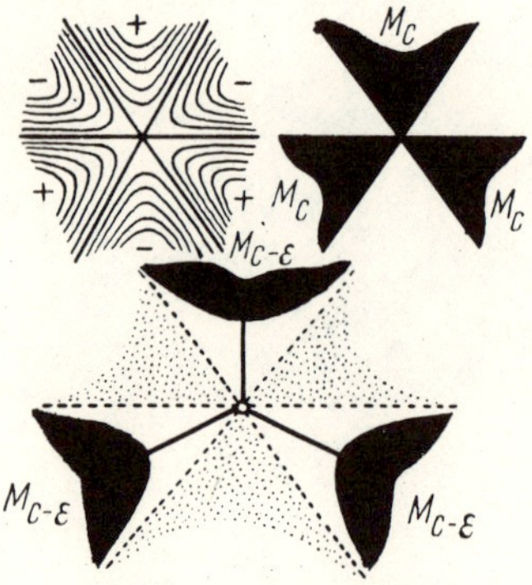

Fig. 89

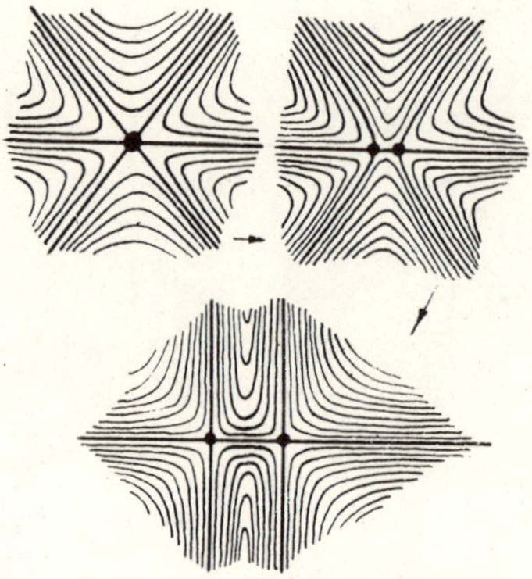

Fig. 90

Consider the following equality that follows from Theorem 9.1: $\sum_k \mu_k t^k = \sum_k \beta_k t^k + (1 + t)\, K\,(t)$. For t = −1, we get

$$\sum_k (-1)^k \mu_k = \sum_k (-1)^k \beta_k,$$

where, on the right-hand side, the value called the Euler characteristic of a manifold (alternated sum of Betti numbers) appears. It is possible to prove that the Euler characteristic is a homotopic invariant of M. It follows that the alternated sum of Morse numbers for an arbitrary smooth function (with isolated singularities) does not depend on the function and is an invariant of a manifold. Further, we expand $(1 + t)^{-1}$ in a power series with respect to t. We get $\sum_{i=0}^{\infty} (-1)^i t^i$. This implies that the series $\left(\sum_k (\mu_k - \beta_k)\, t^k\right) \times \sum_{i=0}^{\infty} (-1)^i t^i$ has, after reduction, nonnegative coefficients of t^k. Fixing some k, we get from here the system of inequalities $(\mu_0 - \beta_0)(-1)^k + (\mu_1 - \beta_1)(-1)^{k-1} + (\mu_2 - \beta_2)(-1)^{k-2} + \ldots + (\mu_k - \beta_k) \geqq 0$; hence $\mu_k - \mu_{k-1} + \mu_{k-2} - \ldots \pm\mu_0 \geqq \beta_k - \beta_{k-1} + \beta_{k-2} - \ldots \pm\beta_0$.

It is useful to know what are polynomials Q(M, f) when bifurcational points are not nondegenerate critical points. Let, for example, x_0 be a degenerate point for f. As a special case, consider $f(x, y) = \mathrm{Re}\, z^n$, where $z = x + iy$. In Fig. 89, level surfaces f_c are depicted. Clearly, $M_{c+\varepsilon}/M_{c-\varepsilon} = S^1 \vee \ldots \vee S^1$ ($n - 1$ times). This implies that at x_0 the polynomial Q is of the form $t + \ldots + t = (n - 1)t$. We see that this polynomial differs from those that appear for nondegenerate singularities. The coefficient $n - 1$ of t is explained in terms of the following geometrical fact. It is known that degenerate critical points may be transformed by small perturbations into the union of nondegenerate critical points (disintegration of degenerate singularities). In our example, a small perturbation of $\mathrm{Re}\,(z^n)$ may be chosen as $\mathrm{Re}\,(z - a_1) \ldots (z - a_n)$, where $a_i \neq a_j$ for $i \neq j$ and $a_i \in \mathbf{R}$. Figure 90 reflects the qualitative picture of perturbation of level surfaces of a function. Clearly, $M_{c+\varepsilon}/M_{c-\varepsilon} = S^1 \vee \vee \ldots \vee S^1$ ($n - 1$ times) and the degenerate singularity of order n splits into the union of $n - 1$ nondegenerate (quadratic) singularities. Since the value of Q at a nondegenerate critical point of index 1 equals t, then $\mu_1 t = (n - 1)t = t + \ldots + t$ ($n - 1$ times). This number shows us how many nondegenerate singularities will appear at the disintegration of one degenerate singular point of the function $\mathrm{Re}\,(z^n)$. In fact, we have just encountered a general property of the polynomial $Q(M, f)$, namely, it does not vary under a sufficiently small perturbation of f. The reason is that this polynomial is expressed in terms of groups of relative homology $H_*(M_{c+\varepsilon}, M_{c-\varepsilon})$, which are evidently independent of small perturbations of f, since a nonsingular level surface remains nonsingular if the perturbation is sufficiently small. Hence, the total polynomial $Q(M, f)$ shows how many nondegenerate critical points of each index appear at the disintegration of degenerate singularities of the function f under a sufficiently small perturbation. Note that Q must not be a homogeneous polynomial at a degenerate critical point as it was in the preceding example. In fact, let $f(x, y, z) = x^3 - 3x(y^2 + z^2)$. Let us compute $H_*(M_{c+\varepsilon}, M_{c-\varepsilon})$. The topological scheme of level surfaces is depicted in Fig. 91. Clearly, $M_{c+\varepsilon}/M_{c-\varepsilon} \approx S^1 \vee S^2$, i.e., Q is $t + t^2$ at x_0.

9.4. Critical Points of Functions on Two-Dimensional Manifolds

Let f be a smooth function on a two-dimensional smooth closed connected orientable manifold. Since each such manifold M_g^2 (a sphere with g handles) may be embedded into $\mathbf{R}^3$, then, among Morse functions, we distinguish the class of

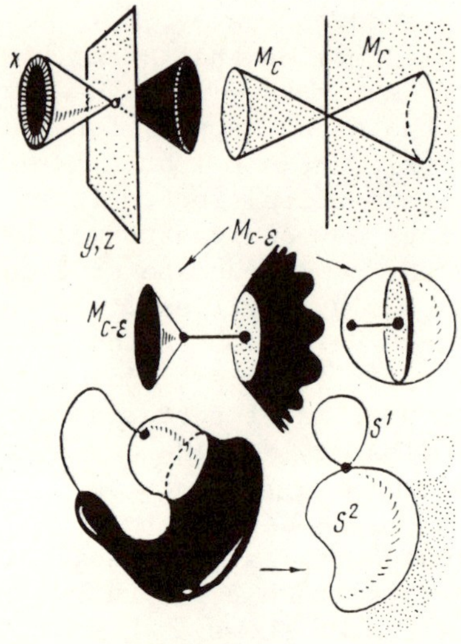

Fig. 91

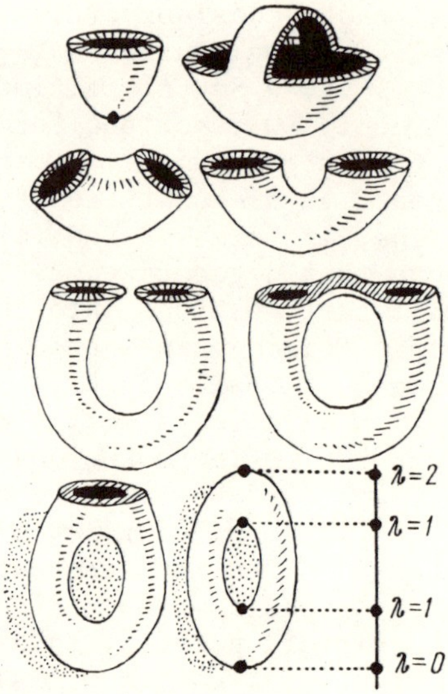

Fig. 92

"height functions" defined as projections of $M_g{}^2$ onto a fixed line in $\mathbf{R}^3$. In [2], pp. 493-494, it is shown that among these "height functions" it is always possible to distinguish a Morse function. In Fig. 92, the process restoring $T^2 = M_1^2 = g$ is depicted. T^2 is embedded naturally into $\mathbf{R}^3$ so that $z = f(P)$, for $P \in T^2$ (the height function), is a Morse function with four nondegenerate critical points: x_1 (minimum), x_2, x_3 (saddle points of index 1), x_4 (maximum). For $g > 1$, a similar height function on $M_g{}^2$ has $2g + 2$ nondegenerate critical points: x_1 (minimum), $x_2, \ldots, x_{2g+1}$ (saddle points of index 1), x_{2g+2} (maximum); see Fig. 93.

However, as we have seen, the relationship between points of bifurcation of a function and the homology of its domain of definition (of a manifold) takes place even when singularities are nondegenerate. The nondegenerate singularities may collapse, forming one degenerate singular point. Hence, the number of degenerate singularities, for example on $M_g{}^2$, may be less than $2g + 2$. Then, evidently, the structure of degenerate points will be more complicated than that of nondegenerate ones (see above). On any $M_g{}^2$, a smooth height function f in $\mathbf{R}^3$ may be constructed with four critical points: minimum, maximum, and two saddle points which will be degenerate for $g > 1$. The desired embedding $M_g{}^2 \to \mathbf{R}^3$ is

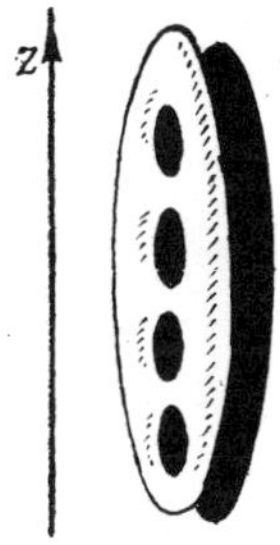

Fig. 93

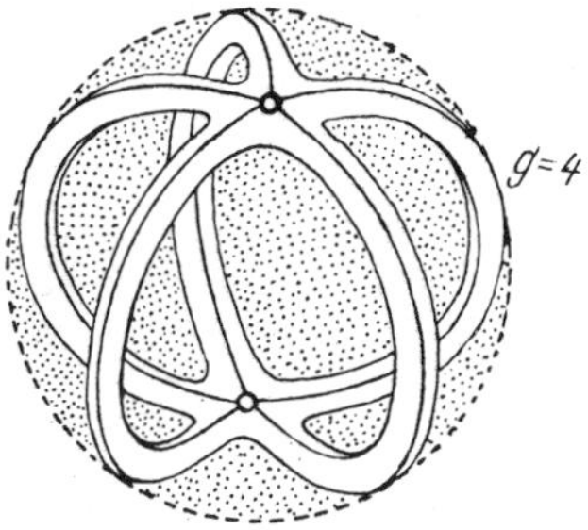

Fig. 94

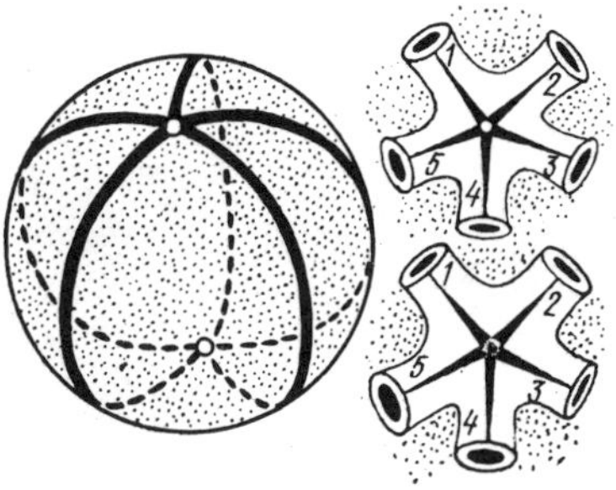

Fig. 95

depicted in Fig. 94. For this, it suffices to glue two constructions from the picture, reproducing the tubular neighborhood of g + 1 meridians that connect northern and southern poles of the sphere (see Fig. 95). Two saddle points x_2, x_3 are degenerate for $g > 1$ and the height function in a neighborhood of these points is locally Re (z^{g+1}) (see Fig. 96). It is possible to show that the minimum number of critical points of the height function on M_g^2 (where $g > 1$) equals 4 (and we have produced such a height function). However, if we neglect the requirement that f is a height function, then the number of critical points may be lessened by 1. Namely, on any two-dimensional smooth compact connected closed manifold (orientable M_g^2 for $g > 1$ or nonorientable M_μ^2), there is also a smooth function with only three critical points: that of the minimum, the maximum, and a degenerate saddle point. We will produce this function f, defining the system of level surfaces of f on the manifold. For this, we will use the classification theorem of two-dimensional surfaces

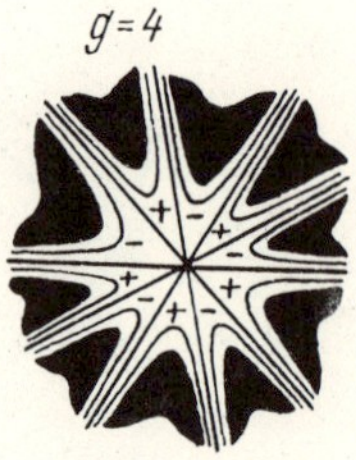

Fig. 96

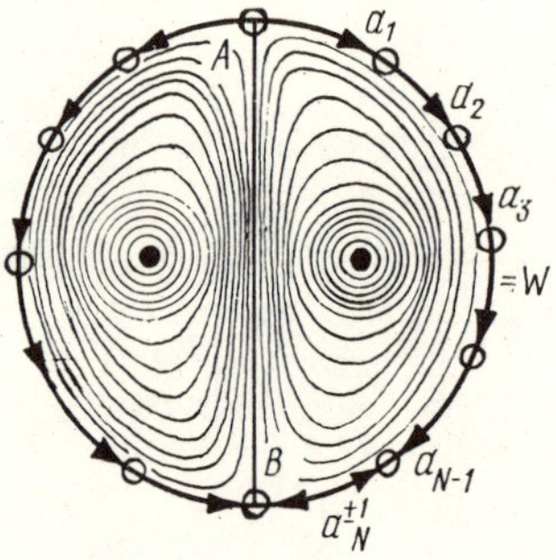

Fig. 97

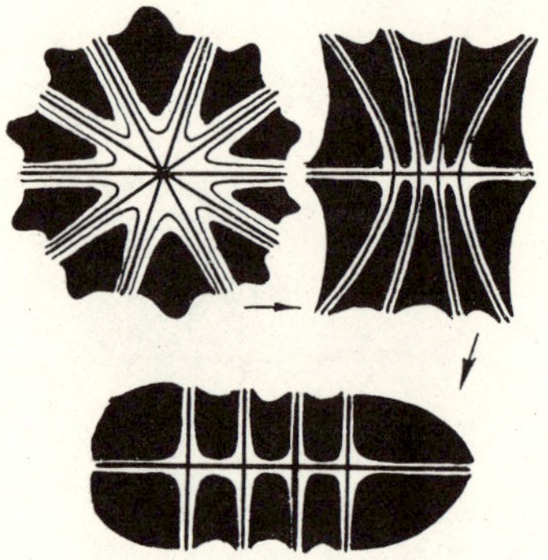

Fig. 98

(see [1]), and present M_g^2 (or M_μ^2 in the nonorientable case, where μ is the number of Moebius bands attached to the sphere) in the form of the glueing of the fundamental polyhedron W, for which we will take the symmetric canonical form, namely, $W = a_1 \ldots a_N a_1^{-1} \ldots a_{N-1} a_N^{\pm 1}$ (see [1], p. 268), where "+1" corresponds to the nonorientable case and "−1" to the orientable case (see Fig. 97). Dividing this polyhedron by halves defines f by its level lines (this definition is ambiguous). To the left of the segment AB (which divides W into halves), we will place the point of the maximum; and to the right of AB, that of the minimum. Then the third critical point — the degenerate saddle point — is at the vertices of the fundamental polyhedron (recall that all these vertices are identified at any point of the surface) (see Fig. 97). Clearly, the constructed function f is of the form Re (z^α) in a small neighborhood of this degenerate saddle point (determine α as a function of g or μ). It is clear where we have made use of the symmetric form of W: dividing W into halves by AB, we have guaranteed that at each inner point of either side a_i the function f has a zero gradient, i.e., either it continues to grow, passing a_i, or continues to decay. If there had been a pair of faces numbered by the same letter from one side of AB, then this face would be entirely filled in by degenerate singularities of the function. Under the small perturbation of the constructed smooth function, the unique degenerate saddle point splits into the union of nondegenerate saddle points, as depicted in Fig. 98.

Problem. Show that the function constructed above cannot be realized as a height function for a smooth embedding of a two-dimensional surface in $\mathbf{R}^3$.

On the other hand, the number of points of bifurcation of a smooth function f on $M_{g>1}^2$ (or on $M_{\mu>0}^2$) cannot be less than 3, i.e., in the two-dimensional case, we have produced a function with the minimum possible number of singularities (equal to 3). In fact, the following statement holds.

Proposition 9.3. Let there be a smooth function f with only two critical points (perhaps degenerate) on a compact connected smooth closed manifold M^n. Then the manifold M is homeomorphic to the sphere S^n.

Proof. We will prove this statement when both points are nondegenerate. Consideration of the degenerate case is more difficult and will be omitted here. Thus, thanks to the compactness of M, one of these points, say x_0, is the minimum and the other, x_1, is the maximum of the function. Let $f(x_0) = 0$, $f(x_1) = 1$. By Proposition 8.2 and Theorem 8.2, for a sufficiently small $\varepsilon > 0$, sets $M_{+\varepsilon}$ and $M \setminus M_{1-\varepsilon} = f^{-1}[1 - \varepsilon, 1]$ are diffeomorphic to disks of dimension n. Consider the vector field $-\text{grad } f$. Applying Lemma 8.1, we get that M^n is homeomorphic to the manifold glued from two disks by their common boundary, i.e., to the sphere. The statement is proved.

In particular, for $n = 2$, we immediately obtain that there are no smooth functions with two singularities on $M_{g>1}^2$ and $M_{\mu>0}^2$, since these manifolds are not homeomorphic to a sphere (see the classification theorem and the computation of homology groups of these surfaces).

Note that, when $n \geqq 7$, the manifold M^n that possesses a function with two singularities (assume even though nondegenerate) must not be diffeomorphic to the standard sphere

S^n, though M^n is homeomorphic to S^n. This is related to the fact that there are, for example, seven-dimensional smooth closed manifolds homeomorphic but not diffeomorphic to the standard sphere S^n. At the same time, there are Morse functions with two nondegenerate critical points on such manifolds. For $n \leqq 6$ and $n \neq 4$, there are no such intuitively inexplicable geometrical phenomena, i.e., the presence of a homeomorphism implies a diffeomorphism. One must not think that these strange "many-dimensional effects" are related to some pathology of manifolds. For example, in dimension 7 there are 28 smooth manifolds which are all homeomorphic to the standard sphere S^7, but are mutually nondiffeomorphic; and these manifolds may be described by simple polynomial equations in the complex space $\mathbf{C}^5(z_1, z_2, z_3, z_4, z_5)$. For this, it suffices to consider the sphere S^9 defined in $\mathbf{C}^5 = \mathbf{R}^{10}$ by the equation $|z_1|^2 + \ldots + |z_5|^2 = 1$, and an eight-dimensional algebraic surface, defined in $\mathbf{C}^5$ by one complex equation $z_1{}^{6k-1} + z_2^3 + z_3^2 + z_4^2 + z_5^2 = 0$, where the integer k may take on any of the following values $k = 1, 2, \ldots, 28$. The intersection of these surfaces gives a seven-dimensional manifold. Variation of k gives 28 seven-dimensional manifolds which are homeomorphic to S^7 but are mutually nondiffeomorphic. The reader interested in the details may refer, for example, to [5]. It is also possible to show that the above polynomial equations that define these "nonstandard" spheres are in a sense simplest ones; in particular, the Riemannian matrices on the manifolds induced by the enveloping Euclidean metric on $\mathbf{C}^5$ are matrices with the maximum number of possible (for these manifolds) isometric groups. In other words, the above embeddings are the most "symmetric" — have the maximum number of motion groups. One of the above manifolds is diffeomorphic to the standard sphere.

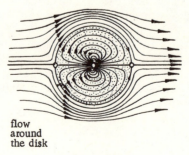

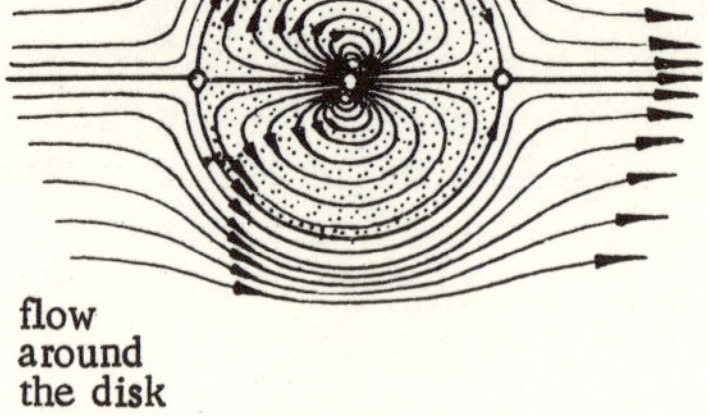

Fig. 99

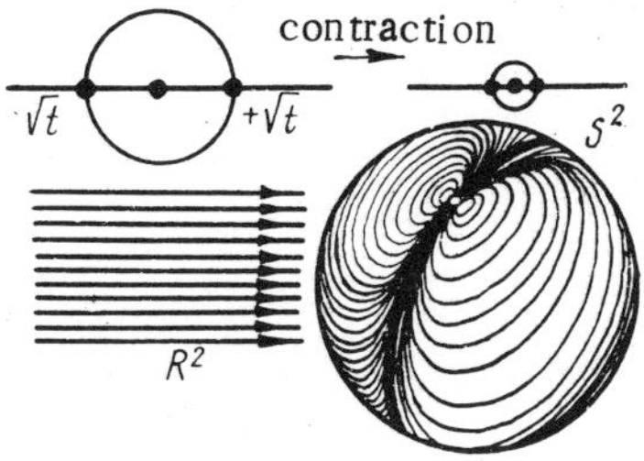

Fig. 100

As we have seen, nondegenerate critical points may sometimes collapse into one point, producing a more complicated degenerate singularity of the function. However, such a collapse may also produce the mutual simplification of singularities. Let us illustrate this with the simplest example. Consider on the sphere the Zhukovsky function $z + 1/z$ and let $f(x, y) = \mathrm{Re}\,(z + 1/z)$ (in the finite part of the plane). Then, integral trajectories of the field grad f are of the form depicted in Fig. 99. There are three singular points: two saddle points (at points ±1) and a degenerate singularity (1-order pole) at 0. The deformation of the function $\varphi_t f = \mathrm{Re}\,(z + t/z)$ leads to the condition that the separatrix diagram of the flow grad $\varphi_t f$ (i.e., the collection of integral trajectories connecting singular points) is deformed along the plane and finally contracts into one point where all three singularities annihilate. As a result (for $t = 0$), we get the smooth function $\mathrm{Re}\,(z) = x$, without any singular points in the finite part of the plane (and having a 1-order pole at infinity, i.e., in the northern pole of the sphere); see Fig. 100. Thus, considering continuous deformations of the function of the space of all smooth functions (with singularities) on M, we may considerably change the picture of distribution of its critical (bifurcational) points. At the same time, as was proved above for smooth functions, there is a clear connection (inequality) between critical points of f on M and the topology of M, namely the connection with homology groups. It is also interesting to determine the minimum number of singularities (perhaps degenerate) that can have a smooth function on a manifold. The following section is devoted to the answer of this question.

10. CRITICAL POINTS OF FUNCTIONS AND THE CATEGORY OF A MANIFOLD

10.1. Definition of the Category

As was shown above, if f is a Morse function on M, then the number of its critical points of index λ is always less than β_λ, where $\beta_\lambda = \dim H_\lambda(M; A)$; and A is one of the following groups: $\mathbf{R}$, $\mathbf{Z}_2$, or $\mathbf{Z}_p$, where $p \neq 2$ and p is prime. For example, if $M = M_g^2$, then any Morse function has no less than $2g + 2$ singularities. However, when we continue on to the study of points of bifurcation, in particular when we allow nondegenerate singularities to collapse into degenerate ones and begin to consider various smooth deformations of the initial function in the space of all smooth functions, the situation immediately becomes very complicated since interactions of singularities at their collapse and decay are very cumbersome. Here the "complicatedness" of points of bifurcation may increase or decrease in these processes. For points of bifurcation, inequalities of the type $\sum_\lambda \pm\mu_\lambda \geqslant \sum_\lambda \pm\beta_\lambda$ are satisfied as before also when f is not a Morse function; but then, numbers μ_λ do not have such a simple meaning as they do for a Morse function (recall that, for a Morse function, μ_λ is equal to the number of critical points of index λ). In general, case numbers μ_λ describe "the degree of complicatedness" of points of bifurcation which is computed in terms of the homology groups of level surfaces close to the bifurcational surface (i.e., containing points of bifurcation). At the same time, μ_λ no longer give direct information on the number of points of bifurcation. Besides, not each degenerate singularity is bifurcational (see above). Thus, due to the nontriviality of the computation of numbers μ_λ, and since in various applications it is only interesting to answer how many points of bifurcation f has, we will only concentrate on the investigation of what is the minimum number of singularities of the function f on the manifold. It turns out that this question is solved in terms of a new topological invariant — the Lusternik–Shnirelman category — connected with an arbitrary CW-complex (space). It is interesting that this invariant appears useful in the analysis of several "infinite-dimensional" variational problems, e.g., in computing the number of closed geodesics on a manifold; cf. [6].

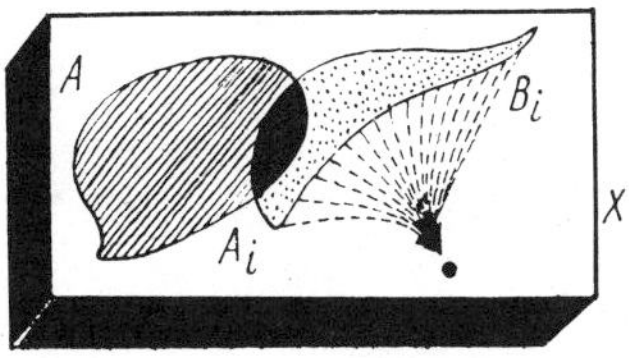

Fig. 101

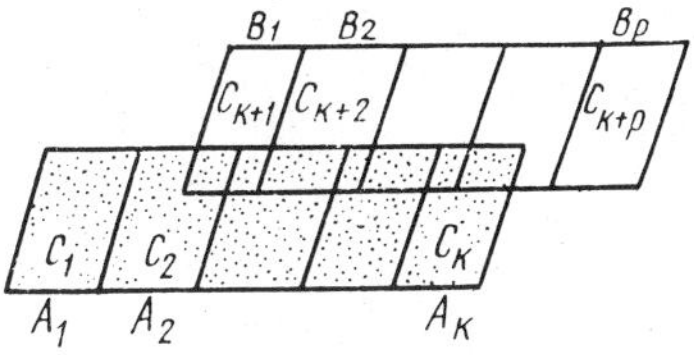

Fig. 102

Definition 10.1. Let X be a topological Hausdorff space and $A \subset X$ an arbitrary closed subset in X. The category $\mathrm{cat}_X A$ of A with respect to X is the minimum number k such that there exist closed subsets $A_1, \ldots, A_k$ in X, such that A is their union $A = \bigcup_{i=1}^{k} A_i$, and each of these subsets is contractible in X into a point. The connectedness of the A_i's is not assumed.

In what follows we will assume for simplicity that X is connected. If A = X, then we will assume that $\mathrm{cat}_X X = \mathrm{cat}\, X$. Thus, $\mathrm{cat}_X A$ is a positive integer. In what follows we will prove several important properties of this number.

10.2. Topological Properties of the Category

Lemma 10.1. If $A \subset B \subset X$ are closed subsets in X, then $\mathrm{cat}_X A \leqq \mathrm{cat}_X B$.

Proof. Let $q = \mathrm{cat}_X B$, i.e., there exist closed subsets $B_1, B_2, \ldots, B_q$ such that $B = \bigcup_{i=1}^{q} B_i$ and each set B_i is contractible in X into a point. Consider new closed subsets $A_i = A \cap B_i$, where $1 \leqq i \leqq q$. Clearly, $A = \bigcup_{i=1}^{q} A_i$ and each set A_i is contractible in X into a point (together with B_i) (see Fig. 101). It follows that $\mathrm{cat}_X A \leqq q = \mathrm{cat}_X B$, as required.

Lemma 10.2. Let A and B be two arbitrary closed subsets in X. Then $\mathrm{cat}_X A \cup B \leqq \mathrm{cat}_X A + \mathrm{cat}_X B$.

Proof. Let $\text{cat}_X A = k$, $\text{cat}_X B = p$, and $A=\bigcup_{i=1}^{k} A_i$, $B=\bigcup_{j=1}^{p} B_j$. Then $A \cup B=\bigcup_{q=1}^{p+k} C_q$, where $C_q = A_q$ for $i \leq q \leq k$, and $C_q = B_{q-k}$ for $k + 1 \leq q \leq k + p$. Since $\{A_i\}$ and $\{B_j\}$ are contractible in X into a point, then C_q is contractible into a point and $\text{cat}_X C \leq k + p = \text{cat}_X A + \text{cat}_X B$, as required; see Fig. 102.

Lemma 10.3. Let $A \subset B$ be closed subsets in X. Then $\text{cat}_X \overline{B\setminus A} \geq \text{cat}_X B - \text{cat}_X A$, where $\overline{B\setminus A}$ stands for the closure of $B\setminus A$ in X.

Proof. Since $B = A \cup (\overline{B\setminus A})$, then by Lemma 10.2 we have $\text{cat}_X B \leq \text{cat}_X A + \text{cat}_X \overline{B\setminus A}$, as required.

Lemma 10.4. Let $A \subset B \subset X$ be closed subsets in X and B be continuously deformable in X into the subset A, i.e., there exists a homotopy φ_t of the embedding $i:B \to X$, in the embedding $\varphi_1:B \to X$, such that $\varphi_1 B \subseteq A$. Then, $\text{cat}_X A \geq \text{cat}_X B$.

Note that the set $\varphi_1 B$ cannot be homeomorphic to B.

Proof. Let $\text{cat}_X A = k$. Consider the covering $A=\bigcup_{i=1}^{k} A_i$, where all sets A_i are contractible in X into a point. Since $\varphi_1 B \subseteq A$, we may consider $H_j = (\varphi_1 B) \cap A_j$, where $1 \leq j \leq k$. By the hypothesis of the lemma, there is a continuous mapping $h:iB \to \varphi_1 B$, where iB is homeomorphic to B. Let $B_j = h^{-1} H_j$, where $1 \leq j \leq k$. Clearly, $B=\bigcup_{j=1}^{k} B_j$. Furthermore, applying the homotopy φ_t to B_j, we will deform B_j along X into $\varphi_1 B_j = H_j \subset A_j$, i.e., H_j is contractible in X into a point; therefore, each B_j is contractible in X into a point. Hence, $\text{cat}_X B \leq k$, as required (see Fig. 103).

Lemma 10.5. Let A be a closed subset in the manifold X. Then there exists an $\varepsilon > 0$ (depending, generally speaking, on A) such that $\text{cat}\, U_\varepsilon A = \text{cat}_X A$, where $U_\varepsilon A$ stands for a closed ε-neighborhood of $A \subset X$.

Proof. Since $A \subset U_\varepsilon A$, then by Lemma 10.1 we have $\text{cat}_X A \leq \text{cat}_X U_\varepsilon A$. It suffices to verify the inverse inequality. Let $\text{cat}_X A = k$ and $A = \bigcup_{i=1}^{k} A_i$, where each set A_i is contractible

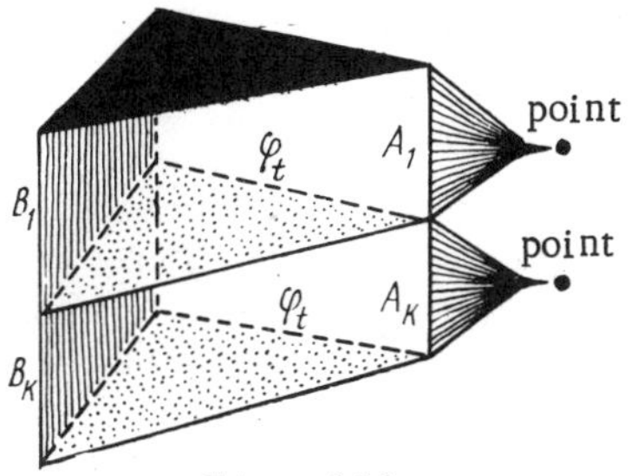

Fig. 103

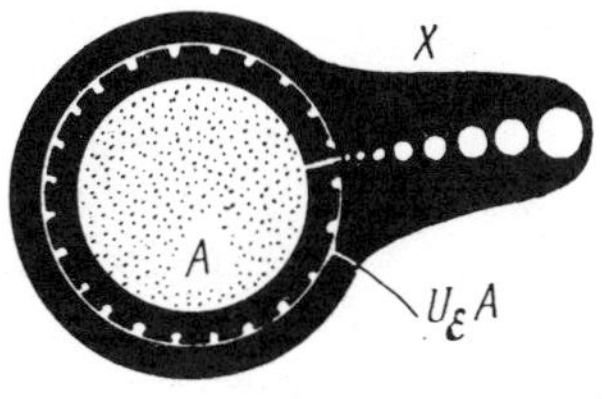

Fig. 104

into a point. Since X is a manifold, then there exists an $\varepsilon > 0$ such that $U_\varepsilon A_i$ is contractible in X into a point after A_i. Since $U_\varepsilon A = \bigcup_{i=1}^{k} U_\varepsilon A_i$, then $\text{cat}_X\ U_\varepsilon A \leqq k = \text{cat}_X A$, as required.

If X is not a manifold, then the statement of Lemma 10.5 is false. This example is depicted in Fig. 104.

If X is a space with a metric ρ, then we may define the distance between any two closed subsets A and B in X putting

$$\rho(A, B) = \sup_{x \in A} \inf_{y \in B} \rho(x, y) + \sup_{y \in B} \inf_{x \in A} \rho(x, y),$$

where $\rho(x, y)$ is the distance between points x and y in X.

Lemma 10.6. Let A, B_n, where n = 1, 2, ..., be closed subsets in the manifold X, endowed with the metric ρ, and $A = \lim_{n\to\infty} B_n$, i.e., $\rho(A, B_n) \to 0$ for $n \to \infty$. If $\text{cat}_X B_n \geqq k$ for all n, then $\text{cat}_X A = \text{cat}(\lim B_n) \geqq k$.

Proof. Lemma 10.5 implies the existence of an $\varepsilon > 0$ such that $\text{cat}_X U_\varepsilon A = \text{cat}_X A$. Since $\rho(A, B_n) \to 0$, there exists a number N such that $B_n \subset U_\varepsilon A$ for all n > N. It follows that $k \leqq \text{cat}_X B_n \leqq \text{cat}_X U_\varepsilon A = \text{cat}_X A$, as required.

10.3. A Formulation of the Theorem on the Lower Boundary of the Number of Points of Bifurcation

Theorem 10.1. Let M^n be a smooth compact connected closed manifold and f be a closed function on M. Then $k \geq \geq \operatorname{cat} M$, where k is the number of different points of bifurcation of f. Similarly, $p \geq \operatorname{cat} M$, where p is the number of different critical points (perhaps degenerate) of f.

For simplicity, we will prove this theorem for the case of critical points since, for points of bifurcation, the considerations can be repeated practically word for word with the replacement of integral trajectories of the vector field grad f by the fibration of neighborhoods U by trajectories—pre-images of I with respect to the homeomorphism $U = f_q \times I$ (see Sec. 9).

The proof of this theorem follows to a great extent a simple analogy that relates the behavior of eigenvalues of a bilinear form in $\mathbf{R}^n$ to the behavior of several numbers related with objects of a fixed category. We will briefly describe this analogy since it makes it easier to follow our subsequent considerations. Let $S^{n-1} \subset \mathbf{R}^n$ be a standard sphere of radius 1 and B(x, y) be a symmetric bilinear real-valued form in $\mathbf{R}^n$. This form defines on a sphere the smooth function $f(x) = B(x, x)$. Let us find the critical points of this function. Let $B:\mathbf{R}^n \to \mathbf{R}^n$ be the symmetric linear operator such that $B(x, x) = \langle Bx, x \rangle$, where $\langle , \rangle$ is the Euclidean scalar product in $\mathbf{R}^n$.

Lemma 10.7. The point $x_0 \in S^{n-1}$ is critical iff $Bx_0 = = \lambda x_0$, where $\lambda \in \mathbf{R}$, i.e., x_0 is an eigenvector of the form.

Proof. Let $a \in T_{x_0}S^{n-1}$ be an arbitrary tangent vector and x(t) be a smooth curve to the sphere passing through $x_0 = = x(0)$ and such that $\dot{x}(0) = a$. Let us express f in the form $\langle Bx, x \rangle$. Then

$$\left.\frac{df}{da}\right|_{x_0} = \frac{d}{dt}\langle B(x(t)), x(t)\rangle|_{t=0} = 2\langle Bx, \dot{x}\rangle|_{t=0} = 2\langle Bx_0, a\rangle.$$

Clearly, $\left.\frac{df}{da}\right|_{x_0} = 0$ iff x_0 is a critical point of f, i.e., grad $f(x_0) = 0$. On the other hand, this holds iff $\langle Bx_0, a \rangle =$

$= 0$ for any tangent vector a, i.e., $Bx_0 \perp T_{x_0}S^{n-1}$. Thus, Bx_0 is proportional to x_0.

It is known that among eigenvectors of B an orthobasis $e_0, e_1, \ldots, e_{n-1}$ can be chosen. The choice of numbers 0, 1, ..., $n - 1$ as indices of vectors will be justified later. Let λ_i be an eigenvalue corresponding to e_i. Let us order all λ_i (and vectors e_i) so that $\lambda_0 \leqq \lambda_1 \leqq \ldots \leqq \lambda_{n-1}$. In S^{n-1}, consider the set E_i of all i-dimensional "equators" S^i, i.e., sections of the sphere cut by planes of dimension $i + 1$ passing through the origin. It is known from the theory of quadratic forms that λ_i can be presented in the form

$$\lambda_i = \inf_{S^i \in E_i} (\max_{x \in S^i} f(x)) \text{ for } 0 \leqslant i \leqslant n-1,$$

i.e., it is equal to the smallest maximum value which f attains on equators S^i. It is clear from the definition of the expression on the right-hand side of the equality that $\lambda_0 \leqq \lambda_1 \leqq \ldots \leqq \lambda_{n-1}$. All equators S^i in the class E_i are obtained from an equator $S_0{}^i$ by a suitable orthogonal transformation $g \in SO_n$. In other words, an orthogonal group SO_n acts transitively on the set E_i. Since the function $f = \langle Bx, x\rangle$ is invariant with respect to the transformation $x \to -x$, then f is, actually, a function on the real projective space $\mathbf{R}P^{n-1}$. This function will be denoted by the same letter f.

Lemma 10.8. Let f be the function on $\mathbf{R}P^{n-1}$ defined above. Then the number of different critical points of f on $\mathbf{R}P^{n-1}$ is no less than the number of classes E_i, i.e., no less than n.

Proof. If all eigenvalues of the form B are different, then critical points of f on the sphere are points $\pm e_i$, where $0 \leqq i \leqq n - 1$, which gives us n points on $\mathbf{R}P^{n-1}$. If for some i, j we have $\lambda_i = \lambda_j$, then B possesses a linear subspace of eigenvectors corresponding to λ_i; hence the sphere S^{j-i} consists of degenerate critical points of f. Since there are finitely many such points, the lemma is proved.

Continuing on to the study of critical points of functions on an arbitrary manifold, we will make the following substitutions in the above construction: 1) the sphere S^{n-1} will be replaced by a manifold M; 2) the form B on the sphere will be replaced by a smooth function f on M; 3) orthogonal transformations preserving each class E_i will be replaced by continuous deformations of closed subsets in M; 4) classes E_i will be replaced by classes M_i, consisting of subsets whose category is no less than i; and, 5) eigenvalues λ_i will be replaced by some analogs defined by a formula similar to the minimax formula given above for λ_i.

Then it turns out that these substitutions enable us to formulate and prove an analog of Lemma 10.8 that will be the required Theorem 10.1.

10.4. Proof of the Theorem

Denote by M_i the class of all closed subsets $X \subset M^n$ such that $\mathrm{cat}_M X \geq i$. Clearly, $M_i \supseteq M_{i+1}$. Let Θ be the space of all closed subsets in M. Then Θ is endowed with a natural structure of the space with the metric $\rho(X, Y)$, whose definition is contained in Sec. 10.2. We will say that $Y = \lim_{q\to\infty} X_q$ if $\rho(X_q, Y) \to 0$ as $q \to \infty$.

Lemma 10.9. Each class M_i is a subset in Θ, closed with respect to two operations: a) the approach to the limit, i.e., $Y \in M_i$ if $Y = \lim_{q\to\infty} X_q$, where $X_q \in M_i$; b) continuous deformations of subsets in the manifold M, i.e., $Y \in M_i$ if $Y = \varphi_1 X$, where $\varphi_t: X \to M$ is a homotopy of the initial embedding.

Proof. Let $X_q \in M_i$ for $q = 1, 2, \ldots$ and $Y = \lim_{q\to\infty} X_q$, where $X_q \in M_i$. It is necessary to prove that $Y \in M_i$. But this follows immediately from Lemma 10.6. Furthermore, let $Y = \varphi_1 X$, where $\varphi_t: X \to M$. Since $\mathrm{cat}_M X \geq i$, then Lemma 10.4 implies that $\mathrm{cat}_M Y \geq i$, as required.

Fix the class M_i and let $X \in M_i$. Put $\lambda_i = \inf_{X \in M_i} (\max_{x \in X} f(x))$. Clearly, this definition stems from the definition of eigenvalues λ_i of B (see Sec. 10.3). Let $N = \mathrm{cat}_M M$. Under the conditions of Theorem 10.1, we have $N < \infty$. We get the chain of inclusions $\Theta = M_0 = M_1 \supseteq M_2 \supseteq \ldots \supseteq M_N$. Here we can assume that $\Theta = M_0 = \{X \in \Theta, \mathrm{cat}_M X \geq 0\}$. Clearly, $\mathrm{cat}_M X \geq 0$ for any $X \in \Theta$. The class M_N contains the manifold M itself. It is also clear that M_N is the last class in the above chain, since there are no subsets whose category is greater than N.

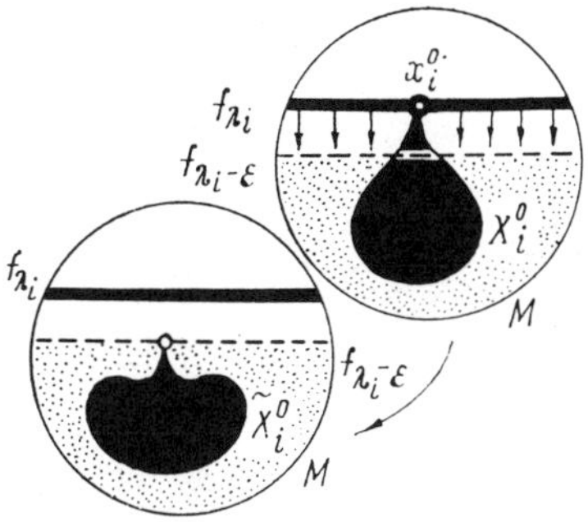

Fig. 105

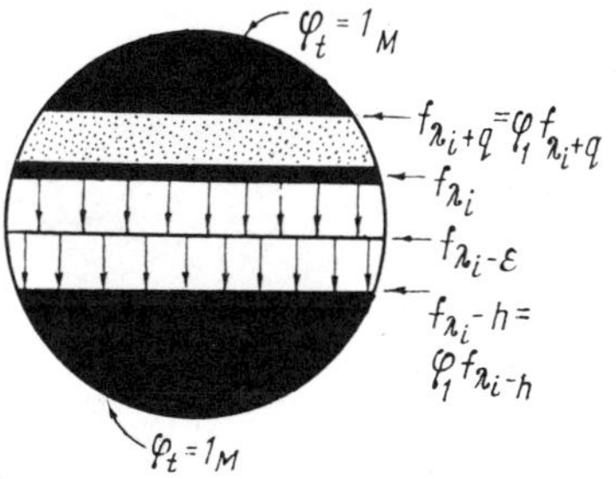

Fig. 106

The function f on M induces functions $f_0, \ldots, f_N$, defined on classes $M_0, \ldots, M_N$, respectively. Namely, $f_i(X) = \max_{x \in X} f(x)$, where $X \in M_i$. Then $\lambda_i = \inf_{X \in M_i} f_i(X)$. Since $M_i \supseteq M_{i+1}$, then numbers λ_i do not decrease as i increases, i.e., $\lambda_0 \leqq \ldots \leqq \lambda_N$. Since classes M_i are closed in Θ upon passage to the limit (see Lemma 10.9), then in each class M_i there is an element X_i^0 (a closed subset in M) such that $f_i(X_i^0) = \lambda_i$. In other words, X_i^0 is a subset in M such that $\lambda_i = \max_{x \in X_i^0} f(x)$.

<u>Lemma 10.10</u>. On the level surface f_{λ_i}, there is at least one critical point of f.

<u>Proof</u>. Assume the contrary, i.e., suppose on f_{λ_i} there are no critical points of f. Consider the case M_i, and let $X_i^0 \in M_i$ be a closed subset in M such that $\max_{x \in X_i^0} f(x) = \lambda_i$, i.e., $f_i(X_i^0) = \lambda_i$. Since X_i^0 is closed in M, there exists a point $x_i^0 \in X_i^0$ such that $x_i^0 \in f_{\lambda_i}$, i.e., $f(x_i^0) = \lambda_i$. Since by assumption $\operatorname{grad} f(x) \neq 0$ for any $x \in f_{\lambda_i}$, then the compactness of M implies the existence of a sufficiently small deformation of the surface f_{λ_i} along integral trajectories of the vector field $-\operatorname{grad} f$ to the domain of eigenvalues of f which are less than λ_i (see Fig. 105). Since M is a compact manifold, then Lemma 8.1 implies the existence of a smooth homotopy $\varphi_t : M \to M$ such that each mapping φ_t is a diffeomorphism of M and φ_t is the identical mapping outside of the domain $\lambda_i - h \leqq f \leqq \lambda_i + q$, where $\lambda_i - h < \lambda_i - \varepsilon$ and $\lambda_i < \lambda_i + q$,

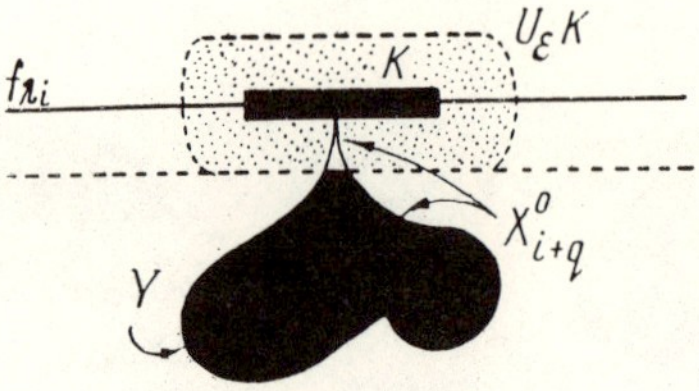

Fig. 107

and φ_1 transforms f_{λ_i} into $f_{\lambda_i-\varepsilon}$ (see Fig. 106). Denote by $\tilde{X}_i{}^0$ the image of $X_i{}^0$ under the deformation φ_1. Since $\tilde{X}_i{}^0$ is obtained from $X_i{}^0$ by homotopy along M, then, due to Lemma 10.4, we have $\mathrm{cat}_M\, \tilde{X}_i{}^0 \geqq \mathrm{cat}_M\, X_i{}^0$. Thus, $\mathrm{cat}_M\, \tilde{X}_i{}^0 \geqq i$, i.e., $\tilde{X}_i{}^0 \in M_i$. But this means that $\sup\limits_{x\in \tilde{X}_i^0} f(x) \leqslant \lambda_i - \varepsilon < \lambda_i$, i.e.,

$$\inf_{X\in M_i} (\sup_{x\in X} f(x)) \leqslant \sup_{x\in \tilde{X}_i^0} f(x) \leqslant \lambda_i - \varepsilon < \lambda_i,$$

which is impossible by definition of the number λ_i. The lemma is proved.

Lemma 10.11. Suppose that the sequence $\lambda_0 \leqq \ldots \leqq \lambda_N$ contains identical numbers. Let $\lambda_i = \lambda_i + q$, where $q > 0$. Let K be the set of all critical points of f on the surface f_{λ_i}. Then $\mathrm{cat}_M\, K \geqq q + 1$.

Proof. Recall that within the framework of the analogy mentioned in Sec. 10.3, this lemma reproduces an algebraic statement on the existence of a subspace of eigenvectors that correspond to a multiple eigenvalue of B. Since K is evidently closed in M, then, by Lemma 10.5, there is an $\varepsilon > 0$ such that $\mathrm{cat}_M K = \mathrm{cat}_M\, U_\varepsilon K$. Assume the contrary, i.e., $\mathrm{cat}_M\, K \leqq q$. Consider the sequence of classes $M_i \supseteq M_{i+1} \supseteq \ldots \supseteq M_{i+q}$. Let $X_{i+q}{}^0 \in M_{i+q}$ be a closed subset such that $\lambda_i = \lambda_{i+q} = \sup\limits_{x\in X^0_{i+q}} f(x)$.

Let us construct the new closed set $Y = \overline{X^0_{i+q} \setminus (X^0_{i+q} \bigcap K)}$ (see Fig. 107). We have

$$\mathrm{cat}_M Y \geqslant \mathrm{cat}_M X^0_{i+q} - \mathrm{cat}_M X^0_{i+q} \bigcap U_\varepsilon K \geqslant \mathrm{cat}_M X^0_{i+q} - \\ - \mathrm{cat}_M U_\varepsilon K = \mathrm{cat}_M X^0_{i+q} - \mathrm{cat}_M K \geqslant i + q - q = i.$$

Thus, $\text{cat}_M Y \geq i$, i.e., $Y \in M_i$. Furthermore,

$$\lambda_i = \lambda_{i+q} = \sup_{x \in X^0_{i+q}} f(x) \geqslant \sup_{x \in Y} f(x) \geqslant \lambda_i = \lambda_{i+q} = \sup_{x \in X^0_{i+q}} f(x).$$

This implies that $\sup_{x \in Y} f(x) = \lambda_i$; hence the set Y can play the role of the compact X_i^0 in the class M_i. At the same time, $Y \cap K = \phi$, contradicting Lemma 10.10, which implies that at least one critical point of f must be contained in Y, since $Y \cap f_{\lambda_i} \neq \phi$. The lemma is proved.

Now we can proceed to the proof of Theorem 10.1.

Consider the sequence of classes $M_0 = M_1 \supseteq M_2 \supseteq \ldots \supseteq \supseteq M_N$. First, suppose that $\lambda_0 = \lambda_1 < \lambda_2 < \ldots < \lambda_N$ (i.e., there are no two coinciding numbers for $i \geq 1$). Then Lemma 10.10 implies that on each surface f_{λ_i} there is at least one critical point of a function; hence the number of different critical points is no less than the number of different critical level surfaces, which is evidently no less than N = cat M. Thus, in the case of "general position," our theorem is proved. Now suppose that some of λ_i coincide, e.g., $\lambda_i = = \lambda_{i+q}$. How many different critical points can we then select on f_{λ_i}? Lemma 10.11 shows that $\text{cat}_M K \geq i$, where K is the set of critical points on f_{λ_i}. But this means that in K we can select at least q + 1 different points. In fact, $K = = \bigcup_{j=1}^{q+1} K_j$, where each K_j is contractible along M into a point; therefore, it suffices to select one point in each K_j. Thus, the "single" value λ_i, i.e., such that $\lambda_{i-1} < \lambda_i < \lambda_{i+1}$, adds at least one critical point, and each (q + 1)-multiple value λ_i, i.e., such that $\lambda_{i-1} < \lambda_i = \ldots = \lambda_q < \lambda_{i+q-1}$, adds at least q + 1 different critical points. Moving upward with the growth of λ_i, we obtain the statement of the theorem.

10.5. Examples of the Computation of Categories

We have obtained an estimation from below of the number of critical (bifurcational) points of a smooth function on M. This estimation is expressed in terms of a topological invariant of M (apparently homeomorphic spaces have the same

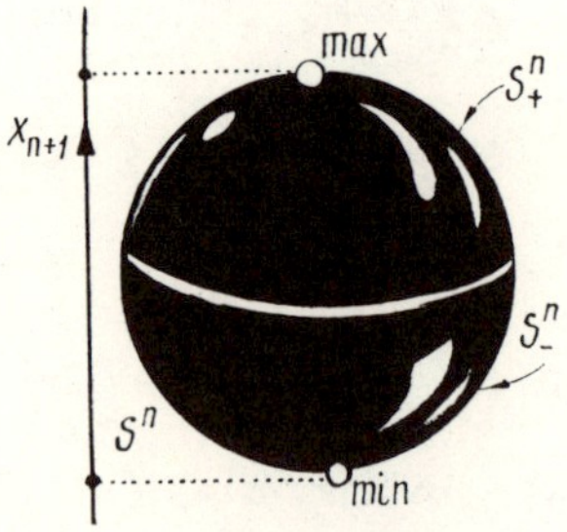

Fig. 108

category). Question: Is this estimation the best one in the general case, i.e., are there examples of functions on M such that they exactly satisfy this estimation? Such examples do exist.

Statement 10.1. The category of the sphere S^n is 2. The category of a two-dimensional smooth closed compact manifold different from a sphere is 3. The minimum number of critical points of a smooth function on the sphere S^n is 2 (and is equal to the category of the sphere). The minimum number of critical points of a smooth function on $M_{g\geqq 1}{}^2$ or $M_{\mu\geqq 1}{}^2$ is equal to 3 (and is equal to the category of this two-dimensional manifold).

Proof. For the sphere S^n, the statement is evident since $S^n = S_+{}^n \cup S_-{}^n$, where $S_\pm{}^n$ are two closed hemispheres and the role of f is played by the standard height function (see Fig. 108). On the other hand, cat $S^n \neq 1$, since S^n is not contractible along itself into a point (nontrivial homology of the sphere is the obstruction). In the case $M_{g>0}{}^2$ or $M_{\mu>0}{}^2$, we have proved in Sec. 10.4 that, on these manifolds, there exists a smooth function with three critical points. It remains to verify that in this case cat $M^2 = 3$. The indices $g > 0$ and $\mu > 0$ will be dropped for the sake of brevity. Clearly, cat $M^2 > 2$, since otherwise the manifold would have been homeomorphic to a sphere. Now, let us produce the partition of M^2 in the union of three closed contractible subsets. For this, it suffices to consider the already known simplest cell partition of M^2. In fact, $M^2 = \sigma^\circ \bigcup \left(\bigcup_{i=1}^{r} \sigma_i^1\right) \bigcup \sigma^2$, i.e., to the bouquet of circles $\bigvee_{i=1}^{r} S_i{}^1$, a two-dimensional

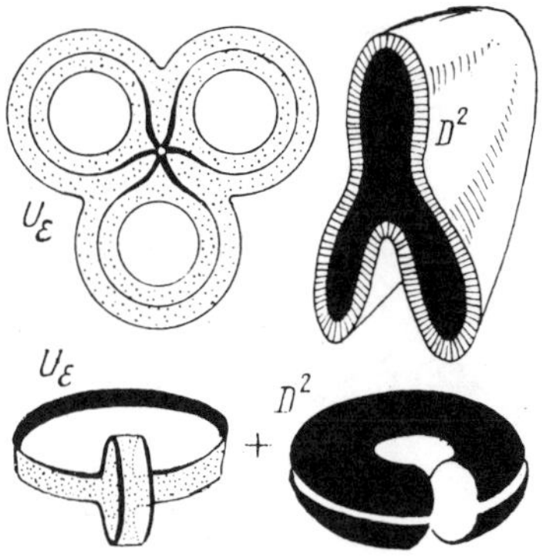

Fig. 109

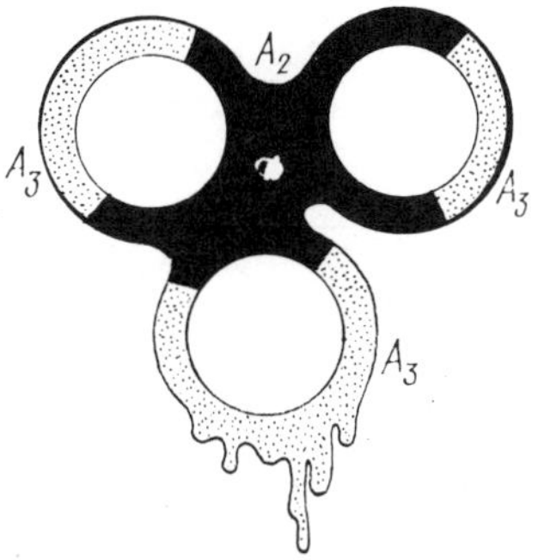

Fig. 110

cell σ^2 is attached which is a model of the fundamental polyhedron W. Let U_ε be a sufficiently small ε-neighborhood of a one-dimensional skeleton (bouquet) $\bigvee_{i=1}^{r} S_i^1$ in the manifold M^2 and let $D^2 = M^2 \setminus U_\varepsilon$ be a closed disk (see Fig. 109). Then M^2 is presentable in the form $A_1 \cup A_2 \cup A_3$, where $A_1 = D^2$ is contractible along itself into a point, and sets A_2 and A_3 are depicted in Fig. 110. Here $A_2 = U_\varepsilon \cap R_\alpha$, where R_α is the disk of radius α with the center in σ^0 (α is sufficiently small), $A_3 = \overline{U_\varepsilon \setminus A_2}$, $U_\varepsilon = A_2 \cup A_3$. In this situation A_2 is contractible along itself into a point and A_3 is contractible along itself into the collection of r points and, therefore (by the connectedness of M^2), A_3 is contractible into a point along M^2. The statement is proved.

At the same time, it is worth mentioning that the category is an invariant of a manifold, and is very difficult to compute. There are no effective algebraic methods for its computation [likewise for spectral sequences for (co)homology]. It is usually not difficult to estimate the category of a manifold in each concrete problem from above. For this, it suffices to use a covering of M by contractible sets. The reader must prove that cat $M^n \leq n + 1$, where $n = \dim M^n$. It is much more difficult to estimate the category from below. We will restrict ourselves to the description of one method based on the notion of a cohomology length of a manifold.

Definition 10.2. Let $H^*(M; A)$ be a ring of cohomology on M, where $A = \mathbf{Z}$ or $\mathbf{R}$ if M is orientable, and $A = \mathbf{Z}_2$ if M is not orientable. Consider all integers q such that there are elements $a_1, \ldots, a_p$ in $H^*(M, A)$, where $\dim a_i > 0$, $1 \leq i \leq p$, whose product $a_1 \cdot \ldots \cdot a_p$ in the cohomology ring $H^*(M; A)$ is nonzero. The cohomology length of a manifold is the maximum of such numbers q.

Theorem 10.2. If K is the cohomology length of the manifold M, then $\operatorname{cat} M \geq k + 1$.

The proof will be given below after we introduce the Poincaré duality. Here we will consider only the corollaries of this theorem that enable us to compute categories of concrete spaces.

Proposition 10.1. $\operatorname{cat} \mathbf{R}P^n = n + 1$.

Proof. First let us prove the estimation from above: $\operatorname{cat} \mathbf{R}P^n \leq n + 1$. Above we have produced the standard presentation of $\mathbf{R}P^n$ in the form of a CW-complex. This gives us the decomposition $\mathbf{R}P^n = \bigcup_{i=1}^{n+1} D_i^n$, where D_i^n is an open n-dimensional disk defined by the formula $D_i^n = \{\lambda(x_1, \ldots, x_{n+1}), \lambda \neq 0, x_i \neq 0\}$, where $\{x_i\}$ are homogeneous coordinates on $\mathbf{R}P^n$. In [1], explicit formulas are contained that define a homeomorphism between D_i^n and the standard open disk D^n. Since $\{D_i^n\}$ is a finite open cover, then in each set D_i^n a less closed disk B_i^n can be inscribed; the union of all B_i^n with respect to all i will still be a cover of $\mathbf{R}P^n$. Since each disk B_i^n is contractible along itself into a point, we have proved the required estimation from above. Let us now prove that $\operatorname{cat} \mathbf{R}P^n \geq n + 1$. By Theorem 10.2, it suffices to prove that the cohomology length of $\mathbf{R}P^n$ (with coefficients in $\mathbf{Z}_2$ in order to include both the orientable and nonorientable cases) equals n. Statement 7.3 implies $H^*(\mathbf{R}P^n; \mathbf{Z}_2) = \mathbf{Z}_2[e]/(e^{n+1})$, where $\deg(e) = 1$. Thus, the product $e^n = e \cdot \ldots \cdot e$ is nonzero, as required.

Lemma 10.12. If $\operatorname{cat} X \geq 2$, then $\operatorname{cat} X \vee S^n = \operatorname{cat} X$, where $X \vee S^n$ is the bouquet of X and S^n, where $n > 0$.

Proof. Let $\operatorname{cat} X = k$ and $X = \bigcup_{i=1}^{k} X_i$, where each X_i is contractible along X into a point. Let $x \in X$ be a point of the attachment of S^n to X. Let us present the sphere in the form of the union of two hemispheres S_+^n and S_-^n, where $x \in S_-^n$ and $x \notin S_+^n$ (see Fig. 111). Let X_j be a subset such

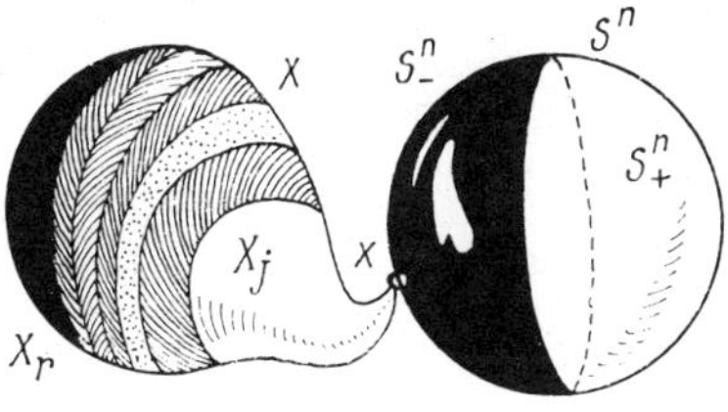

Fig. 111

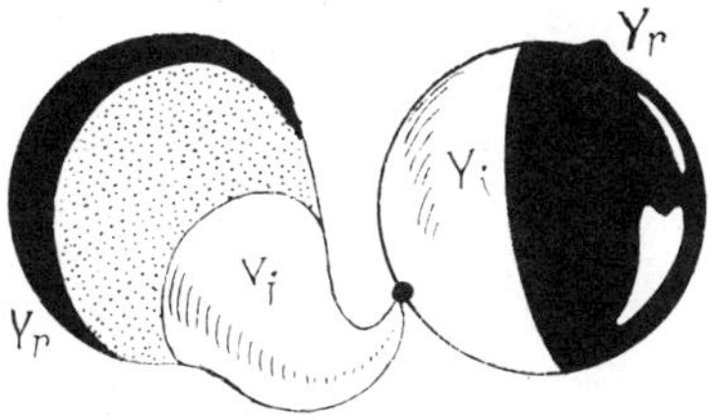

Fig. 112

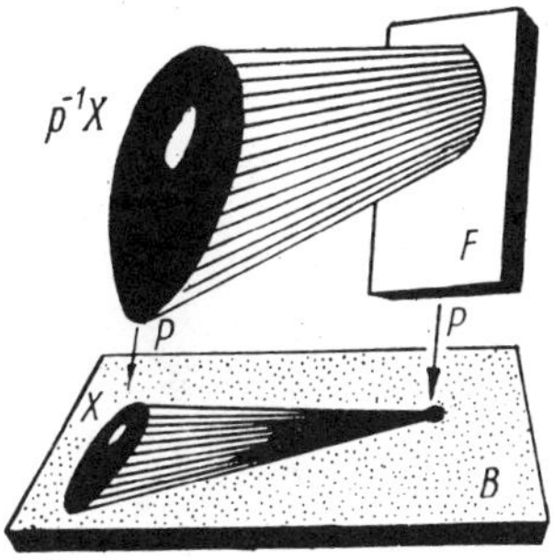

Fig. 113

that $x \in X_j$. Set $Y_i = X_i$, where $i \neq j$ and $i \neq r$, and where r is any index different from j. Finally, set $Y_j = X_j \cup S_-^n$, $Y_r = X_r \cup S_+^n$ (see Fig. 112). In particular, Y_r can be non-connected. Thus, $X \vee S^n = \bigcup_{s=1}^{k} Y_s$, where each set Y_s is contractible along $X \vee S^n$ into a point. Therefore, cat $X \vee S^n =$ = cat X, as required.

The proof of the following statement is quite similar: cat $X \vee Y$ = max (cat X, cat Y) for any spaces X, Y.

Proposition 10.2. If T^n is a torus, then cat $T^n = n + 1$.

Proof. Since T^n is the direct product of circles, then $H^*(T^n, \mathbf{Z}) = \wedge(x_1, \ldots, x_n)$ (the exterior algebra in one-dimensional generators $x_1, \ldots, x_n$). Hence, the product $x \cdot \ldots \cdot x_n$ is nonzero and cat $T^n \geq n + 1$. Let us prove that cat $T^n \leq n + 1$. Since $T^n = S^1 \times T^{n-1}$, then T^n is presentable in the form $(S^1 \vee T^{n-1}) \cup \sigma^n$ (see the n-dimensional analog in Fig. 109), hence cat T^n = cat $(S^1 \vee T^{n-1}) \cup \sigma^n \leq$ cat $(S^1 \vee T^{n-1}) + 1$. Since cat $T^{n-1} \vee S^1$ = cat T^{n-1} and cat $T^2 = 3$ (see the case of two-dimensional manifolds considered above), then cat $T^{n-1} \leq n - 1$.

Proposition 10.3. Let $p: E \xrightarrow{F} B$ be a locally trivial bundle. Then cat $E \leq \mathrm{cat}_E F \cdot \mathrm{cat}\, B$, where $F \subset E$ is the fiber of the bundle.

Proof. Even a more general statement holds. If $X \subset B$ is a closed subset in the base and $p^{-1}X \subset E$ is its inverse image, then $\mathrm{cat}_E\, p^{-1}(X) \leq \mathrm{cat}_E F \cdot \mathrm{cat}_B X$. Setting $X = B$, we get the required statement. First, consider the special case $\mathrm{cat}_B X = 1$. Hence, we must verify the inequality $\mathrm{cat}_E\, p^{-1}X \leq \mathrm{cat}_E F$. Contracting X along the base B into a point and covering this deformation by a continuous deformation of the pre-image $p^{-1}X$ along the space E into the fiber F (the existence of a covering homotopy is proved, for example, in [2], Chapter 5), we get, due to Lemma 10.4, that $\mathrm{cat}_E\, p^{-1}X \leq \mathrm{cat}_E F$, as required (see Fig. 113).

Let us proceed to the general case by applying induction. Let $\mathrm{cat}_B X = k$, i.e., $X = \bigcup_{i=1}^{k} X_i$, where each X_i is contractible along B into a point. Set $Y = \bigcup_{i=1}^{k-1} X_i$, where $Z = X_k$; then $X = Y \cup Z$, where $\mathrm{cat}_B Y \leq k - 1$, $\mathrm{cat}_B Z = 1$. It is necessary to prove that $\mathrm{cat}_E F \cdot \mathrm{cat}_B Y \cup Z \geq \mathrm{cat}_E\, p^{-1}(Y \cup Z)$. We have $\mathrm{cat}_E\, p^{-1}(Y \cup Z) = \mathrm{cat}_E\, p^{-1}Y \cup p^{-1}Z \leq \mathrm{cat}_E\, p^{-1}Y + \mathrm{cat}_E\, p^{-1}Z$. The required inequality follows from $\mathrm{cat}_E\, p^{-1}Y + \mathrm{cat}_E\, p^{-1}Z \leq \mathrm{cat}_E F \cdot \mathrm{cat}_B Y \cup Z = k \cdot \mathrm{cat}_E F$. Since $\mathrm{cat}_E\, p^{-1}Z \leq \mathrm{cat}_E F$ (because Z is contractible along B into a point), it suffices to prove that $\mathrm{cat}_E\, p^{-1}Y + \mathrm{cat}_E F \leq k\, \mathrm{cat}_E F$, i.e., $\mathrm{cat}_E\, p^{-1}Y \leq (k - 1)\, \mathrm{cat}_E F$. But this inequality follows from the stronger statement $\mathrm{cat}_E\, p^{-1}Y \leq \mathrm{cat}_E F \cdot \mathrm{cat}_B Y$, since $\mathrm{cat}_B Y \leq k - 1$. The latter statement holds by the inductive hypothesis in which induction is carried out in $\mathrm{cat}_B X$. Recall that the first step of induction $\mathrm{cat}_B X = 1$ was already considered above. Our statement is completely proved.

The obtained estimation of the category of the space of a bundle is, in a sense, the exact one, i.e., there exist bundles such that this inequality turns into an equality, i.e., such that cat $E = \mathrm{cat}_E F \cdot \mathrm{cat}\, B$. Among these bundles

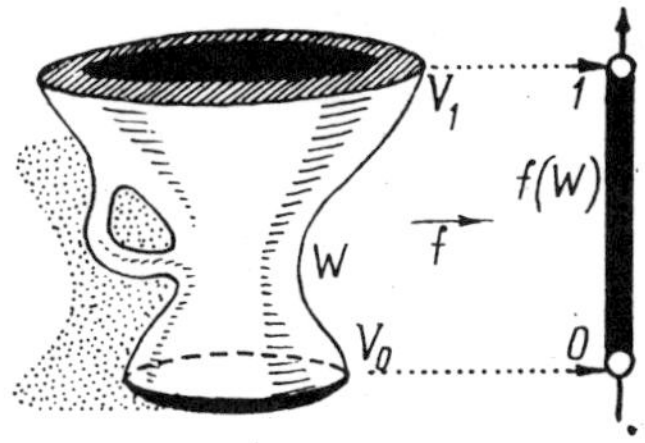

Fig. 114

are nontrivial ones; for example, let $p: S^3 \xrightarrow{S^1} S^2$ be the Hopf bundle. Then cat S^3 = cat S^2 = 2 and $\mathrm{cat}_{S^3} S^1 = 1$ (since the fiber S^1 is contractible along the sphere S^3 into a point), i.e., $2 = 1 \cdot 2$.

11. ADMISSIBLE MORSE FUNCTIONS AND BORDISMS

11.1. Bordisms

This section contains results that play an extremely important role in the study of the topology of manifolds. In particular, we will prove here the existence on any compact manifold of the so-called "admissible Morse functions," i.e., functions f such that $f(x) \geqq f(y)$, where x and y are critical points and ind $(x) \geqq$ ind (y). From the existence of such functions, we will derive several important topological consequences. The main objects of study in this section are orientable smooth compact manifolds W^n, with the boundary ∂W^n, which is a nonconnected union of two orientable closed, i.e., without boundary (but not necessarily connected), manifolds V_0^{n-1} and V_1^{n-1}. We will assume that $\partial W = V_0 \cup (-V_1)$ with regard to the induced orientation, where the minus sign indicates orientation opposite to the initial one (see Fig. 114). Manifolds V_0 and V_1 are sometimes called bordant ones and the manifold W is called a bordism. Sometimes we will use the notation (W, V_0, V_1). For instance, any two-dimensional closed orientable manifold M_g^2 is bordant to a two-dimensional sphere; therefore, all such manifolds are bordant to each other. To prove this, it suffices to consider the standard embedding of M_g^2 into $\mathbf{R}^3$, then consider a three-dimensional manifold with M_g^2 as a boundary in $\mathbf{R}^3$ (i.e., the solidification of M_g^2) and extract from it a small ball.

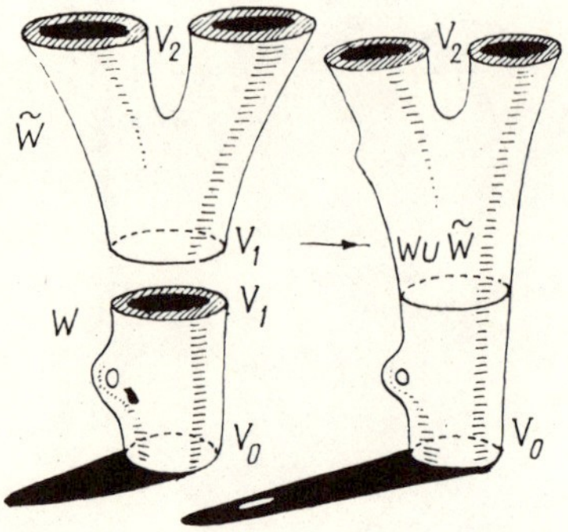

Fig. 115

Morse functions on W will be of special interest to us. We will assume that f satisfies the following conditions (see Fig. 114): $f|V_0 = 0$, $f|V_1 = 1$, $0 \leqq f \leqq 1$ on W. Such Morse functions always exist (see, for example, [7]).

Definition 11.1. We will say that f is a Morse function on W and $0 \leqq f \leqq 1$ if $f|V_0 = 0$, $f|V_1 = 1$, and the boundary ∂W has no critical points of f, i.e., they are situated inside of the bordism.

As in the case of closed manifolds, the following theorem holds: The set of Morse functions is dense in the space of all smooth functions g on W such that $0 \leqq g \leqq 1$, $g|V_0 = 0$, $g|V_1 = 1$.

11.2. A Decomposition of a Bordism into a Composition of Elementary Bordisms

Since we consider pairs (W, f), it is natural to raise the question: What is the simplest structure of such a pair?

Definition 11.2. The bordism (W, f) is elementary if the Morse function f on W has only one critical point.

Suppose two bordisms (W, V_0, V_1) and $(\tilde{W}, V_1, V_2)$ are given; then we can make a new bordism $W \circ \tilde{W} = (W \cup \tilde{W}, V_0, V_1)$ attaching the boundary V_1 of W to the boundary V_1 of $\tilde{W}$ (see Fig. 115). For the attachment, we can use any fixed diffeomorphism between $V_1 \subset \partial W$ and $V_1 \subset \partial\tilde{W}$. Surely, we must trace those "edges" that will not appear at the place of attachment;

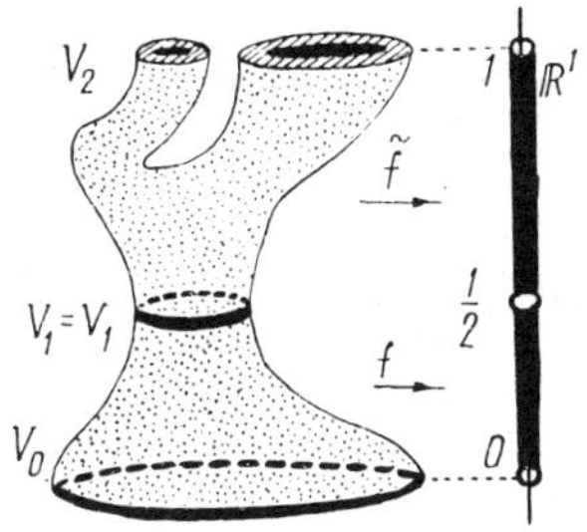

Fig. 116

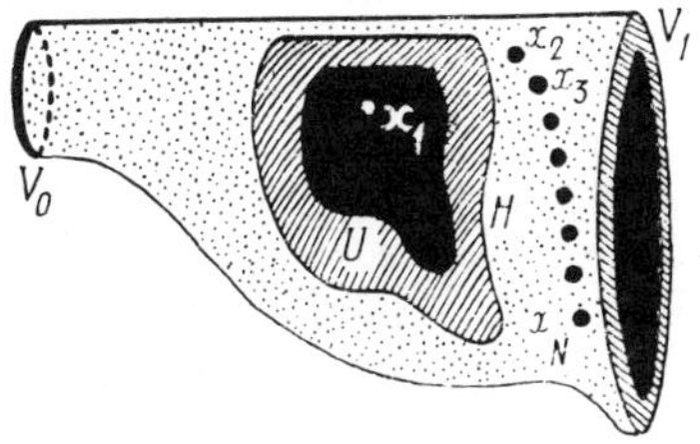

Fig. 117

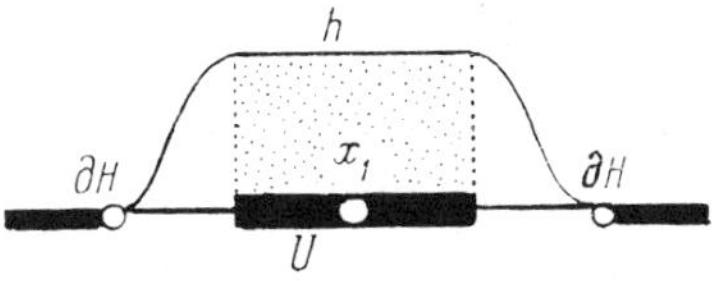

Fig. 118

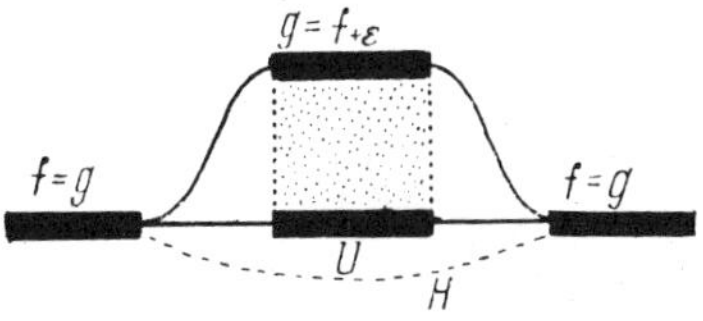

Fig. 119

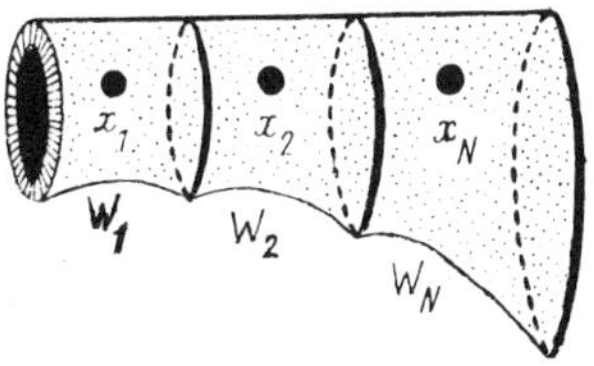

Fig. 120

however, we can avoid them by "smoothening," as we have smoothed "angles" of manifolds that appear at the attachment of a handle (see above). We will say that the bordism $W \circ \tilde{W}$ is the composition of bordisms W and $\tilde{W}$. It is also clear that if f and $\tilde{f}$ are Morse functions on W and $\tilde{W}$, respectively, the composition $W \circ \tilde{W}$ enables us to define the new Morse function on $W \circ \tilde{W}$ obtained by glueing these functions (see Fig. 116).

Proposition 11.1. Let (W, f) be an arbitrary bordism with a Morse function on it. Then it admits a presentation in the form of a composition of elementary bordisms $W = W_1 \circ \circ \ldots \circ W_N$ and $(W, f) = (W_1, g_1) \circ \ldots \circ (W_N, g_N)$, where g_i is the restriction of a Morse function g onto W_i, defined on the whole W, and the function g can be chosen however-close to the initial function f.

The proof follows immediately from the following lemma.

Lemma 11.1. Let (W, f) be a bordism with a Morse function. Let $x_1, \ldots, x_N$ be critical points of f. Then the Morse function f can be approximated however-close by a Morse function g with the same critical points, and such that $g(x_i) \neq g(x_j)$ for $i \neq j$.

Proof of the Lemma. Fix a point x_1 and let U and H be two of its open neighborhoods such that $x_1 \in U \subset \bar{U} \subset H$ and $x_j \notin \bar{H}$ for $j \neq 1$. Since critical points are discontinuous, such neighborhoods exist (see Fig. 117). Let us construct then on W a smooth function h such that $h \equiv 1$ in U and $h \equiv 0$ outside of H (see Fig. 118). After this, consider the new function $f_0 = f + \varepsilon_1 h$, where $\varepsilon_1 > 0$ is such that $0 \leqq f_0 \leqq 1$ on W and $f_0(x_1) \neq f_0(x_j)$ for $j \neq 1$. Consider the "ring" $H \setminus \bar{U} = K$, and assume that on W there is a Riemannian metric (which exists because W is compact). Clearly, on the closure of K, the module of the gradient of f is separated from zero from below, i.e., there exists a constant c such that $0 < c \leqq |\text{grad } f|$ and we can also assume that $|\text{grad } f| \leqq c < \infty$ on K. We consider an ε such that $0 < \varepsilon < \min(\varepsilon_1, c/c_1)$. Finally, take $g = f + \varepsilon h$ (see Fig. 119). We claim that g is a Morse function with the same critical points as f. In fact, if $\text{grad } g(x') = 0$, where $x' \in K$, then $\text{grad } f(x') + \varepsilon \times \text{grad } h(x') = 0$, whence $c \leqq |\text{grad } f(x')| = \varepsilon |\text{grad } h(x')| < c$. The obtained contradiction proves the lemma.

Thus, if a bordism W with a Morse function f is given, then, by a little manipulating of f, we can obtain the result that all critical points of the new function will be situated on different critical levels (see Fig. 120).

11.3. Gradient-like Fields and Separatrix-like Disks

Let ξ be a vector field on W and let f be a smooth function. Let $\xi(f)$ be a generating function of f along the field ξ.

Definition 11.3. A smooth vector field ξ on W is called gradient-like for the function f if

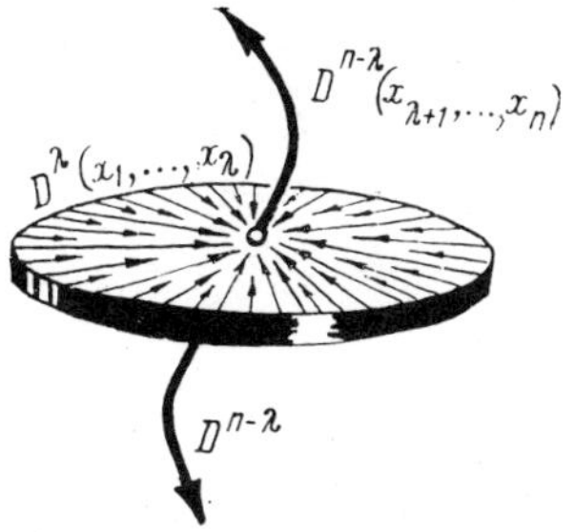

Fig. 121

1) $\xi(f) > 0$ on $W\setminus(y_1 \cup \ldots \cup y_N)$, where $y_1, \ldots, y_N$ are critical points of f,

2) for any critical point y_i, there exists an open neighborhood U_i such that, in any coordinate system where

$$f(x)|_{U_i} = f(y_i) - \sum_{j=1}^{\lambda} x_j^2 + \sum_{r=\lambda+1}^{n} x_r^2,$$

the field ξ is of the form $\xi(x) = (-x_1, \ldots, -x_\lambda, x_{\lambda+1}, \ldots, x_n)$.

For any Morse function f on W, such fields surely exist. It suffices, for example, to take for ξ the field grad f. Note that if there is a Morse function f without critical points on W, then W is diffeomorphic to $V_0 \times I$; in particular, V_0 and V_1 are homeomorphic. Such bordisms are sometimes called trivial ones (or cylinders). For the proof, it suffices to consider the foliation of W by integral trajectories of the field ξ, gradient-like for the function f. Now let x_0 be a critical point of index λ for f on W and ξ a gradient-like field for f.

Definition 11.4. A separatrix diagram of the critical point $x_0 \in W$ is the collection of all integral trajectories of the field ξ with x_0 as the target or with x_0 as the source.

This diagram is depicted in Fig. 121. Entering integral trajectories fill in the λ-dimensional disk D^λ, corresponding to local coordinates $x_1, \ldots, x_\lambda$, and leaving trajectories fill in the disk $D^{n-\lambda}$, corresponding to other co-

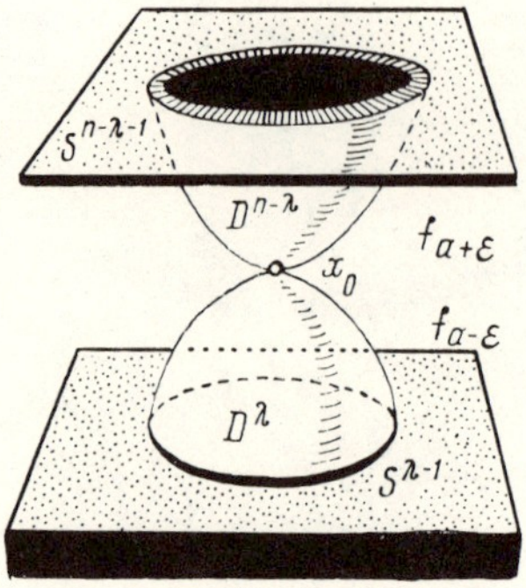

Fig. 122

ordinates and transverse to D^λ at the point of their intersection x_0. Note that x_0 is the unique point of intersection of these disks. If we interpret the vector field ξ as the field of velocities of the fluid that flows along W, then along D^λ the fluid flows into x_0, and along $D^{n-\lambda}$ the fluid flows out of x_0. Now consider two spheres obtained by the intersection of separatrix disks with two level surfaces of the function, one above and one below a critical point. The disk D^λ situated in the domain of smaller values of f [i.e., in the domain $f(x) \leqq f(x_0)$] is sometimes called a left one and the disk $D^{n-\lambda}$ situated in the domain of larger values is called a right one. Let $f(x_0) = a$. Now consider two spheres $S^{\lambda-1} = D^\lambda \cap f_{a-\varepsilon}$ and $S^{n-\lambda-1} = D^{n-\lambda} \cap f_{a-\varepsilon}$, where $\varepsilon > 0$ is a sufficiently small number. We can assume that

$$S^{\lambda-1} = \partial D^\lambda,\ S^{n-\lambda-1} = \partial D^{n-\lambda-1}$$

in $U(x_0)$ (see Fig. 122). Consider the extensions of the integral trajectories of the field outside of the limits of the segment $a - \varepsilon \leqq f \leqq a + \varepsilon$. This extension generates a movement of left and right spheres which begin to move along W; at the same time, disks D^λ and $D^{n-\lambda}$ begin to expand and move along integral trajectories. Two right (or two left) disks can intersect under such an extension only at a critical point of f (outside of critical points, integral trajectories do not intersect). Thus, we have invariantly connected a pair of disks D^λ and $D^{n-\lambda}$ with each critical point. It turns out that interaction of these disks, issuing from different critical points, defines to a great extent the topology of W.

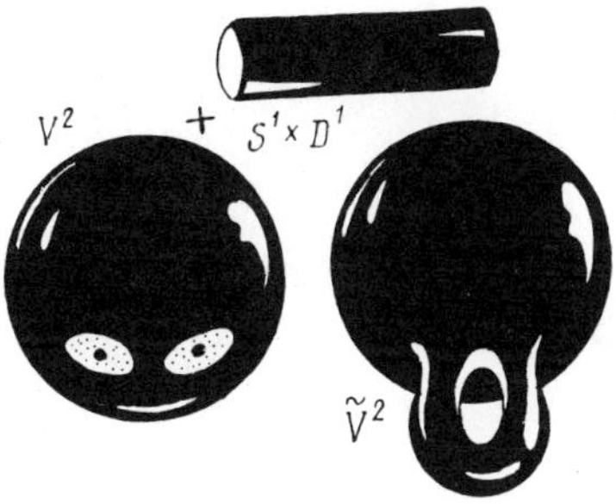

Fig. 123

11.4. Reconstructions of Level Surfaces of a Smooth Function

Let V^{n-1} be a smooth compact manifold and $S^{\lambda-1} \subset V^{n-1}$ be a smooth embedding of $S^{\lambda-1}$ into V^{n-1} such that a sufficiently small tubular neighborhood $N_\varepsilon S^{\lambda-1}$ of the sphere in V^{n-1} is diffeomorphic to $S^{\lambda-1} \times D^{n-\lambda}$. Clearly, $\partial(S^{\lambda-1} \times D^{n-\lambda}) = S^{\lambda-1} \times S^{n-\lambda-1}$ (see Fig. 72). Consider the manifold $D^\lambda \times S^{n-\lambda-1}$ with the boundary $\partial(D^\lambda \times S^{n-\lambda-1}) = S^{\lambda-1} \times S^{n-\lambda-1}$. Let us construct the new manifold $\tilde{V}^{n-1} = (V^{n-1} \setminus (S^{\lambda-1} \times D^{n-\lambda})) \cup (S^{n-\lambda-1} \times D^\lambda)$, i.e., let us reject from V^{n-1} the tubular neighborhood $N_\varepsilon S^{\lambda-1}$ and attach $S^{n-\lambda-1} \times D^\lambda$ instead. In other words, we interchange the sphere and the disk in the direct product: $S^{\lambda-1} \times D^{n-\lambda}$ is replaced by $D^\lambda \times S^{n-\lambda-1}$. We make use of the fact that both these manifolds have homeomorphic boundaries. We will say that $\tilde{V}^{n-1}$ is obtained from V^{n-1} by the Morse reconstruction of index λ. Consider the simplest examples. Let $n = 3$, $\lambda = 1$, then $\lambda - 1 = 1$ (see Fig. 123). Here,

$$U_\varepsilon S^{\lambda-1} = D^2 \cup D^2 = S^0 \times D^2, \quad \partial(U_\varepsilon S^{\lambda-1}) = S^1 \times S^0, \quad D^\lambda \times S^{n-\lambda-1} = D^1 \times S^1,$$

$$\partial(D^1 \times S^1) = S^0 \times S^1.$$

As a result of this reconstruction of index 1 to the initial two-dimensional manifold V^2, a "handle" in the sense of the theory of two-dimensional surfaces is attached (see Fig. 123). For example, from a sphere we obtain a torus.

<u>Problem</u>. Make a reconstruction of index 2 for a two-dimensional manifold.

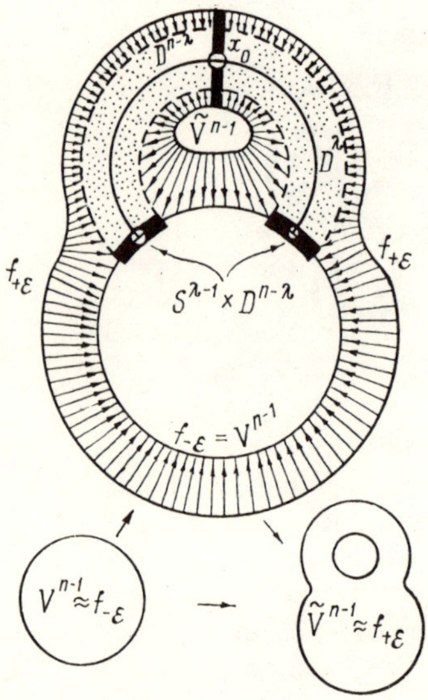

Fig. 124

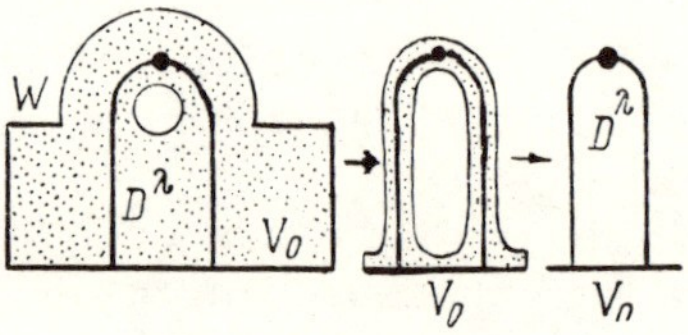

Fig. 125

The reader has already guessed, perhaps, that this operation of Morse reconstruction of index λ is actually known: we have encountered it during the study of reconstructions of level surfaces of a smooth function during passage through a critical point of index λ.

Proposition 11.2. Let f be a Morse function on W and x_0 be the unique critical point of index λ in the segment $-\varepsilon \leqq f \leqq +\varepsilon$. Then the level surface $f_{+\varepsilon}$ is obtained from the level surface $f_{-\varepsilon}$ by a Morse reconstruction of index λ.

Proof. Consider a left separatrix disk D^λ issuing from x_0 (see Fig. 124). This disk intersects $f_{-\varepsilon}$ through its boundary $S^{\lambda-1}$. Clearly, a sufficiently small tubular neighborhood $U_\varepsilon S^{\lambda-1}$ of this sphere in $f_{-\varepsilon} = V^{n-1}$ is diffeomorphic to $S^{\lambda-1} \times D^{n-\lambda}$. Therefore, we can perform a Morse reconstruction of index λ. From Fig. 124 it is clear that $f_{+\varepsilon}$ is smoothly deformed by a diffeomorphism into a reconstructed (by Morse) new manifold $\tilde{V}^{n-1}$. This deformation can be fulfilled along integral trajectories of the field $-\mathrm{grad}\, f$. The statement is proved.

The converse statement is also true: if $\tilde{V}^{n-1}$ is obtained from V^{n-1} by a Morse reconstruction of index λ, then there is a bordism W^n, with boundaries $\tilde{V}^{n-1}$ and V^{n-1}, and a Morse function f on W^n, with the unique critical point of index λ, such that $f_{-\varepsilon} = V^{n-1}$ and $f_{+\varepsilon} = \tilde{V}^{n-1}$. In fact, it suffices to refer to the already proven theorem on recovering Morse functions from the decomposition of a manifold into the sum of handles (see Sec. 8). Thus, models of Morse reconstructions of index λ can be as follows: We must consider the initial manifold V^{n-1}, multiply it by a segment, then attach to $W^n = V^{n-1} \times 1$ the handle H_λ^n of index λ, according to the embedding of the sphere $S^{\lambda-1}$ in $V^{n-1} \times 1$ with a tubular neighborhood which is a direct product (see Sec. 8). We will obtain the new manifold $\tilde{W}^n$ whose boundary consists of two components of the initial manifold V^{n-1} and the new $\tilde{V}^{n-1}$, obtained from V^{n-1} by a Morse reconstruction of index λ.

Corollary 11.1. Let (W, V_0, V_1) be a bordism with the Morse function f with only one critical point of index λ. Then W is contractible onto the CW-complex $V_0 \cup D^\lambda$ obtained from V by attaching a cell of dimension λ. In particular, W is homotopically equivalent to $V_0 \cup \sigma^\lambda$.

Proof. It suffices to consider the left separatrix disk issuing from the critical point and extend it up to the intersection with V_0. The necessary statement follows from Proposition 11.2 (see Fig. 125).

11.5. Construction of Admissible Morse Functions

The existence on compact manifolds of Morse functions whose critical values are equal to indices of a critical point, i.e., such that f(x) = ind (x) for any critical point, is a remarkable geometric fact. It means, in particular, that critical points are ordered by the value of their index, i.e., the level f(x) of the critical point x is an ascending function in index of this point. First prove an auxiliary statement.

Proposition 11.3. Let the Morse function f have only two critical points x_0 and y_0 on the bordism (W, V_0, V_1) and $a = f(x_0) < f(y_0) = b$. Assume that, for a gradient-like vector field ξ, the function f satisfies the following condition: separatrix diagrams of critical points x_0 and y_0 do not intersect, i.e.,

$$(D^{\lambda}(x_0) \cup D^{n-\lambda}(x_0)) \cap (D^{\lambda'}(y_0) \cup D^{n-\lambda'}(y_0)) = \varnothing,$$

where λ and λ' are indices of x_0 and y_0, respectively. Then, on W, there is a new Morse function g such that 1) the field ξ remains gradient-like for g; 2) g has the same critical points as f; 3) $g(x_0) > g(y_0)$, i.e., critical levels "are interchanged;" 4) g coincides with f in a neighborhood of the boundary $\partial W = V_0 \cup V_1$ and equals f + const in a neighborhood of x_0 and in a neighborhood of y_0.

Proof. In Fig. 126, the main condition of the theorem is depicted: separatrix diagrams of two critical points do not intersect. For the sake of brevity, let

$$X = D^{\lambda}(x_0) \cup D^{n-\lambda}(x_0), \quad Y = D^{\lambda'}(y_0) \cup D^{n-\lambda'}(y_0).$$

Evidently, the complement to $X \cup Y$ in W is the direct product of a manifold by a segment, namely,

$$W \setminus (X \cup Y) = (V_0 \setminus (V_0 \cap (X \cup Y)) \times I = (V_1 \setminus (V_1 \cap (X \cup Y)) \times I.$$

Since each separatrix disk intersects with V_0 (or V_1) by its separatrix sphere, the above direct product can be expressed as follows:

$$W \setminus (X \cup Y) = (V_0 \setminus (S^{\lambda-1}(x_0) \cup S^{\lambda'-1}(y_0)) \times I =$$
$$= (V_1 \setminus (S^{n-\lambda-1}(x_0) \cup S^{n-\lambda'-1}(y_0))) \times I$$

(see Fig. 126). In particular, manifolds

$$V_0 \setminus (S^{\lambda-1} \cup S^{\lambda'-1}) \text{ and } V_1 \setminus (S^{n-\lambda-1} \cup S^{n-\lambda'-1})$$

are diffeomorphic and this diffeomorphism is established along integral trajectories of ξ that glide past x_0 and y_0. On the lower (left) base of V_0, consider a smooth function α such that $\alpha(x) = 0$ in a sufficiently small neighborhood of $V_0 \cap X$ and $\alpha(x) = 1$ in a sufficiently small neighborhood of $V_0 \cap Y$. Since $X \cap Y = \phi$, then such a function exists. Then, continue the function α from the base of V_0 up to a smooth function α on the whole W. For this, let us extend α, defined on V_0, by constant values along integral trajectories of ξ.

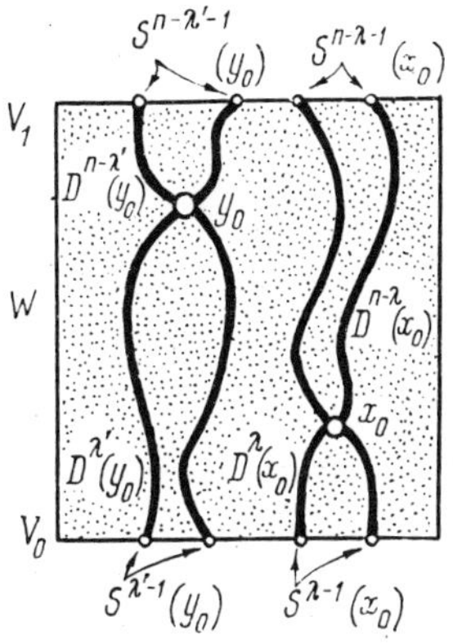

Fig. 126

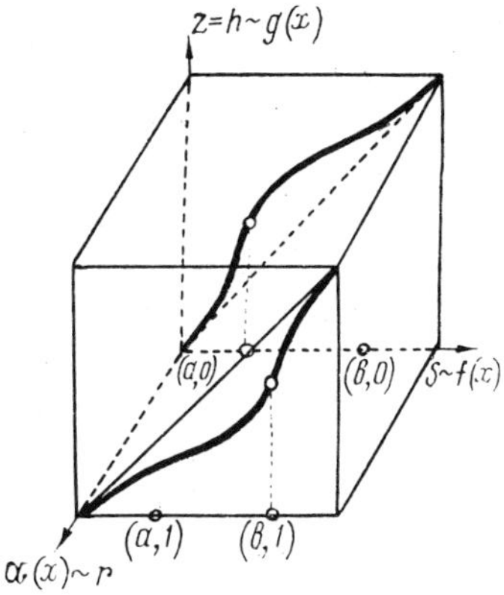

Fig. 127

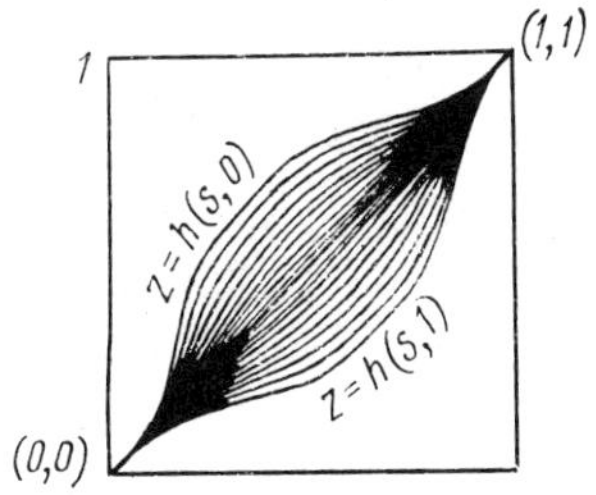

Fig. 128

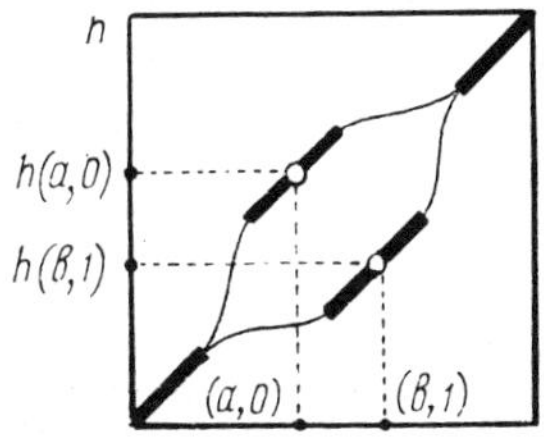

Fig. 129

Evidently this extension is well defined, since integral trajectories ξ outside of $X \cup Y$ do not intersect. As a result, we obtain the function α on W which vanishes in a neighborhood of X and is equal to 1 in a neighborhood of Y. Now consider a smooth function $z = h(s, r)$ defined by the graph of Fig. 127. In Fig. 128, the evolution of lines of intersection of this graph $z = h(s, r)$ with the plane $r = t = \text{const}$ is shown as t ranges from 0 up to 1. Let us take the function h such that the following conditions will be satisfied:

1) $\partial h/\partial s > 0$ for all (s, r), and $h(s, r)$ increases from 0 to 1 as s increases from 0 to 1;

2) $h(a, 0) = b$, $h(b, 1) = a$;

3) $\partial h(s, 0)/\partial s \equiv 1$ for all s in a neighborhood of a, and $\partial h(s, 1)/\partial s \equiv 1$ for all s in a neighborhood of b (see Fig. 129).

Now define the required function $g(x) = h(f(x), \alpha(x))$. Then we have

$$g(x_0) = h(f(x_0),\ \alpha(x_0)) = h(a, 0) = b > a = h(b, 1) = h(f(y_0),\ \alpha(y_0)) = g(y_0).$$

Thus, $g(x_0) > g(y_0)$. Conditions 1-3 imposed on h imply that g satisfies all requirements of the theorem. The proof is completed.

Now we can formulate and prove the main theorem.

Theorem 11.1 (theorem on the regrouping of critical points). Let the Morse function f have only two critical points x_0, y_0 on (W, V_0, V_1); let $a = f(x_0) < f(y_0) = b$. Suppose that the index of x_0 is no less than the index of y_0. Then there is a new Morse function g such that its critical points are the same as those of f, $g(x_0) > g(y_0)$, and g satisfies all the conditions listed in Proposition 11.3.

Proof. In the case when $X \cap Y = \phi$, where X, Y are separatrix diagrams of x_0 and y_0, respectively, the statement follows immediately from Proposition 11.3. Now, consider the general case $X \cap Y \neq \phi$. It turns out that this situation can be reduced to the case $X \cap Y = \phi$. For this, fix a surface $f_{1/2} = V$ and let λ be the index of x_0 and λ' the index of y_0. Then $\lambda \geqq \lambda'$. Since $X \cap Y \neq \phi$, then right and left separatrix spheres of points x_0 and y_0 intersect on V, i.e., $S^{n-\lambda-1}(x_0) \cap S^{\lambda-1}(y_0) \neq \phi$ (see Fig. 130). In fact, if these spheres do not intersect, then neither do the extended separatrix disks, meaning that $X \cap Y = \phi$. Since 1/2 is not a critical value for f, then the surface V is a smooth $(n - 1)$-dimensional submanifold in W and spheres $S^{n-\lambda-1}$ and $S^{\lambda'-1}$ are smooth submanifolds in V. Thus, we obtain in V two submanifolds and the sum of their dimensions is strictly less than $\dim V = n - 1$. In fact,

$$\dim S^{n-\lambda-1} + \dim S^{\lambda'-1} = n - \lambda - 1 + \lambda' - 1 =$$
$$= n - 2 - (\lambda - \lambda') < n - 1,$$

since $\lambda - \lambda' \geqq 0$. From the theorems for "general position," namely from the theorem on a transversal regularity (see [1],

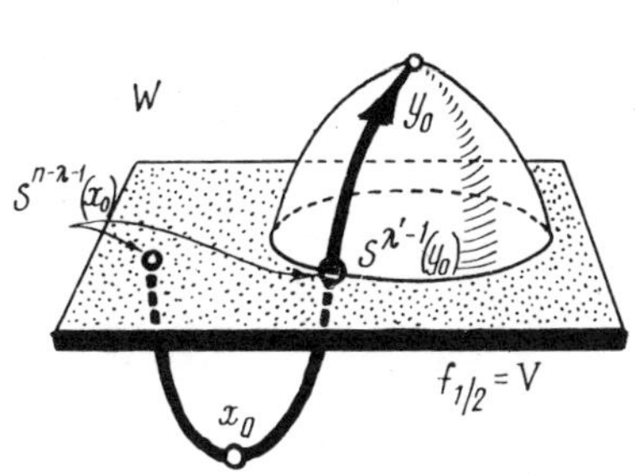

Fig. 130

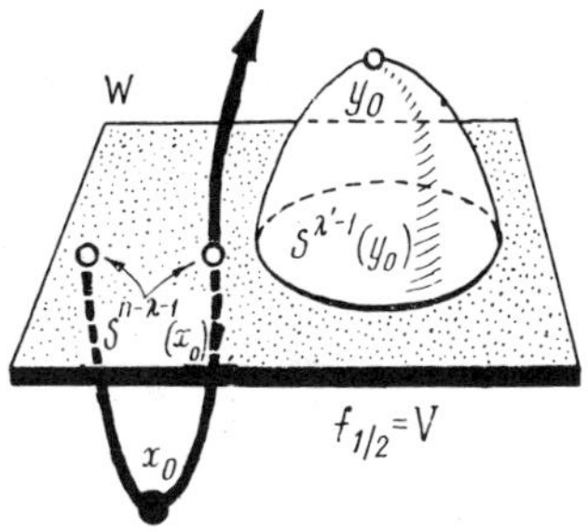

Fig. 131

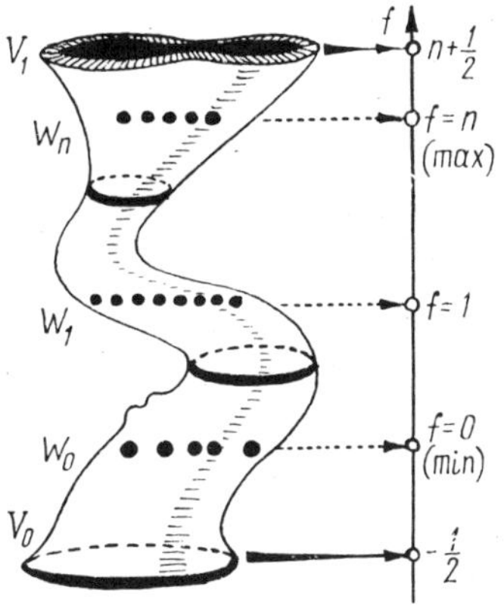

Fig. 132

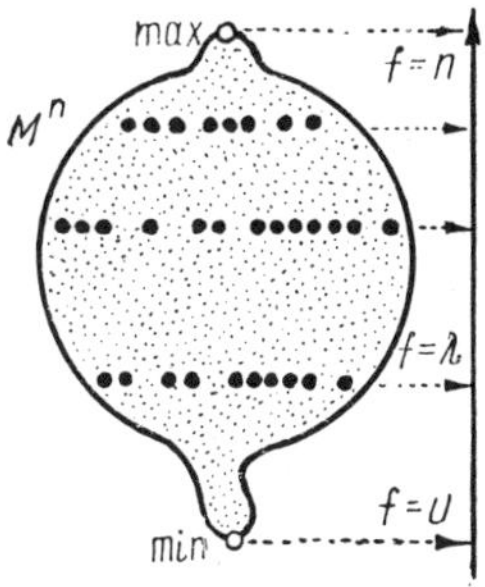

Fig. 133

p. 433 and following), it follows that there is a however-small smooth deformation φ_t of the initial embedding $S^{\lambda'-1} \to V$ to the new closed embedding which has a nonzero intersection with $S^{n-\lambda-1}$, and all mappings φ_t are also embeddings. This deformation is extendable onto a small neighborhood of V so that outside of this neighborhood all the φ_t's are identical mappings. Applying this deformation (defined now on the whole W) to the initial field ξ, we obtain the new field ξ' with the empty intersection of separatrix diagrams of x_0 and y_0. Hence, we have reduced the problem to the case $X \cap Y = \phi$ (see Fig. 131). The theorem is proved.

<u>Theorem 11.2</u> (existence of an admissible Morse function). Let (W, V_0, V_1) be an arbitrary bordism. Then on it there is a Morse function f such that $f|V_0 = -1/2$, $f|V_1 = n + 1/2$,

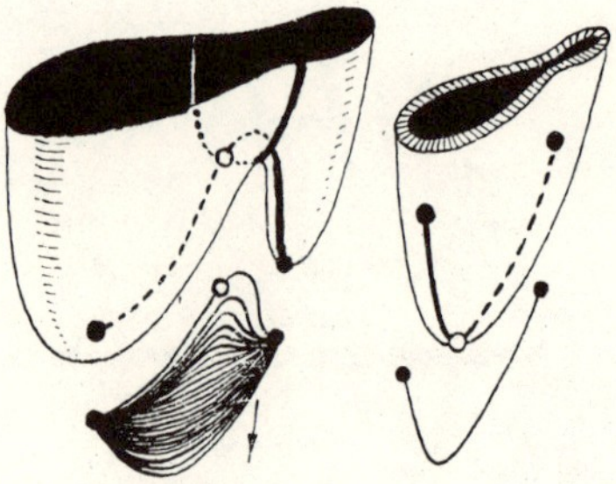

Fig. 134

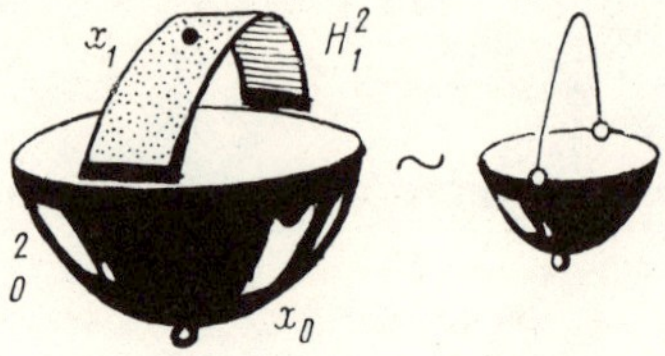

Fig. 135

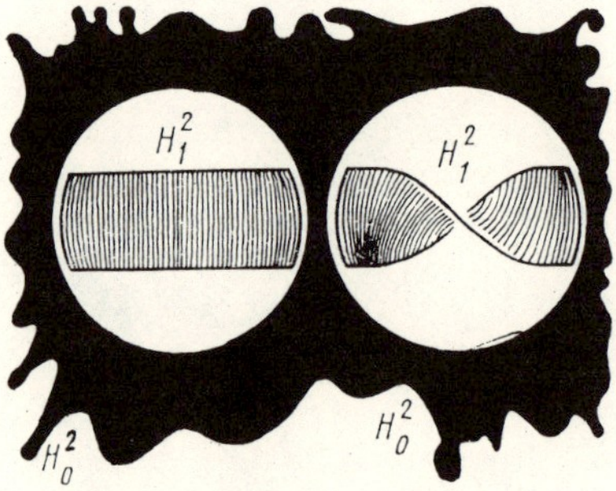

Fig. 136

where n = dim W, and the value of f at each critical point is equal to the index of this point. In particular, all critical points of the same index λ are situated on the same level surface f_λ. Thus, the initial bordism is presented as a composition of elementary bordisms $W = W_0 \circ \ldots \circ W_n$ and, in each of them, f has only one critical value λ for W_λ (see Fig. 132).

Thus, critical points of greater index of an admissible function are situated higher than points of smaller index (to the right with respect to the horizontal depiction of a bordism). Note that these functions do not constitute a dense subset in the space of all smooth functions of W.

The proof follows immediately from Theorem 11.1. If two critical points have equal indices, then we separate them into different level surfaces so that the point of smaller index will be "below." If indices of these points coincide, we stop the process of regrouping of these points (see proof of Theorem 11.1) when they are situated on the same level surface.

Theorem 11.3. Let M^n be a smooth compact connected closed manifold. Then, on M, there is an admissible Morse function f, where $0 \leqq f \leqq n$, such that it has only one minimum (point of index 0), only one maximum (point of index n), and each critical point of index λ is situated on the level surface f_λ (see Fig. 133).

Proof. Let the number of local minima of f be greater than 1. Take two points of minimum and join them by the shortest geodesic γ. Let U_ε be a sufficiently small tubular neighborhood of γ. The geodesic contains the "saddle point," i.e., the critical point of a positive index. Let us make a deformation of f in the neighborhood of U_ε as depicted in Fig. 134. The mentioned singularities of f are glued into one local minimum and no new minima appear during this process. Repeating this operation, we obtain a function with one minimum. Let us repeat a similar consideration for points of maximum. Let x and x' be the points of unique minimum and maximum, respectively. Cutting out their small neighborhoods, we obtain a bordism with a Morse function without points of index 0 and n. After that, we apply Theorem 11.2 and, reconstructing this function, we terminate the proof.

In particular, on any compact manifold, there is a Morse function f such that it has only one minimum, one maximum, and only one critical point on each critical level, so that $f(x) > f(x')$ if ind $(x) >$ ind (x'). Consider several simple corollaries.

On a two-dimensional smooth compact closed manifold, take an admissible Morse function f. Let x_0 be its point of minimum, $x_1, \ldots, x_r$ points of index 1, x_{r+1} a point of maximum, and $f(x_i) < f(x_{i+1})$, where $0 \leqq i \leqq r$. Let $0 \leqq f \leqq r + 1$ and $f(x_i) = i$. The set $0 \leqq f \leqq \varepsilon$ is diffeomorphic for small ε to a two-dimensional disk, i.e., to a handle H_0^2. During passage through the critical point x_1, a handle of index 1 is attached to this disk (see Fig. 135). In this situation, in the two-dimensional case, there are only two ways to attach this handle, shown in Fig. 136. From a homotopical point of view, they lead to the same result (a segment is attached). But from a differential point of view, we obtain different manifolds with boundary. In the first case, it is the cylinder $S^1 \times D^1$, while in the second case, it is the Moebius band, i.e., orientable and nonorientable surfaces. In particular, they are not homeomorphic. Continuing this process of recovering a manifold from an admissible Morse function and passing through $x_2, \ldots, x_r$, we attach at each step (in terms of handles) either $S^1 \times D^1$ or the Moebius band. After passage through the last point of index 1, we obtain from the homotopic viewpoint a bouquet of circles, where each circle corresponds to a critical point of index 1. The last step, i.e., passage through the point of maximum x_{r+1}, is the attachment of a handle of index 2, i.e., of a two-dimensional disk. Thus, M is diffeomorphic to $H_0^2 \cup \underbrace{H_1^2 \cup \cdots \cup H_1^2}_{r} \cup H_2^2$.

The attachment of the last handle (cell) to the manifold with boundary S^1, obtained in the (r + 1)-th step, is done uniquely: by the identity mapping of the boundary of the disk. The disk can be identified with the fundamental polyhedron W that defines the code of M^2 (see [1]); and the bouquet $\bigvee_{i=1}^{r} S_i^r$ is identified with the boundary of this polyhedron, where vertices are already identified into a point. Thus, we have obtained a new proof of the classification theorem of two-dimensional surfaces based on the existence of an admissible Morse function.

11.6. The Poincaré Duality

In geometry, an important role is played by the duality between homology and cohomology, called the Poincaré duality. The simplest variant of this statement is that, for a closed orientable connected compact smooth manifold M^n, an isomorphism $H_k(M^n, \mathbf{R}) = H^{n-k}(M^n, \mathbf{R}) = H_{n-k}(M^n, \mathbf{R})$ holds. If M^n is not orientable, then $H_k(M^n, \mathbf{Z}_2) = H^{n-k}(M^n, \mathbf{Z}_2) = H_{n-k}(M^n, \mathbf{Z}_2)$. For definiteness, we restrict ourselves to the study of the orientable case, since the nonorientable case may be considered by the same scheme.

Theorem 11.4. If M^n is an orientable closed compact smooth connected manifold, then $H_k(M^n, \mathbf{Z}) = H^{n-k}(M^n, \mathbf{Z})$. In particular, $H_k(M^n, \mathbf{R}) = H^{n-k}(M^n, \mathbf{R}) = H_{n-k}(M^n, \mathbf{R})$.

Proof. There are several different proofs of this fact. Our proof is based on the existence of admissible Morse functions since it seems to us that, in this case, the geometric nature of the Poincaré duality is clearest. It turns out that, on orientable manifolds, there are two cell decompositions that are dual to each other in the sense that each cell of one decomposition is in one-to-one correspondence with a cell of a complementary dimension of the other decomposition, and this correspondence preserves (up to a sign) incidence coefficients of cells.

To prove the theorem we must construct such a correspondence in geometric terms. Let f be an admissible Morse function on M with only one minimum and only one maximum. Consider also $-f$. Clearly, if x is the critical point of index λ for f, then the same point is also critical for $-f$, but its index is $n - \lambda$. Due to Sec. 8.4, the function f defines a cell decomposition K of the manifold in the sum of cells (handles). The function $-f$ also defines a cell decomposition $\tilde{K}$ of this manifold.

We can establish a simple geometric correspondence between these cell decompositions. For this, fix a critical point x of f and consider a small neighborhood U of this point. This neighborhood admits two decompositions (in the sense discussed in Sec. 8): with respect to f and with respect to $-f$ (see Fig. 137). These decompositions are different and define the required complexes K and $\tilde{K}$. Cells that constitute, for example, the complex K, are realized in M as left-separatrix disks D^λ, corresponding to the field grad f. Cells that constitute $\tilde{K}$ are realized as left-separatrix disks corresponding to grad $(-f)$, i.e., as right disks $D^{n-\lambda}$ for grad f (see Fig. 137). Now we may construct the Poincaré duality operator $P: K \to \tilde{K}$. Let $P(\sigma^\lambda) = \tilde{\sigma}^{n-\lambda}$ (see Fig. 138). To follow the action of this operator on (co)homology, we must establish the relationship between the incidence coefficients $[\sigma^\lambda : \sigma^{\lambda-1}]$ and $[\tilde{\sigma}^{n-\lambda-1} : \tilde{\sigma}^{n-\lambda}]$. For this, we will make use of the cell homology of complexes (see Sec. 4). Consider the cell σ_i^λ (where i is the number of a cell) and the cell $\sigma_j^{\lambda-1}$. Due to Sec. 4, $[\sigma_i^\lambda : \sigma_j^{\lambda-1}]$ is the degree of the mapping $q_{ij}^\lambda : S_i^{\lambda-1} \to S_j^{\lambda-1}$, where $S_i^{\lambda-1} = \partial\sigma_i^\lambda$ (the boundary of σ_i^λ) and the mapping q_{ij}^λ is the composition of the characteristic mapping $\partial\sigma_i^\lambda \to K^{\lambda-1}$ restricted from σ_i^λ onto its boundary $\partial\sigma_i^\lambda$ and the subsequent projection of the quotient complex $K^{\lambda-1}/K^{\lambda-2} = \bigvee S_j^{\lambda-1}$ onto the j-th summand of this bouquet, i.e., on $S_j^{\lambda-1}$ (see Fig. 139). The obtained number

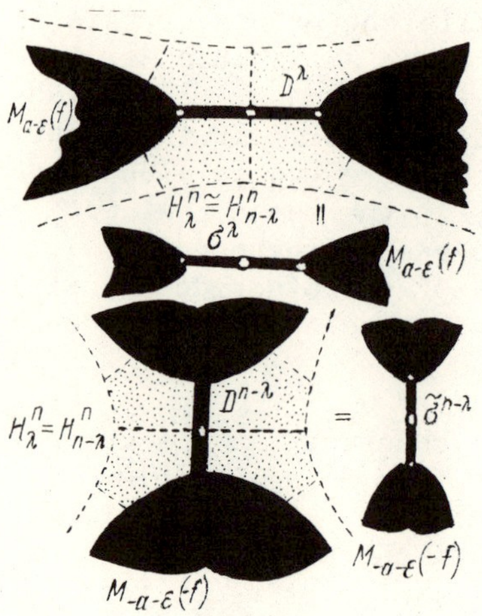

Fig. 137

Fig. 138

coincides with the index of intersection of $S_i^{\lambda-1} = \partial\sigma_i^\lambda$ with $\tilde{\sigma}_j^{n-\lambda+1}$.

Recall the definition of the index of intersection. Let N^m and Q^q be smooth orientable submanifolds in the orientable manifold M^n, whose dimension equals the sum of dimensions of N and Q, i.e., $n = m + q$. Suppose that N and Q are "in general position," i.e., their intersection consists of a finite number of points and, in these points, tangent planes T_xN and T_xQ generate (with respect to the sum) all T_xM (i.e., N and Q intersect transversally) (see Fig. 140). Fix orientations on M, N, Q. Then at each point x of intersection of N and Q, the number +1 or −1 accrues, which shows whether the fixed orientation of M coincides with the orientation induced on M by the tangent frame at $x \in N \cap Q$, consisting of two frames $e_1, \ldots, e_m \in T_xN^m$ and $a_1, \ldots, a_q \in T_xQ^q$, each of them being positively oriented from the points of view of N and Q, respectively (see Fig. 141). Adding these numbers ±1 with respect to all points of intersection, we get an integer called the index of intersection of N and Q.

The index of intersection $S_i^{\lambda-1}$ with $\tilde{\sigma}_j^{n-\lambda+1}$ can be considered as the linking coefficient (see [2], p. 528) of $S_i^{\lambda-1}$ with the boundary of this cell, i.e., with $\tilde{S}_j^{n-\lambda}$ (see Fig. 142). On the other hand, by similar considerations, this incidence coefficient $\pm\alpha$ can be presented as the index of intersection of $\tilde{S}_j^{n-\lambda}$ with σ_i^λ. Hence, $[\sigma_i^\lambda:\sigma_j^{\lambda-1}] =$

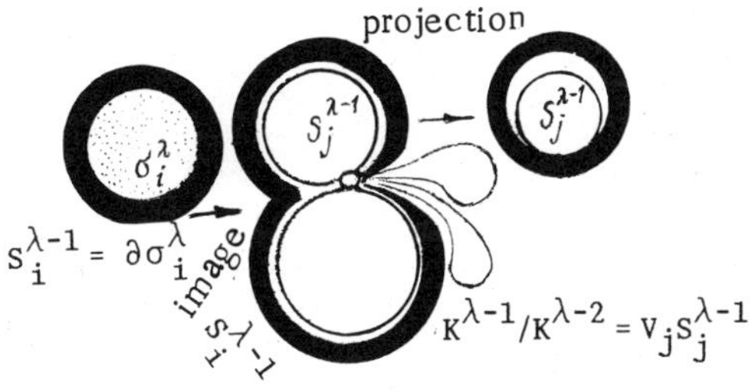

Fig. 139

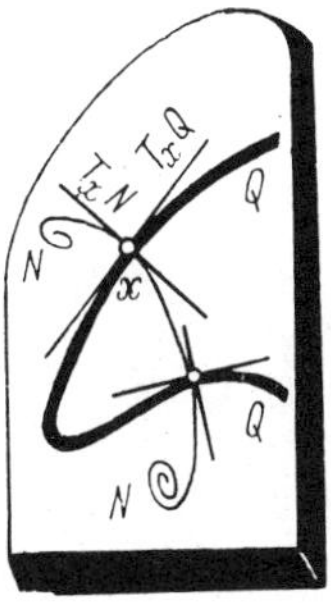

Fig. 140

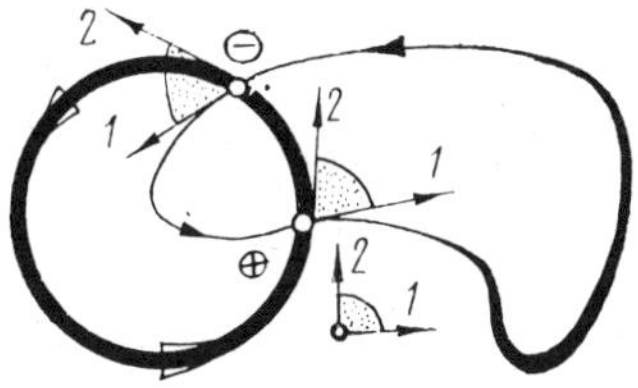

Fig. 141

$= \pm[\tilde{\sigma}^{n-\lambda} : \tilde{\sigma}^{n-\lambda+1}]$. But this means that P preserves incidence coefficients up to a sign that depends only on dimension, i.e., CW-complexes K and $\tilde{K}$ are dual. Thus, dual cells σ_j^{λ} and $\tilde{\sigma}_j^{n-\lambda} = P(\sigma_j^{\lambda})$ intersect only at one inner point (and this intersection is transversal). As we know, cells of a complex can be taken as a basis of groups of integer chains; hence, P establishes a one-to-one correspondence between bases of groups of chains $C_{\lambda}(K)$ and $C_{n-\lambda}(\tilde{K})$. Here it is important that, between groups of chains, we have obtained a nondegenerate pairing generated by the index of intersection. Choosing in each dimension the sign in an appropriate way, we may assume that

$$(a, b) = \sum_{i,j} a_i b_j (\sigma_i^{\lambda}, \tilde{\sigma}_j^{n-\lambda}), \text{ where } (\sigma_i^{\lambda}, \tilde{\sigma}_j^{n-\lambda})$$

is the index of intersection (see above) for

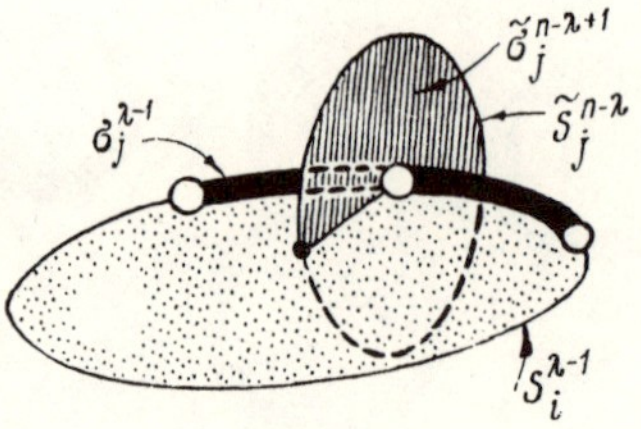

Fig. 142

$$a=\sum_i a_i\,\sigma_i^\lambda\in C_\lambda(K)\text{ and }b=\sum_j b_j\,\tilde{\sigma}_j^{n-\lambda}\in C_{n-\lambda}(\tilde{K}).$$

In particular, $(\sigma_i{}^\lambda, P\sigma_j{}^\lambda) = \delta_{ij}$, i.e., $\{\sigma_i{}^\lambda\}$ and $\{P\sigma_j{}^\lambda\}$ constitute a dual basis in groups of chains. In the nonorientable case, we ought to perform these constructions "modulo 2," i.e., with coefficients in $\mathbf{Z}_2$. In terms of the above pairing, the proven property of P can be expressed as $(\partial a, b) = (a, \partial b)$, where $a\in C_\lambda(K)$, $b\in C_{n-\lambda-1}(\tilde{K})$, since ∂ is defined by incidence coefficients and $[\sigma_i{}^\lambda:\sigma_j^{\lambda-1}] = [P\sigma_i{}^\lambda:P\sigma_j{}^{\lambda-1}]$. Here ∂ is the boundary operator. Thus, the complex (K, ∂) is dual to the complex $(\tilde{K}, \partial)$; hence $H_k(M^n, \mathbf{R}) = H_{n-k}(M^n, \mathbf{R})$. Since all CW-complexes K and $\tilde{K}$ are homotopically equivalent to the same manifold M, then $H_k(M^n, \mathbf{Z}) = H^{n-k}(M^n, \mathbf{Z})$. The theorem is proved.

An important corollary is the existence of a nondegenerate pairing between the homology and cohomology of complementary dimensions, i.e., H_k, H^{n-k}, and between the homology H_k and H_{n-k} generated by the index of intersections of cycles (see above). If $n = 2k$, then $n - k = k$, and we obtain on $H_k(M^{2k}, \mathbf{Z})$ a nondegenerate bilinear form $(a, b) = (-1)^k(b, a)$. Invariants of this form (e.g., the signature) play an important role in the topology of manifolds.

For any orientable connected closed manifold we have $H_0 = H_n = \mathbf{Z}$. We advise the reader to perform the following useful exercises:

a) Let $X\supset Y$, where X and Y are finite CW-complexes and $X\setminus Y$ is an open smooth orientable manifold. Then

$$H_i(X, Y, \mathbf{Z})=H_i(X/Y, \mathbf{Z})=H^{n-i}(X\setminus Y, \mathbf{Z}),\ i>0;$$

$$H^i(X, Y, \mathbf{Z})=H_i(X/Y, \mathbf{Z})=H_{n-i}(X\setminus Y, \mathbf{Z}),\ i>0.$$

b) Let $X^m \subset S^n$ be a finite subcomplex in S^n, $m < n$. Then

$$H_i(X, \mathbf{Z}) = H^{n-i-1}(S^n \setminus X, \mathbf{Z}), \quad i>0;$$

$$H^i(X, \mathbf{Z}) = H_{n-i}(S^n \setminus X, \mathbf{Z}), \quad i>0.$$

c) Let M be a compact closed orientable manifold and $H_k(M; \mathbf{Z}) = R_k \oplus T_k$ the decomposition of H_k into the direct sum of a free Abelian group R_k and a finite-order Abelian group T_k. Then $R_k = R_{n-k}$, $R_k = T_{n-k-1}$.

d) For any finite CW-complex, we have $R_k = R^k$, $T_k = T^{k+1}$, where $H^k = R^k \oplus T^k$.

e) The Euler characteristic of a manifold is $\chi(M^n) = \sum_{i=0}^{n} (-1)^i \beta_i$, where $\beta_i = \dim H_i(M^n; F)$ and F is a field. The Poincaré duality (for $F = \mathbf{R}$ or $\mathbf{Z}_2$ in the orientable case and for $F = \mathbf{Z}_2$ in the nonorientable case) implies $\beta_i = \beta_{n-i}$. In particular, for odd-dimensional manifolds M^{2k+1}, we have $\chi(M^{2k+1}) = 0$.

As an application of the Poincaré duality, let us prove Theorem 10.2 by estimating a category in terms of the cohomology length of a manifold.

Proof of Theorem 10.2. Let $D: H^k(M; \mathbf{Z}) \to H_{n-k}(M; \mathbf{Z})$ be the constructed isomorphism of Poincaré duality. In the ring $H^*(M; Z)$, consider elements $a_1, \ldots, a_k$ such that $a_1 \cdot \ldots \cdot a_k \neq 0$, and let $\gamma_1, \ldots, \gamma_k$ be cycles corresponding to these cocycles with respect to the duality. Then the construction of D easily implies that $D(a_1 \cdot \ldots \cdot a_k) = \gamma_1 \cap \ldots \cap \gamma_k = \gamma$, where $\gamma_1 \ \ldots \cap \gamma_k$ stands for the intersection of all cycles that can be realized in M as subcomplexes. Since $a_1 \cdot \ldots \cdot a_k \neq 0$, then γ is not homologic to zero. Now suppose that cat M = $s \leqq k$. This means that in M there are closed subsets $A_1, \ldots, A_s$, such that $M = \bigcup_{i=1}^{s} A_i$, and each A_i is contractible along M into a point. Without loss of generality, we may assume that s = k and $M = \bigcup_{i=1}^{k} A_i$, where all A_i are contractible into a point. It suffices to add k − s points to $\{A_i\}$ if $s < k$. Assign to any cycle γ_i a subset A_i. Since A_i is contractible into a point along M, then this implies that each cycle γ_i is homologic to the cycle $\tilde{\gamma}_i \subset M \setminus A_i$, i.e., the cycle γ_i can be "removed" from A_i. But then

$$\gamma = \gamma_1 \cap \ldots \cap \gamma_k \sim \bigcap_{i=1}^{k} \tilde{\gamma}_i \subset \bigcap_{i=1}^{k} (M \setminus A_i) = M \setminus \bigcup_{i=1}^{k} A_i = \varnothing,$$

i.e., γ is homologic to zero. The obtained contradiction proves the theorem.

Chapter 3

TOPOLOGY OF THREE-DIMENSIONAL MANIFOLDS

12. THE CANONICAL PRESENTATION OF THREE-DIMENSIONAL MANIFOLDS

12.1. Admissible Morse Functions and Heegaard Splittings

In a course of differential geometry and topology (see [1]), the classification theorem for two-dimensional closed manifolds M^2 is proved. It turns out that the result of the classification is simple: any M^2 is defined (nonuniquely) by a fundamental polyhedron with certain identifications on its boundary. The set of such polyhedrons defines the set of codes of two-dimensional manifolds, and there is a simple algorithm that answers the question whether two such codes define the same manifold or they define different (nonhomeomorphic) manifolds. For this, it suffices to reduce codes (i.e., polyhedrons) to the canonical form. The problem of classification of three-dimensional manifolds is much more difficult and has not been solved as yet. This problem is interesting not only in itself, but is also related to various important problems of algebra, analysis, and geometry. In this chapter, we shall briefly discuss some of these questions.

Consider two copies of a two-dimensional sphere with g handles with its standard embedding in $\mathbf{R}^3$ (see Fig. 143). Each of these two-dimensional manifolds $M_g{}^2$ is a boundary of a three-dimensional manifold in $\mathbf{R}^3$ that can be considered as a "solidification" of $M_g{}^2$ in $\mathbf{R}^3$. Denote these two-dimensional manifolds by Π_1 and Π_2. Clearly, $\partial\Pi_i = M_i$, where M_1 and M_2

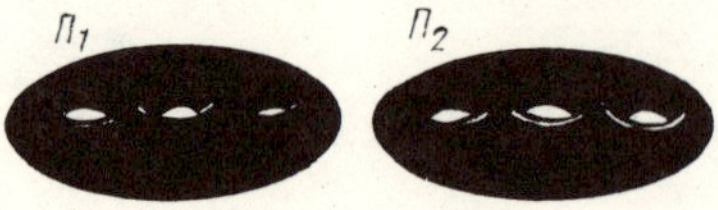

Fig. 143

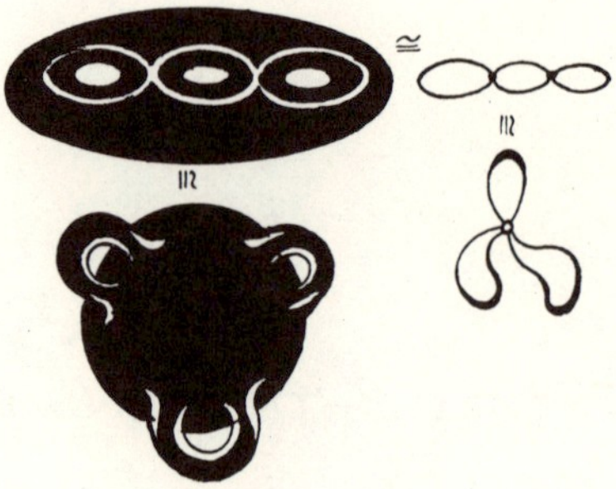

Fig. 144

denote the two copies of M_g^2. Each of Π_i is homotopically equivalent to the bouquet of g circles (see Fig. 144.) Let $\alpha: M_1 \to M_2$ be an arbitrary diffeomorphism of M_1 onto M_2. Then we may construct a three-dimensional manifold without a boundary (i.e., closed), attaching Π_2 to Π_2 with respect to this mapping of their boundaries. Denote the obtained manifold by $M^3(\alpha)$.

Theorem 12.1 (the canonical presentation of three-dimensional manifolds). Any three-dimensional smooth compact connected closed manifold can be (nonuniquely) presented for a certain integer $g \geq 0$ as $M^3(\alpha)$, i.e., as the attachment of two three-dimensional manifolds Π_1 and Π_2 with respect to a diffeomorphism (homeomorphism) of their boundaries M_1 and M_2. Both Π_1 and Π_2 are homeomorphic to a three-dimensional ball with g handles.

Proof. On a three-dimensional manifold M^3, consider an admissible Morse function (see 11.5) with one minimum, one maximum, and points $x_1, \ldots, x_g$ of index 1, and $y_1, \ldots, y_r$ of index 2. Let us assume that the critical points are ordered, i.e., $0 \leq f \leq 1$, $f_0 = \min$, $f_1 = \max$. All critical points of index 1 are situated on the level surface $f_{1/3}$ and all points of index 2 are situated on the level surface $f_{2/3}$. The Poincaré duality implies that $g = r$, i.e., the number of critical points of indices 1 and 2 is the same (recall that the Euler characteristic of M^3 is zero; hence $0 = 1 - g + r - 1$, i.e., $g = r$). Consider the surface $f_{1/2}$. Since it has no critical points of f, then $f_{1/2}$ is homeomorphic to a two-dimensional compact smooth connected orientable manifold M^2, i.e., M^2 is homeomorphic to the sphere with a certain number of handles (see Fig. 145). Clearly, $M^2 = (f \leq \frac{1}{2}) \cup (f \geq \frac{1}{2})$, where manifolds $(f \leq \frac{1}{2})$ and $(f \geq \frac{1}{2})$ have M^2 as a boundary and

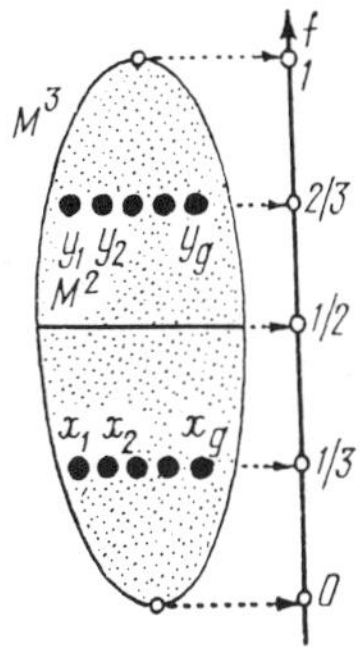

Fig. 145

each of them is therefore homeomorphic to a three-dimensional ball with g handles. Thus, we may set $\Pi_1 = (f \leqq \frac{1}{2})$, $\Pi_2 = (f \geqq \frac{1}{2})$. The theorem is proved.

A presentation of M^3 in the form $M^3(\alpha)$ is sometimes called the Heegaard splitting. Thus, this diagram is defined by a certain diffeomorphism $\alpha: M_1 \to M_2$.

Lemma 12.1. If two diffeomorphisms α, $\beta: M_1 \to M_2$ are homotopic in the class of diffeomorphisms, then the corresponding manifolds $M^3(\alpha)$ and $M^3(\beta)$ are diffeomorphic. If $M^3 = M^3(\alpha)$ is a presentation of the manifold corresponding to a Heegaard diagram, then, on M^3, there is an admissible Morse function that defines the decomposition of M^3 into the union of Π_1 and Π_2, which coincides with the initial Heegaard diagram.

Proof. The first part of the lemma is evident since a smooth homotopy in the class of diffeomorphisms is extendable into a small tubular neighborhood of M^2 in M^3. The second statement holds since, on Π_1 and Π_2, the standard Morse functions f_1 and f_2 with critical points of index 1 for Π_1 and index 2 for Π_2 are defined, as constructed above in Sec. 8 (see Fig. 146). Recall that these functions are constant on boundaries $\partial\Pi_1$ and $\partial\Pi_2$, which enables us to construct the general Morse function f on the whole M^3. The lemma is proved.

12.2. Examples of Heegaard Splittings

The number g, equal to the number of handles of $M^2 = f_{1/2} = \partial\Pi_i$, is sometimes called the genus of the Heegaard

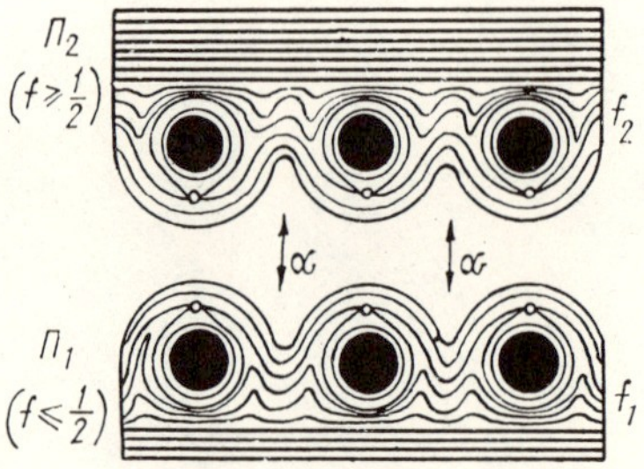

Fig. 146

splitting. Consider the simplest examples that clarify the process of construction of diagrams $M^3(\alpha)$. The unique manifold that admits the Heegaard splitting of genus g = 0 is a three-dimensional sphere. In fact, since g = 0, then Π_1 and Π_2 are homeomorphic to the three-dimensional ball. Therefore, a necessary statement follows from the fact that the corresponding Morse function constructed in Lemma 12.1 has only two critical points (of minimum and maximum); by Proposition 9.3, M^3 is homeomorphic in the standard sphere. There are many more manifolds M^3 with Heegaard splitting of genus 1. Here $\Pi_1 \approx \Pi_2 = S^1 \times D^2$. First, the sphere S^3 admits such a splitting. We are actually already acquainted with the presentation of S^3 as the attachment of two "solid tori" (g = = 1) by their common boundary. Such an attachment was analyzed in Sec. 6.3 during the study of the geometry of the Hopf bundle. In Fig. 31 (see Sec. 6), a diffeomorphism α: $T_1^2 \to T_2^2$ is depicted and the corresponding attachment is as follows: parallels of $T_1^2 = \partial\Pi_1$ are identified with meridians of $T_2 = \partial\Pi_2$, and vice versa. The result of the attachment is depicted in Fig. 30. This Heegaard splitting of genus 1 is realized also as follows: Let us embed S^3 in $\mathbf{C}^2(z, w)$; then $\Pi_1 = S^3 \cap \{|z| \geq |w|\}$, $\Pi_2 = S^3 \cap \{|z| \leq |w|\}$ (see Sec. 6). The orthogonal transformation $(z, w) \to (w, z)$ transforms Π_1 into Π_2 (and vice versa), transforming the boundary of the solid tori into itself. The obtained mapping is the required attaching diffeomorphism α in the splitting $S^3 = M^3(\alpha)$.

The next example is the manifold $S^1 \times S^2$, admitting the Heegaard splitting of genus 1. It suffices to take for $\alpha: T^2 \to$ $\to T^2$ the identity mapping of the torus. Clearly, the attachment of two solid tori $S^1 \times D^2$ with respect to such an identification of their boundaries gives us $S^1 \times D^2$. To clarify this process, it is useful to consider its analog in the two-dimensional case (see Fig. 147). Attaching two rings $S^1 \times D^1$

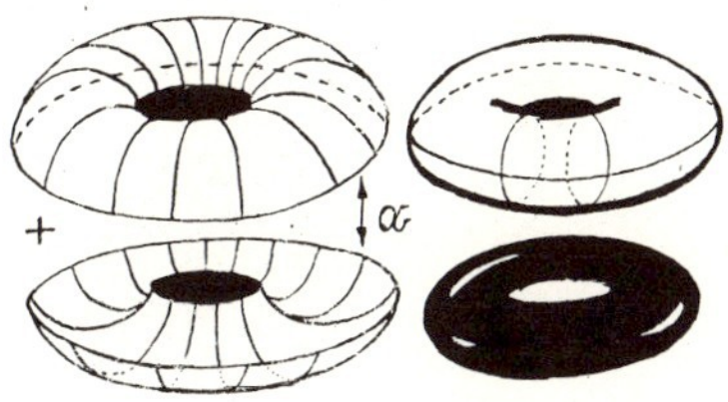

Fig. 147

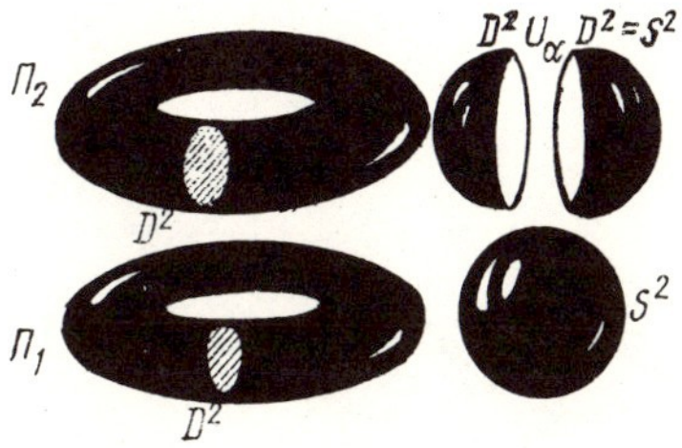

Fig. 148

and $S^1 \times D^1$ with respect to the identity mapping of their boundaries, we obviously obtain a torus, i.e., $S^1 \times S^1$. The same occurs in the three-dimensional case (see Fig. 148). Let D^2 be an arbitrary disk-fiber in the direct product $\Pi_1 = S^1 \times D^2$. This disk, being attached to the second disk in Π_2 with respect to the identity mapping of their boundaries-cycles, gives us a two-dimensional sphere (see Fig. 148).

Thus, a general principle of construction of manifolds admitting the Heegaard splitting of genus 1 is clear. The diffeomorphism $\alpha: T^2 \to T^2$ defines the induced mapping α_*: $H_1(T^2; \mathbf{Z}) \to H_1(T^2; \mathbf{Z})$, i.e., $\alpha_*: \mathbf{Z} \oplus \mathbf{Z} \to \mathbf{Z} \oplus \mathbf{Z}$. Hence, the homomorphism α_*, which is an automorphism of $\mathbf{Z} \oplus \mathbf{Z}$, is defined by an integer matrix $\begin{pmatrix} a & b \\ c & d \end{pmatrix}$ such that $ad - bc = \pm 1$ (+1 if α preserves the orientation and -1 otherwise). Thus, for example, in the above examples we have: a) $\alpha_* = \begin{pmatrix} 0 & 1 \\ -1 & 0 \end{pmatrix}$ in the case of a sphere and, b) $\alpha_* = \begin{pmatrix} 1 & 0 \\ 0 & 1 \end{pmatrix}$ in the case $S^1 \times S^2$. Prove that diffeomorphisms with the matrix $\begin{pmatrix} 1 & b \\ 0 & 1 \end{pmatrix}$ also define $S^1 \times S^2$. We leave the proof of the following statement to the reader.

Proposition 12.1. Any three-dimensional smooth compact closed connected manifold admitting the Heegaard splitting of genus 1 is homeomorphic (and diffeomorphic) to one of the following manifolds: 1) the standard sphere S^3, 2) $S^1 \times S^2$, 3) the so-called lens space obtained after factorization of the standard sphere S^3 with respect to the smooth $\mathbf{Z}_p$-action, defined by the formula $(z, w) \to (e^{2\pi i/p} \cdot z, e^{2\pi ik/p} \cdot w)$, where (z, w) are complex coordinates in $\mathbf{C}^2(z, w)$, $S^3 = \{|z|^2 + |w|^2 = 1\}$. In particular, $S^3/\mathbf{Z}_2$ (for $p = 2$) is diffeomorphic to the projective space $\mathbf{R}P^3$.

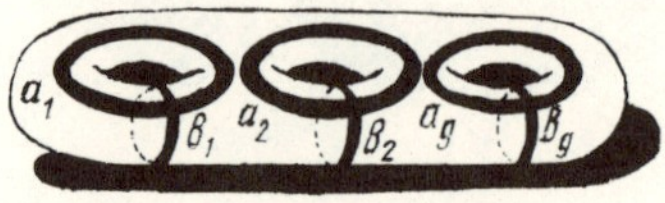

Fig. 149

Fig. 150

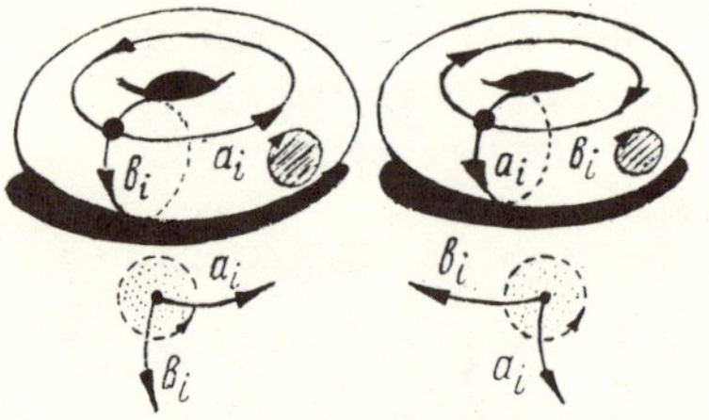

Fig. 151

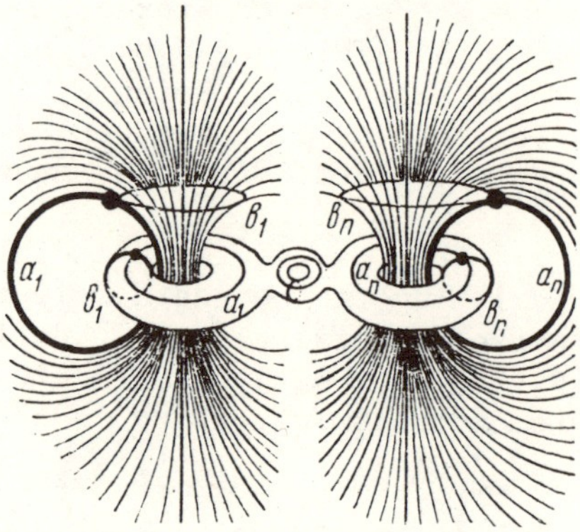

Fig. 152

Thus, we have a complete description of all manifolds M^3 admitting the Heegaard splitting of genus 1. However, passage to the case $g > 1$ highly complicates the situation and there are as yet no similar classification theorems. Let us give the simplest example of the Heegaard splitting of genus $g > 1$. Let a_i and b_i, where $1 \leqq i \leqq g$, be standard parallels and meridians on M_g^2 (see Fig. 149). Consider the diffeomorphism $\alpha: M_1 \to M_2$ that interchanges parallels and meridians, i.e., $\alpha(a_i) = b_i$, $\alpha(b_i) = -a_i$, $1 \leqq i \leqq g$. This diffeomorphism may be constructed as follows. Present M_g^2 in the form of a sphere with g handles (see Fig. 150). We may assume that, to the sphere S^2 with g handles, g copies of the torus T^2 with a hole are attached. Define the diffeomorphism α on the sphere with holes as follows. On each torus with a hole, consider a diffeomorphism that interchanges a parallel and a meridian (see Fig. 151). We may assume that the

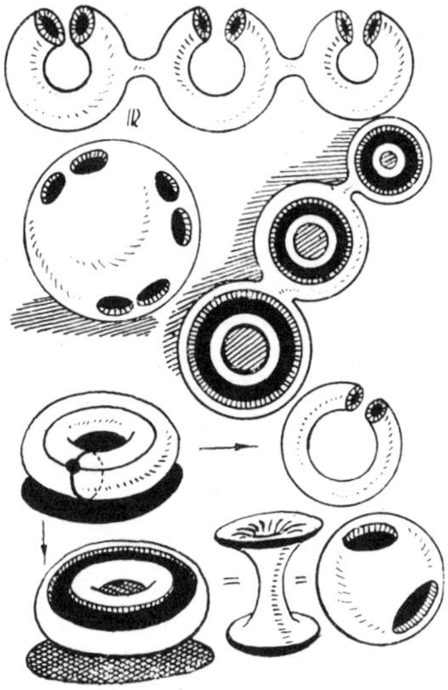

Fig. 153

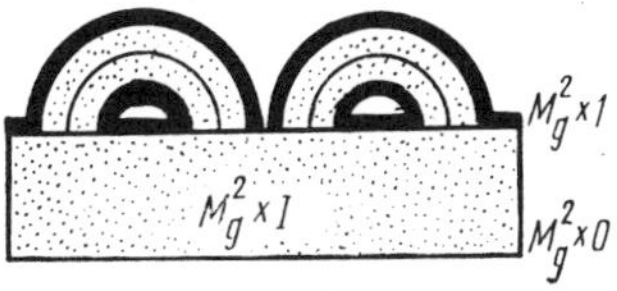

Fig. 154

hole passes into itself. Since $a_i \to b_i$ and $b_i \to -a_i$, then, as is clear from Fig. 151, the boundary of the hole (circle) is mapped identically. Hence, glueing the constructed diffeomorphisms on handles and on the sphere with holes, we obtain α. Attaching Π_1 and Π_2 with respect to this diffeomorphism, we get the standard sphere S^3. This statement is proved exactly as it is for the similar fact when g = 1. In fact, in $S^3 = \mathbf{R}^3 \cup \infty$, consider the three-dimensional manifold with boundary M_g^2 in its standard embedding. Then, the complement to Π_1 in $S^3 = \mathbf{R}^3 \cup \infty$ is, evidently, homeomorphic to the second copy of Π_2, where the role of parallels is played now by meridians, and vice versa (see Fig. 152).

12.3. The Coding of Three-Dimensional Manifolds in Terms of Nets

Heegaard splittings may be defined by an equivalent but clearer method. Consider the surface of a genus g, i.e., M_g^2, and define on it two systems of smooth circles $S_1, \ldots, S_g$ and $\bar{S}_1, \ldots, \bar{S}_g$, without self-intersections, which will be called circles of index 1 and index 2, respectively. Furthermore, suppose that 1) circles of the same index do not intersect; 2) if we cut M_g^2 along all circles of index 1 or along all circles of index 2, the result in both cases will be the sphere S^2 with 2g holes. An example of such a system of circles is already known (see Fig. 149). Clearly, cutting M_g^2 along all $a_1, \ldots, a_g$ or along all $b_1, \ldots, b_g$, we get the two-dimensional sphere with 2g holes (see Fig. 153). The system of circles that satisfies all the above-listed conditions will be denoted by (α_1, α_2).

Definition 12.1. Two systems (α_1, α_2), (α_1', α_2') of circles described above will be called equivalent if there is a diffeomorphism $\alpha: M_g^2 \to M_g^2$ such that it transforms one system into another, i.e., $\alpha(S_i) = \bar{S}_i$ for $1 \leq i \leq g$. The class of equivalent systems (α_1, α_2) on M_g^2 will be called the net $[\alpha_1, \alpha_2]$. The number g will be called the genus of the net.

Lemma 12.2. Each net $[\alpha_1, \alpha_2]$ defines a three-dimensional smooth compact connected closed manifold that will be denoted by $M^3[\alpha_1, \alpha_2]$. Conversely, any three-dimensional closed compact smooth connected manifold is presentable in the form $M^3[\alpha_1, \alpha_2]$ for a certain integer g.

Proof. Let us recover from a net a three-dimensional manifold. For this, consider the direct product of M_g^2 and the segment I. In the upper boundary of this direct product, let us distinguish the system of circles of index 2, i.e., $\bar{S}_1, \ldots, \bar{S}_g$. Since M_g^2 is orientable, then a sufficiently small tubular neighborhood of each of the circles $\bar{S}_i$ is homeomorphic to the direct product of $\bar{S}_i$ by the segment. Hence, the scheme of Sec. 8.5 allows us to attach g handles H_2^2 of index 2, i.e., to make a reconstruction of the upper boundary of $M_g^2 \times I$ using circles $\bar{S}_1, \ldots, \bar{S}_g$. We will obtain the new three-dimensional manifold $(M_g^2 \times I) \bigcup \left(\bigcup_1^g H_2^2\right)$, whose upper boundary is the result of the reconstruction of M_g^2 (of the lower boundary) (see Fig. 154). We claim that this upper boundary is homeomorphic to S^2. In fact, the definition of a net implies that $M_g^2 \setminus \left(\bigcup_{i=1}^g \bar{S}_i\right)$ is homeomorphic to the sphere with 2g holes. Fix a circle $\bar{S}_i$ generating (after cutting) two holes D_i and D_i'. The reconstruction of index 2 has been studied earlier (see Fig. 75). Clearly, it is equivalent to the attachment of two disks to boundaries of the holes obtained after cutting the surface along $\bar{S}_i$. Since this is true for any $\bar{S}_i$, we will finally close off all holes in the sphere and

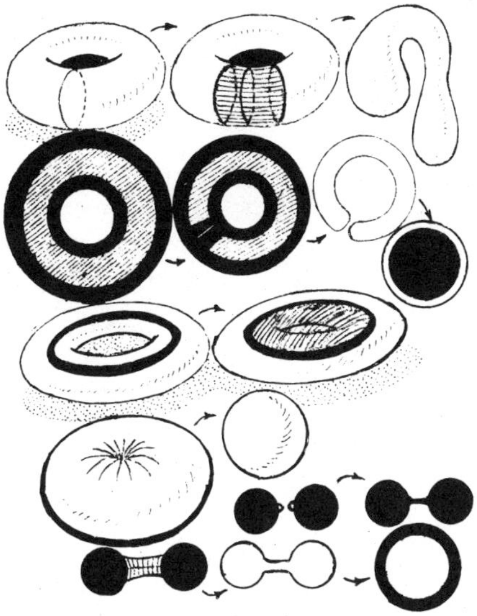

Fig. 155

Fig. 156

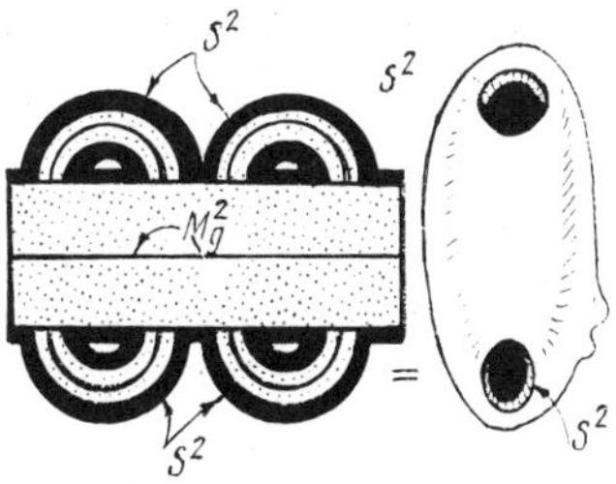

Fig. 157

obtain the sphere S^2. For g = 1 and g = 2, this operation is depicted in Figs. 155 and 156. Now let us consider another system of circles $S_1, \ldots, S_g$ of index 1 on M_g^2 and repeat the above operation for the system $S_1, \ldots, S_g$ with $M_g^2 \times \times I$. Attaching two reconstructed direct products by their common boundary M_g^2, we obtain a three-dimensional manifold whose boundary consists of two two-dimensional spheres. One of them is obtained by reconstruction with respect to circles of index 2, while the other is obtained by reconstruction with respect to circles of index 1 (see Fig. 157). Here both spheres are depicted by thick black curves. The last step is the attachment of three-dimensional disks to these spheres, resulting in the three-dimensional manifold $M^3[\alpha_1, \alpha_2]$. Conversely, let M^3 be an arbitrary manifold of the type indicated in the theorem. Let us present it in the form $M^3[\alpha_1, \alpha_2]$. On M^3, consider an admissible Morse function f with one point of minimum (f = 0), one point of maximum (f = 1), critical points $x_1, \ldots, x_g$ of index 1 situated on $f_{1/3}$, and points $y_1, \ldots, y_g$ of index 2 situated on $f_{2/3}$ (see Fig. 145). Then $M^3 = \Pi_1 \bigcup_{\alpha} \Pi_2$, where Π_1, Π_2 are "solidifications" of $M_g{}^2$ in $\mathbf{R}^3$ (see above). Consider left separatrix disks $D^2(y_1), \ldots, D^2(y_q)$ for points $y_1, \ldots, y_q$ and right separatrix disks $D^2(x_1), \ldots, D^2(x_g)$. These disks intersect with the point $f_{1/2} = M_g{}^2$ along two systems of circles which evidently are right circles $S^1(x_1), \ldots, S^1(x_g)$ and left circles $S^1(y_1), \ldots, S^1(y_g)$. Clearly, circles $S^1(x_i)$ are identified with circles S_i of index 1, and circles $S^1(y_i)$ with circles $\bar{S}_i$ of index 2. In particular, the notion "circles of indices 1 or 2," accepted in Definition 12.1, now becomes clear. The definition of separatrix disks immediately implies that these two systems of circles on $M_g{}^2$ define a net in the sense of Definition 12.1. The lemma is proved.

Thus, a genus of a net is the number of critical points of the corresponding admissible Morse function on a three-dimensional manifold. It is clear from the proof of Lemma 12.2 that the definition of a net $[\alpha_1, \alpha_2]$ defines a diffeomorphism $\alpha: M_g{}^2 \to M_g{}^2$ and, conversely, each such diffeomorphism defines a net. Therefore, in what follows we will denote the net $[\alpha_1, \alpha_2]$ by the same symbol α as the corresponding diffeomorphism (more exactly, the class of homotopic diffeomorphisms). Now, we may give one more interpretation of systems $\alpha_1 = \{S_i\}$ and $\alpha_2 = \{\bar{S}_i\}$. Lemma 12.2 implies that if α is a diffeomorphism corresponding to the net $[\alpha_1, \alpha_2]$, then for $\bar{S}_j$ we may take the image of S_j with respect to this diffeomorphism (or vice versa). Thus, the definition of the Heegaard splitting is equivalent to that of a net.

12.4. Nets and Separatrix Diagrams

Seeking to find still simpler methods of coding three-dimensional manifolds, we will now introduce an extremely simple method of defining M^3. Consider an arbitrary net

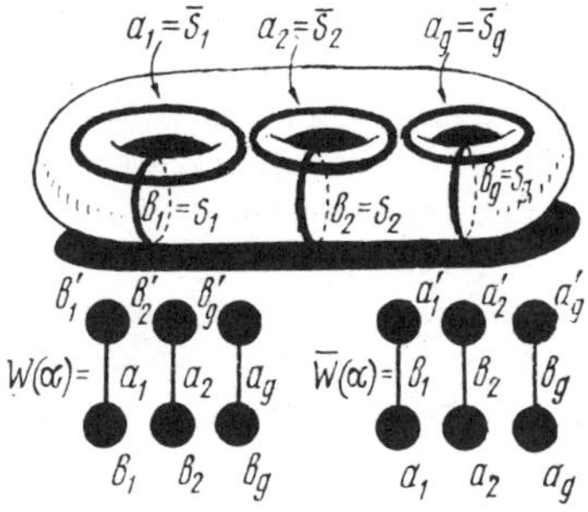

Fig. 158

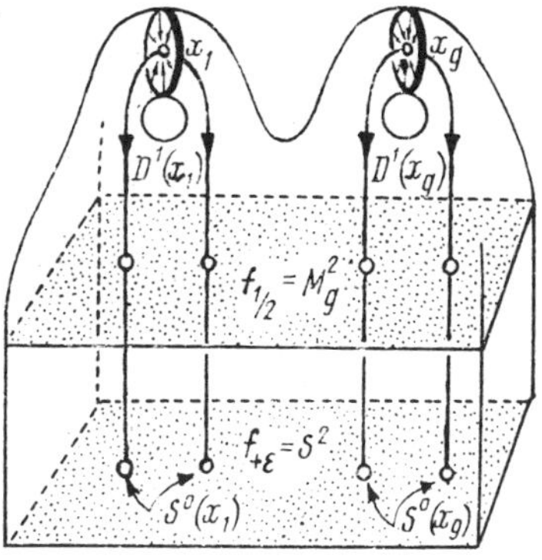

Fig. 159

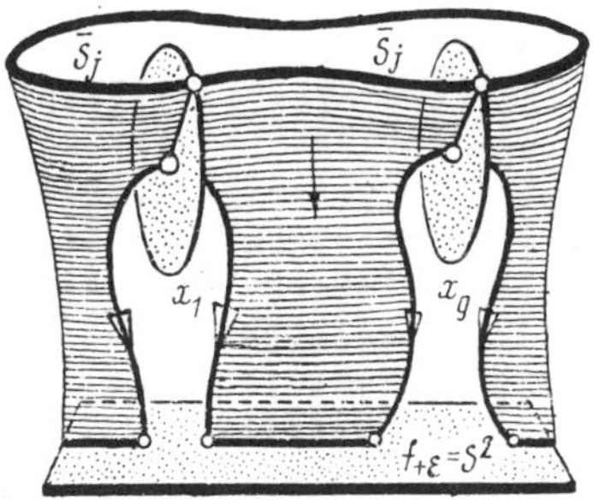

Fig. 160

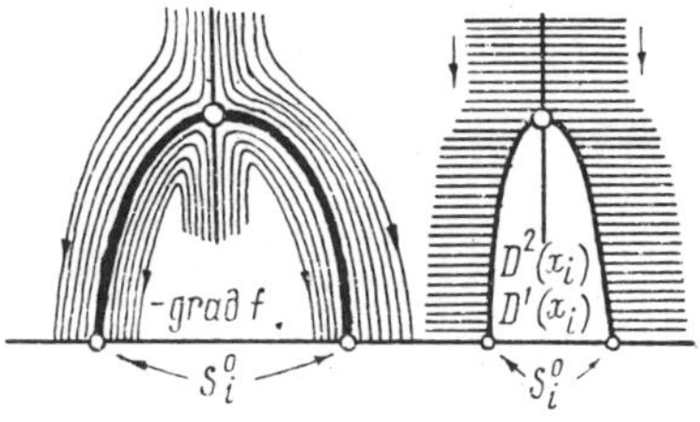

Fig. 161

$\alpha = [\alpha_1, \alpha_2]$ and let (α_1, α_2) be a representative, i.e., two systems of circles. Fix, for instance, the system $\{\bar{S}_j\}$ and let us cut $M_g{}^2$ along all circles $\{S_i\}$. We will obtain a sphere with 2g holes. Then circles $\bar{S}_j$ will also be cut, and convert S^2 into the collection of segments that connect holes. Considering each hole as a point on a sphere (i.e., we puncture a point instead of rejecting a disk), we obtain on S^2 a planar graph that will be denoted by $W(\alpha)$, where α is a net. The second graph $\bar{W}(\alpha)$ is similarly obtained by interchanging $\{S_i\}$ and $\{\bar{S}_i\}$. Thus, each net is uniquely defined by graphs $W(\alpha)$ and $\bar{W}(\alpha)$ that can be considered, if needed, as linear graphs by rejecting a point (which is not a vertex of graphs) from the sphere. In Fig. 158 these graphs are shown for the simplest net that defines S^3, i.e., the net of parallels and meridians on $M_g{}^2$. It turns out that both these graphs are uniquely defined by the separatrix diagram of f on M^3, whose separatrix circles coincide with $\{S_i\}$ and $\{\bar{S}_j\}$ (see above). In fact, consider points $x_1, \ldots, x_g$ and separatrices that emerge from them. The right disk $D^2(x_i)$ is two-dimensional, while the left disk $D^1(x_i)$ is one-dimensional. In Fig. 159 the interaction of these disks is depicted. Arrows show the direction of the vector field $-$grad f. Two separatrices that form the one-dimensional disk $D^1(x_i)$ emerge from x_i and, descending, reach $f+\varepsilon$, homeomorphic to the sphere S^2 of small radius which encircles the point of minimum of $f+\varepsilon$. Reaching this sphere, $D^1(x_i)$ cuts on it two points, a zero-dimensional sphere $S^0(x_i)$. Now consider a circle $\bar{S}_j$, i.e., the boundary of the left disk $D^2(y_j)$. This circle is situated on $M_g^2 = f_{1/2}$ and is also led by integral trajectories of the field $-$grad f downward toward $S^2 = f_{1/2}$. Then $\bar{S}_j$ breaks, meeting with separatrix disks $\{D^2(x_i)\}$. This situation is described in Fig. 160. The circle $\bar{S}_j$ meeting $D^2(x_i)$ is cut by this disk into halves, and points of the cut begin to glide downward, descending along the one-dimensional separatrix until they reach $S^2 = f+\varepsilon$. In Fig. 161, integral trajectories of $-$grad f are shown in a neighborhood of a separatrix diagram. This flow leads pieces of the circle $\{\bar{S}_j\}$ downward. Thus, on S^2, a graph arises whose vertices are separatrix zero-dimensional spheres (pairs of points) $\{S^0(x_i)\}$ and whose edges are parts of the circles $\{\bar{S}_j\}$. In exactly the same way, the graph $\bar{W}(\alpha)$ arises on the upper sphere $S^2 = f_{1-\varepsilon}$ when $\{S^0(y_j)\}$ and pieces of circles $\{S^1(x_i)\}$ are considered. In this case, parts of circles $S^1(x_i)$ are led upward by the flow grad f.

13. THE PROBLEM OF RECOGNITION OF A THREE-DIMENSIONAL SPHERE

13.1. Homological Spheres

Above we have produced "the list of all three-dimensional manifolds" (but not their classification), i.e., we have shown the countably many codes, for example, graphs $W(\alpha)$, $\bar{W}(\alpha)$, each of which defines a three-dimensional manifold. Besides, any three-dimensional manifold is necessarily contained in this list (with infinite multiplicity). But the existence of such a list does not mean in the least that we have obtained the classification of three-dimensional manifolds in the same sense as we have classified two-dimensional manifolds.

The situation is that each manifold M^3 is presented in this list not by a unique canonical code but by an infinite number of codes. The following classification problem arises: Is there an algorithm that, acting by a "uniform program," determines whether two codes define the same manifold or different ones, i.e., nondiffeomorphic ones? We require that this algorithm be defined by a sequence of operations (a program) whose definition does not depend on the pair of codes entering the input of the algorithm. The situation is that, dealing with two concrete codes, it is sometimes possible to answer the question because of some specific considerations applicable only for these codes, and impossible in other cases. We will not dwell here on the exact algorithmic formulation of the problem, but restrict ourselves to the intuitive notion of an algorithm as a device that acts according to a given problem and processes signals that enter its input. In this case, it is pairs of codes of manifolds.

The simplest question in this vein is that of the recognition, among sets of codes of all three-dimensional manifolds, of codes of the simplest manifold, i.e., of the standard sphere. In other words, how can one recognize Heegaard splittings corresponding to S^3? How can one single out these diagrams from those of other manifolds not diffeomorphic to S^3? This question is sometimes called the algorithmic Poincaré problem of recognition of a sphere. There are several natural candidates for "the characteristic property of the standard sphere." The simplest conjecture is that, if the one-dimensional homology of M^3 is trivial, then this manifold is diffeomorphic to the sphere. This conjecture seems plausible at first glance because the following simple statement holds.

Proposition 13.1. Let M^3 be a three-dimensional closed orientable connected manifold and $H_1(M^3; \mathbf{Z}) = 0$. Then M^3 has the same integer homology as the sphere, i.e., $H_*(M^3; \mathbf{Z}) = H_*(S^3; \mathbf{Z})$.

Such manifolds are sometimes called homological spheres (cf. [8]).

Proof. Since $H_1(M; \mathbf{Z}) = 0$, then by the Poincaré duality $H^2(M; \mathbf{Z}) = 0$. Since $R_1 = R^1$ and $T_0 = T^1$, then $H^1(M; \mathbf{Z}) = 0$. Since $H^3(M; \mathbf{Z}) = \mathbf{Z}$, then $T_2 = T^3 = 0$ and $R^2 = R_2 = 0$, i.e., $H_2(M; \mathbf{Z}) = 0$. The statement is proved.

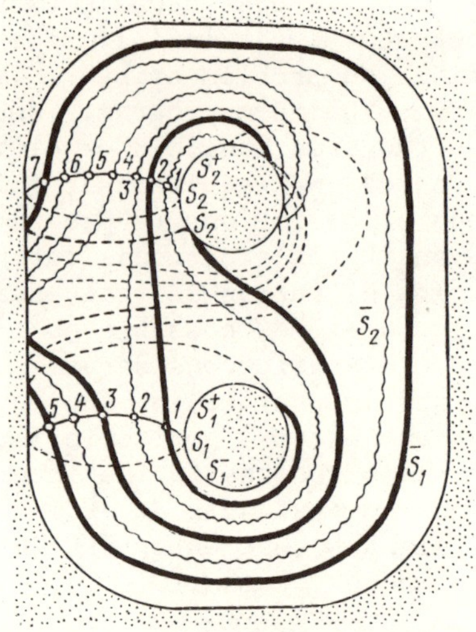

Fig. 162

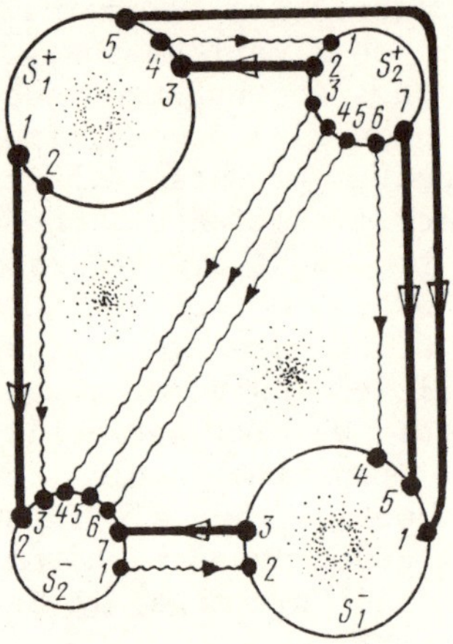

Fig. 163

However, the above conjecture is false.

Proposition 13.2. There are three-dimensional homological spheres that are not homotopically equivalent (and, moreover, not diffeomorphic) to the standard three-dimensional sphere.

We will prove this statement by producing a three-dimensional manifold which is a homological sphere but with a nonzero fundamental group, and hence is not homotopically equivalent to the sphere. For this, we will produce a Heegaard diagram of genus 2, i.e., on M_g^2 (on a pretzel*), the net $[\alpha_1, \alpha_2]$. This net consists of two families of circles S_1, S_2 and $\bar{S}_1$, $\bar{S}_2$ (see Definition 12.1). This net is shown in Fig. 162. Here circles S_1 and S_2 (of index 1) are standard meridians, the circle of index 2, i.e., $\bar{S}_1$, is depicted by a continuous smooth trajectory on the pretzel, and the circle $\bar{S}_2$ of index 2 is depicted by the wavy line. We leave it to the reader to verify that this is a net, i.e., that by cutting the pretzel along $\bar{S}_1$ and $\bar{S}_2$ we obtain a sphere with four holes. This statement is evident if we cut along S_1 and S_2.

*In Russia pretzels are made in the form of a genus 2 solid body — Translator.

In Fig. 163 this net is shown in the other equivalent model: in the form of the graph $W(\alpha)$ on S^2 with four deleted points. However, we have depicted these deleted points as rejected disks to preserve all information on the identification of $S_1 = S_1^+ \cup S_1^-$ and $S_2 = S_2^+ \cup S_2^-$. In other words, the graph from Fig. 163 is obtained from the scheme in Fig. 162 by cutting the pretzel along circles S_1 and S_2. Thus, we have produced a three-dimensional manifold. We claim that the fundamental group of this manifold is nonzero. To prove this, we need a more general statement of the structure of the fundamental group of the three-dimensional manifold $M^3(\alpha)$ corresponding to α. Consider the net $\alpha = (S_1, \ldots, S_g; \bar{S}_1, \ldots, \bar{S}_g)$. On all circles that form the net, fix an orientation. Then consider an arbitrary circle of index 2, say $\bar{S}_j$, and moving along it, in accordance with its orientation, we will subsequently label all points of its intersection with circles of index 1. The point of its intersection with the circle S_i of index 1 will be labeled by S_i^{+1} if the index of intersection at this point of $\bar{S}_j$ and S_i is positive and by S_i^{-1} otherwise. Thus, following along the entire circle $\bar{S}_j$, we write a word $W_j = S_{i_1}^{\pm 1} S_{i_2}^{\pm 1} \ldots$ and simultaneously define with which circles of index 1 $\bar{S}_j$ intersects. Thus, we uniquely define the family of g words $W_1, \ldots, W_g$.

Theorem 13.1. Let $M^3 = M^3(\alpha)$, where α is a net. Consider a free group F_g with generators $S_1, \ldots, S_g$ and the set of elements $\{W\} = (W_1, \ldots, W_g)$ constructed during the above procedure. Let $N\{W\}$ be the unique minimal normal subgroup F_g that contains $W_1, \ldots, W_g$, i.e., generated by $W_1, \ldots, W_g$ and their various conjugations by arbitrary elements of F_g. Then $\pi_1(M^3)$ is isomorphic to $F_g/N\{W\}$.

Proof. Recall a general method of computation of the fundamental group of the CW-complex K with one vertex (zero-dimensional cell); see, for example, [9, 10]. Let $\sigma_1, \ldots, \sigma_g$ be one-dimensional cells of the complex K, and $e_1, \ldots, e_r$ its two-dimensional cells. Then the characteristic mappings of the two-dimensional cells being restricted onto boundaries of the cells are defined up to the conjugacy elements $W_1, \ldots, W_r$ of the free group $F_g(\sigma_1, \ldots, \sigma_g)$. Then the fundamental group of the complex is isomorphic to the quotient group of the free group by the unique minimum normal subgroup that contains $W_1, \ldots, W_r$. This normal subgroup is generated by elements of the form $T_i W_i T_i^{-1}$, where T_i is an arbitrary element of F_g.

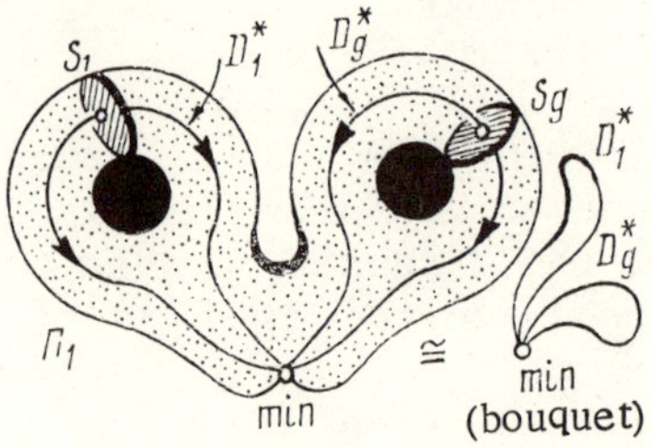

Fig. 164

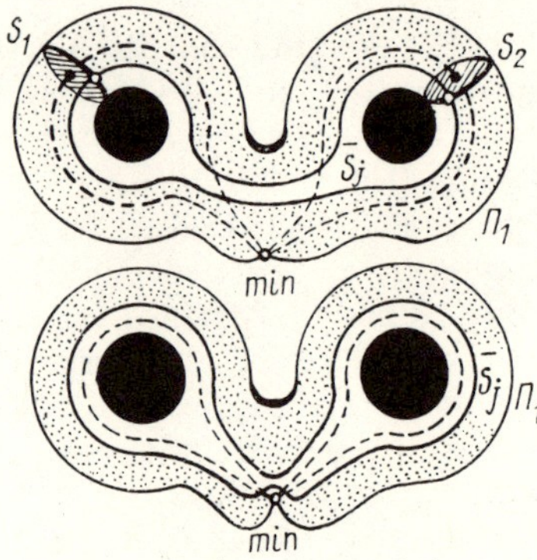

Fig. 165

Let us apply this theorem to our situation. Theorem 12.1 implies that M^3 is obtained by attachment of Π_1 and Π_2. The complex Π_1 is homotopically equivalent to the bouquet of circles $\bigvee_{i=1}^{g} D_i^*$, each of which is "the axis of a handle" of index 1 and is obtained from the left disk that emerges from a point of index 1. Each handle of this kind (see Fig. 159) is obtained by direct multiplication of the "axis"-circle and the right separatrix disk whose boundary is the circle S_i of index 1 that belongs to $f_{1/2}$. Let two-dimensional disks that emerge from points of index 2 descend onto $f_{1/2}$, and define on it circles $\bar{S}_j$ of index 2. Clearly, each intersection of $\bar{S}_j$ with the circle S_i means that the boundary of a two-dimensional cell (i.e., $\bar{S}_j$) belongs to the handle corresponding to S_i. Hence, words $W_1, \ldots, W_g$, recovered above from circles of index 2, define characteristic mappings of boundaries of two-dimensional cells (of index 2) that transform these circles S_j into the one-dimensional skeleton of Π_1. The theorem is proved.

Figure 164 shows the complex Π_1, circles that participate in the proof of the theorem, and the contraction of Π_1 onto $\bigvee_{i=1}^{g} D_i^*$. In Fig. 165 is shown the example of Π_1 of genus 2 and the circle $\bar{S}_j$ that forms the word $W_j = S_1S_2$ when it intersects S_1 and S_2. The figure also shows the deformation of $\bar{S}_j$ onto the bouquet of two disks $D_1^* \vee D_2^*$. With respect to this deformation, $\bar{S}_j$ realizes, evidently, the element of the free group $F(S_1, S_2)$ equal to $S_1 \cdot S_2$, i.e., coinciding with $W_j = S_1 \cdot S_2$. Now, apply Theorem 13.1 to the study of the concrete net described above. In this case, the genus of Π_1 equals 2. Therefore, we obtain two words W_1 and W_2 in $F_2(S_1, S_2)$. For simplicity of notation, substitute a and b for S_1 and S_2. Then, circles $\bar{S}_1$ and $\bar{S}_2$ of index 2 define the following elements of $F_2(a, b)$ (see Fig. 162): $aba^{-1}ba = W_1$ and $b^{-1}ab^4a = W_2$. Hence, $F_2(a, b)/N(W_1, W_2) = \pi_1(M^3)$. Let us prove that this group does not consist only of the unit (in multiplicative notation). For this, let us convert the relations $W_1 = 1$, $W_2 = 1$ in F_2 into a simpler form. Making use of the fact that these relations are defined up to conjugation of an arbitrary element of F_2, we obtain the chain of equivalencies $W_1 = aba^{-1}ba \sim ba^{-1}ba^2 \sim a^{-2}b^{-1}ab^{-1} \sim a^{-2}cac$, where $c = b^{-1}$, and $W_2 = b^{-1}ab^4a \sim b^4ab^{-1}a = c^{-4}aca$. These relations can also be rewritten as follows: $a^2 = cac$, $c^4 = aca$. Thus, $\pi_1(M^3) = F_2(a, c)/N(a^2 = cac,\ c^4 = aca)$. This implies that in $\pi_1(M^3)$ there is a relation $c^5 = a^3$ between a, c. In fact, $c^5 = caca = (ca)^2$, $a^3 = caca = (ca)^2 = c^5$. This group is known in the theory of the symmetry groups of perfect polyhedrons. The above relations are verified in the icosahedron group if, for c, we take the rotation around a vertex by $2\pi/5$ and, for a, we take a rotation in the same direction by the angle $2\pi/3$ around the center of a triangle adjacent to this vertex. This implies that the icosahedron group is either $\pi_1(M^3)$ or its quotient group; in particular, $\pi_1(M^3)$ is, surely, nontrivial since the icosahedron group is different from the unit.

Problem: Prove that $\pi_1(M^3)$ is a group of 120th order.

Thus, we have proved that M^3 is not homotopically equivalent to the sphere. On the other hand, it turns out that $H(M^3; Z) = 0$. In fact, it is known (see, for example, [2]) that $H_1 = \pi_1/[\pi_1, \pi_1]$ for any CW-complex where $[\pi_1, \pi_1]$ is the commutant of π_1. Therefore, in our example, to compute $H_1(M^3; Z)$, we must assume that a, b commute (which is equivalent to the factorization of a group by its commutant) and, after shifting (for convenience) to the additive notation, we obtain $W_1^* = a + b - a + b + a = a + 2b$, $W_2^* = -b + a + 4b + a = 2a + 3b$. Hence, elements W_1^*, W_2^* define the integral transformation of the Abelian group $\mathbf{Z} \oplus \mathbf{Z}$ with the matrix $\begin{pmatrix} 1 & 2 \\ 2 & 3 \end{pmatrix}$. Since the determinant of this transformation is -1, then elements W_1^*, W_2^* generate the whole $\mathbf{Z} \oplus \mathbf{Z}$, as required. Thus, $\pi_1(M^3)$ is a simple group, coinciding with its commutant, and M^3 is a homological sphere. The theorem is completely proved.

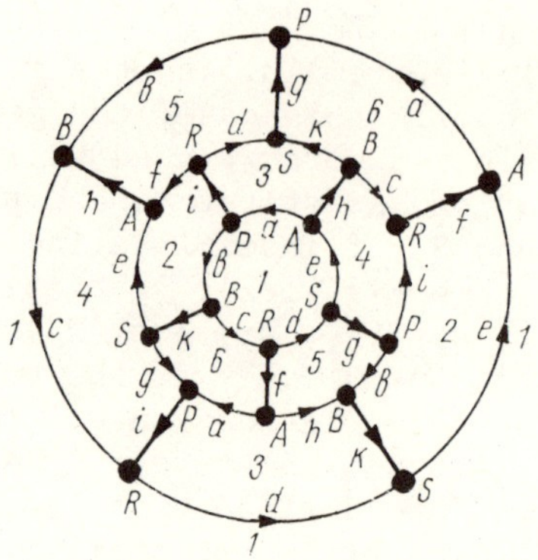

Fig. 166

The manifold M^3 that we have described is, as it turns out, diffeomorphic to the so-called "Poincaré sphere" (see [8]), whose definition is usually given in other terms. This manifold is obtained from the dodecahedron after identification of its opposite faces – pentagons – turned with respect to each other by the angle $\pi/5$. Figure 166 shows the graph of the dodecahedron with the required identifications of its faces. We leave to the reader as a useful exercise the identification of this manifold with the one studied above.

Let us return to the problem of recognition of a standard three-dimensional sphere. Let α' be an arbitrary diffeomorphism of a surface of genus g. Let us consider the corresponding net α. The Whitehead graph $W_i(\alpha)$ of index i is the graph obtained on the plane after we cut the surface along circles of index i, as described above. Thus, each diffeomorphism defines two Whitehead graphs. Defining any of these graphs (with indicated orientations and numberings of edges and vertices), we completely define the initial net. On the surface of genus g, consider two circles of index 1, i.e., S_i and S_j. Let γ be a smooth path that connects a point of S_i with a point of S_j without other intersection points with circles of index 1. Let us replace the pair S_i, S_j by the new pair $\tilde{S}_i$, $\tilde{S}_j$, where $\tilde{S}_j = S_j$; the new circle $\tilde{S}_i$ is depicted in Fig. 168. In other words, from the circle S_i, a thin tongue protrudes which glides along γ; then, coming closer to S_j, bifurcates and begins to embrace S_j. Finally, the circle S_j is swallowed up and the final result is shown in Fig. 168. The circle S_j is not affected by this process.

Fig. 167

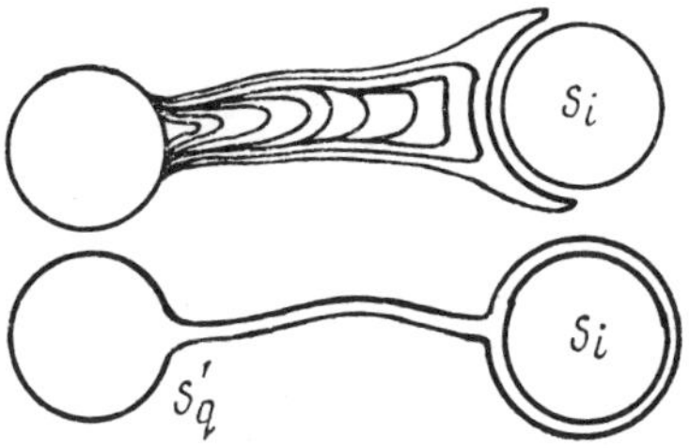

Fig. 168

The operation described above is called the operation of index 1. The operation of index 2 is similarly described. Note that the path γ from the definition of the operation of index 1 could intersect circles of index 2. As always, we assume all intersections to be transversal, i.e., in general position. The following lemma holds.

Lemma. Suppose to some net s that defines the manifold M operations of indices 1 and 2 are applied (in arbitrary number and order). Then we get a new net of the same kind as the initial one. This new net defines the manifold diffeomorphic to the initial one.

Thus, operations of indices 1 and 2 do not change the topological type of a three-dimensional manifold. The proof is carried out by methods of the Morse theory and is contained, for example, in [11, 12]. Thus, it is possible to start from a fixed net and to construct infinitely many nets that define the same manifold. Operations of indices 1 and 2 can be applied in random order. The following statement holds.

Let α_0 be the simplest net of genus g of the standard sphere depicted in Fig. 158. Then any other net on the standard sphere of the same genus can be obtained from α_0 after several operations of indices 1 and 2.

Thus, the set of nets obtained from the simplest net by operations of indices 1 and 2 coincides exactly with the set of all nets corresponding to the standard sphere. The important circumstance is also that any net of a fixed genus can be obtained by the above operations from the simplest net of the same genus. Note that operations of indices 1 and 2 do not affect the genus of the net. This statement gives the description of all codes of the standard sphere in the set of codes of all three-dimensional manifolds. However, the reader must not think that it provides an algorithm of recognition of codes of the sphere. The reason is that, in addition to the aforementioned, we must now learn to recognize nets obtained from the simplest ones by the above operations.

Let us give the definition of the splitting vertex and the operation of simplification of Whitehead graphs. Consider a net on the surface. Recall that we consider only oriented nets. The net defines a splitting of the pretzel into several domains.

Let us say (see [11]) that the domain U is fixed if, among edges that define its boundary, there exist two edges, β_1 and β_2, that belong to the same circle and whose orientations are compatible with one of the bypasses of the boundary of the domain. Edges β_1 and β_2 will be called fixed. The wave τ is a path inside of the fixed domain that connects inner points of fixed edges without other intersection points to circles of indices 1 and 2. The properties of the wave are studied in [11]. This notion became useful in the construction of the recognition algorithm for Heegaard splittings of the sphere (for genus 2). Let us reformulate the notion of the wave in terms of Whitehead's graphs. It turns out that the presence of a wave in a net develops most clearly in terms of these graphs.

Let, for definiteness, fixed edges belong to the circle S_i of index 1. Let us cut the surface along circles of index 1, obtaining the graph $W_1(\alpha)$. The circle S_i will become two vertices S_i^{+1} and S_i^{-1} of $W_1(\alpha)$. The wave τ turns into a smooth path that begins and terminates at one of these vertices (depending on the relationship of orientations). Let, for definiteness, τ begin and terminate in S_i^{+1}. The important property of τ is that it intersects $W_1(\alpha)$ only at the point S_i^{+1}. Hence, ejecting S_i^{+1} from the graph, we split it onto at least two connected components. One of these is situated in the domain A bounded on the plane by the wave τ, and the other is situated outside of this domain. Let us assume that $S_i^{-1} \notin A$.

Let us give the following definition. The vertex of the Whitehead graph, where τ begins and terminates, will be called the splitting vertex if there is at least one edge of this graph that belongs to A and contains the vertex S_i^{+1} [11].

Such vertices first appeared in the remarkable paper by Whitehead [13], where the problem of recognition of generators of a free group was solved.

The role of the splitting vertex is that its presence in one of the Whitehead graphs enables one to simplify both Whitehead graphs. Equivalently, the presence of a wave in a net enables one to simplify a net and skip over, using operations of indices 1 and 2 to the new net with a smaller number of edges. Let us describe the procedure of simplification of the Whitehead graph.

Let S_i be a splitting vertex on $W_1(\alpha)$. The wave τ splits the plane into two domains A and B, one of which contains the vertex S_i^{-1}. Recall that each circle S_i generates two vertices S_i^{+1} and S_i^{-1} when the surface is cut. Let A be the domain that does not contain S_i^{-1}. By definition of the splitting vertex, there is at least one edge of the graph that belongs to A and contains S_i^{+1}. Let us say that the edge of the Whitehead graph is incident on a vertex if the edge enters the vertex and emerges from it. Now consider the set of all edges of the Whitehead graph incident on S_i^{+1} that belong to the domain A encircled by the wave. These edges split into three classes.

1) Edges of the first class are characterized by the fact that they are parts of a circle (of index 2) that interacts with the vertex S_i^{+1} as follows. This part of the circle enters S_i^{+1}, then jumps over into S_i^{-1}, emerges from it, makes a loop, and returns also into S_i^{-1} without passing other vertices of the graph. Furthermore, returning into S_i^{-1}, the part of the circle jumps backward into S_i^{+1} and, emerging from it, returns into the part of the Whitehead graph encircled by the wave.

2) Edges of the second class are characterized by the fact that they are the part of the circle that interacts with S_i^{+1} as follows. This part of the circle comes into S_i^{+1}, jumps over into S_i^{-1}, then enters another vertex of the graph different from S_i^{-1} and situated outside of A.

3) Edges of the third class are characterized by the fact that they are parts of the circle that interact with S_i^{+1} as follows. This part of the circle emerges from a vertex of the graph situated outside of A, comes into S_i^{-1}, jumps over into S_i^{+1}, then makes a loop and returns again into S_i^{+1}

without passing other vertices of the graph. After this, the part of the circle jumps backward into S_i^{-1} and, emerging from it, enters a vertex of the graph situated outside of A.

Now let us describe the reconstruction – the operation of simplification of the Whitehead graph when it contains at least one splitting vertex S_i^{+1}. This operation transforms the Whitehead graph as follows:

1) In the case of edges of the first class, the loop that begins and terminates in S_i^{-1} annihilates and the two edges incident with S_i^{+1} annihilate and are replaced by the one edge that joins vertices of the graph different from S_i^{+1} that were initially incident with these edges.

2) In the case of edges of the second class, both edges incident with vertices S_i^{+1} and S_i^{-1}, respectively (see above), are annihilated and replaced by the one edge that, escaping S_i^{+1} and S_i^{-1}, directly joins the vertices of the graph different from S_i^{+1} and S_i^{-1}, which were initially incident with these edges.

3) In the case of edges of the third class, the operation coincides with the operation described above for edges of the first class. The unique difference is that vertices S_i^{+1} and S_i^{-1} should be interchanged.

The operation of reconstruction of the Whitehead graph is completely described. However, it requires further elucidation. This operation was described abstractly, without taking into account that the possibility of realizing the described reconstructions in the plane is not evident a priori. The following important theorem, proved in [11], holds.

Theorem. (a) Let $\tilde{W}$ be the graph obtained from the Whitehead graph $W_i(\alpha)$ by the reconstruction described above in the presence of at least one splitting vertex. Then $\tilde{W}$ is a planar graph (i.e., admits a realization on a plane) and, moreover, all above reconstructions could be realized on the plane from the very beginning.

(b) The graph $\tilde{W}$ is the Whitehead graph of the new net $\tilde{\alpha}$, i.e., $\tilde{W} = W_i(\tilde{\alpha})$.

(c) The new net $\tilde{\alpha}$ is obtained from the initial net α (corresponding to the initial Whitehead graph) after application of operations of indices 1 and 2 (operations of index 1 are actually sufficient).

(d) The net $\tilde{\alpha}$ defines the same manifold (up to a diffeomorphism) as the initial net α, i.e., the described reconstruction preserves the topological type of the manifold.

(e) The new Whitehead group $W_i(\tilde{\alpha})$ has fewer edges than the initial graph (with the same number of vertices).

Earlier we called the reconstruction "the simplification of the Whitehead graph operation" due to statement (e) of this theorem. The new graph $W_i(\tilde{\alpha})$ is simpler when compared with the initial one in the sense that the number of its edges is strictly less than that of the initial graph. The latter statement (e) is obvious since reconstruction makes the number of edges less by at least one (see above). The statement (d) follows from the statement (c) and the above lemma. Statements (a) and (b) are more difficult to prove (see [11]).

Thus, if the net α of a manifold M^3 contains a wave [i.e., at least one of the Whitehead groups $W_i(\alpha)$ has a splitting vertex], then there is an extremely simple algorithm that enables us to transform the net α into the new net $\tilde{\alpha}$ that defines the same manifold, but is simpler than the initial net.

Now let us describe the algorithm of recognition of the standard sphere in the class of Heegaard splittings of genus 2. It turns out that on nontrivial nets of genus 2 that correspond to the standard sphere, a wave (i.e., the splitting vertex in one of the Whitehead graphs) always exists. The simplest net α_0 corresponding to the standard sphere (Fig. 158) does not contain any splitting vertex. However, for other nets that correspond to the sphere, a splitting vertex always exists. After studying the fundamental paper by Whitehead [13], I. A. Volodin and A. T. Fomenko formulated the following conjecture in 1974: Any net that corresponds to the standard sphere and differs from the simplest one contains at least one wave [11, 12]. The conjecture was tested using a BÉSM-6 computer, where the process of net construction was modeled by operations of indices 1 and 2 [11]. It turned out that all 10^6 nets constructed at random by the computer that define the standard sphere actually contain at least one wave. However, later, it was proved [14, 15] that for nets of genus 3 and higher the conjecture fails. Thus, the situation was unclear for nets of genus 2. (The answer is trivial for nets of genus 1.) As a consequence, the computational experiment was repeated by the author, especially for nets of genus 2, geometrical properties that sometimes annihilate waves on nets of genus 3 and higher being taken into account. It turned out that, despite the limitations of the computer, this extended experiment did not show any counterexamples to the conjecture on nets of genus 2 of the sphere. Finally, in 1980, in [16], this conjecture by Volodin and Fomenko was proved for all nets of genus 2 of the sphere. Namely, the following statement holds.

Theorem. Each net of genus 2 of the standard sphere, different from the simplest one, contains at least one wave. In particular, at least one of the Whitehead graphs contains at least one splitting vertex.

The proof of this fact is quite nontrivial and therefore is omitted here. In [16] there is obtained an interesting byproduct that describes the structure of an arbitrary Whitehead graph of genus 2, which corresponds to the three-dimensional manifold.

Thus, the effectiveness of the following extremely simple and clear algorithm of recognition of the standard sphere in the class of Heegaard splittings of genus 2, formulated in [11] and first verified with the aid of a computer (for genus 2), is established and justified. Let us take an arbitrary net α and determine whether it contains at least one wave, i.e., whether at least one of its Whitehead graphs contains a splitting vertex. If there are no splitting vertices on both graphs, then the initial net defines the three-dimensional manifold that is not diffeomorphic to the standard sphere and we obtain an answer at the very first step. If there is at least one splitting vertex, then let us apply a simplifying operation to the net (the Whitehead graph). As a result, we obtain the new net α' corresponding to the same manifold. Let us determine again whether there is a splitting vertex. If there is none, then the initial manifold is not a sphere. If there is one, then let us simplify the net, and so forth. Thus, after a finite number of steps (which do not exceed the number of edges of the initial net), we either obtain the net without any waves (in this case, the initial manifold is not a sphere) or get the simplest net α_0 corresponding to a sphere (in this case, the initial manifold is diffeomorphic to the sphere).

13.2. Homotopic Spheres

Thus, the vanishing of the one-dimensional homology group H_1 is not sufficient for M^3 to be diffeomorphic to the standard sphere. Therefore, the next natural conjecture regards still further restriction of the topology of the manifold.

The Poincaré Conjecture. Any three-dimensional smooth closed compact connected and simply connected (!) manifold is diffeomorphic to the standard sphere.

Now we suppose that $\pi_1(M^3)$ vanishes, which is, evidently, a more restrictive condition than the initial one. Since $\pi_1 = 0$, such a manifold is automatically an orientable one; otherwise, π_1 would have contained the subgroup of index 2. Since vanishing of π_1 implies vanishing of H_1, then any such manifold is a homological sphere (see 13.1). However, a stronger statement holds.

Proposition 13.3. Any three-dimensional smooth closed compact connected and simply connected manifold is homotopically equivalent to the standard sphere.

Proof. Since $H_1 = H_2 = 0$ and M is simply connected, by the Gurevich theorem (see, e.g., [3], p. 153) we have $\pi_1 = \pi_2 = 0$, $\pi_3(M^3) = H_3(M^3, \mathbf{Z}) = \mathbf{Z}$. Consider M^3 as a CW-complex and construct a continuous mapping $f: M^3 \to M^3$ retracting the two-dimensional skeleton of M^3 into a point. The mapping f induces an isomorphism of homotopic groups and is a homotopic equivalence.

It is unknown as yet whether any homotopic sphere (i.e., a manifold homotopically equivalent to the three-dimensional sphere) is the standard sphere.

14. ON ALGORITHMIC CLASSIFICATION OF MANIFOLDS

14.1. Fundamental Groups of Three-Dimensional Manifolds

We have actually described fundamental groups of two-dimensional manifolds. It turns out that these groups are "sufficiently small," all of them defined explicitly by a simple set of generating relations. As we have already seen in the three-dimensional case, fundamentally new groups appear which are $\pi_1(M^3)$ (see Sec. 13). Nevertheless, there are still finitely generated groups that cannot be fundamental groups of three-dimensional manifolds.

Proposition 14.1. The group $\mathbf{Z}^4 = \mathbf{Z} \oplus \mathbf{Z} \oplus \mathbf{Z} \oplus \mathbf{Z} \oplus \mathbf{Z}$ cannot be the fundamental group of a closed connected compact three-dimensional manifold.

Proof. Assume the contrary; suppose there is an M^3 such that $\pi_1(M^3) = \mathbf{Z}^4$. The group π_1 is defined by the set of generators $a_1, \ldots, a_p$ and relations $W_1, \ldots, W_g$ whose geometric meaning was investigated in Sec. 13. Generators $a_1, \ldots, a_p$ are defined by left separatrix disks of points of index 1, and relations $W_1, \ldots, W_q$ are defined by left two-dimensional disks of points of index 2. Since, thanks to Theorem 12.1, on M^3 there is a Morse function with only one minimum, one maximum, and an equal number of critical points of indices 1 and 2, then there is a presentation of $\pi_1(M^3)$ such that the number of generators is equal to the number of relations. Thus, let $\pi_1(M^3) = F_g(a_1, \ldots, a_g)/N(W_1, \ldots, W_g)$, where g is the genus of $f_{1/2}$. Let us embed M^3 in the CW-complex $K(\mathbf{Z}^4, 1)$ (see [3]), i.e., in the complex such that $\pi_1(K(\mathbf{Z}^4, 1)) = \mathbf{Z}^4$, $\pi_i(K(\mathbf{Z}^4, 1)) = 0$ for $i > 1$. For this it suffices to close off all groups $\pi_i(M)$ for $i \geqq 2$ by attaching cells of dimension no less than 3. In particular, one- and two-dimensional skeletons of M do not change. Since any two spaces of type $K(\pi, 1)$ with the same π are homotopically equivalent (see [3]), we can assume that $K(\pi, 1)$ is homotopically equivalent to T^4. Thus, for T^4 we have obtained a presentation in the form of a complex with one vertex, such that the algebraic complex (over $\mathbf{R}$) of chains of this complex is of the form $\ldots \xrightarrow{\partial_2} \xrightarrow{\partial_2} P_2 \xrightarrow{\partial_1} P_1 \xrightarrow{\partial_0} 0$ where $\dim_{\mathbf{R}} P_2 = \dim_{\mathbf{R}} P_1 = g$ (the genus of $f_{1/2}$). Consider $H_1(T^4, \mathbf{R}) = \mathbf{R}^4$. Since $H_1 = \operatorname{Ker} \partial_0 / \operatorname{Im} \partial_1 = P_1/\operatorname{Im} \partial_1$, then $\operatorname{codim} \operatorname{Im} \partial_1 = 4$. Since $\operatorname{Im} \partial_1 = P_2/\operatorname{Ker} \partial_1$, then ∂_1 is nonzero on $g - 4$ linearly independent vectors in P_2; hence $\dim \operatorname{Ker} \partial_1 = 4$. Therefore, $\dim H_2 \leqq 4$ since $H_2 = \operatorname{Ker} \partial_1 / \operatorname{Im} \partial_2$. But, on the other hand, $H_2(T^4, \mathbf{R}) = \mathbf{R}^6$. This contradiction proves our statement.

Problem. Find all Abelian groups which are fundamental groups of three-dimensional closed connected compact manifolds. (Answer: $\mathbf{Z}$, $\mathbf{Z} \oplus \mathbf{Z} \oplus \mathbf{Z}$, $\mathbf{Z}_p$, $\mathbf{Z} \oplus \mathbf{Z}_2$).

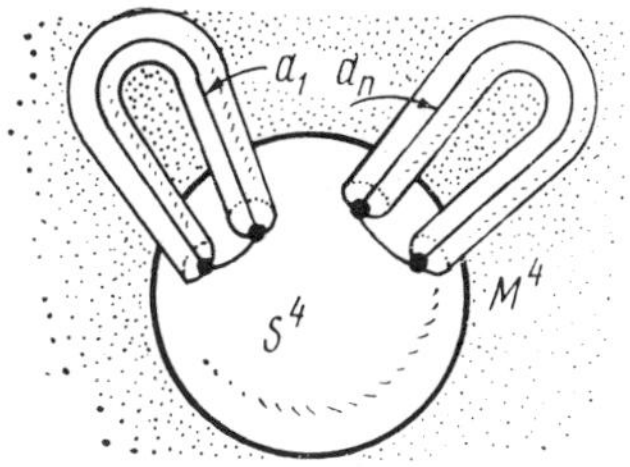

Fig. 169

Consider the class H of all finitely generated groups, i.e., groups presentable by a finite set of generators and defining relations. Let Q be a subclass of groups that are of the form $\pi_1(M^3)$ for a certain M^3. As we have seen, $Q \neq H$ since, for example, $\mathbf{Z}^4 \not\in Q$. It turns out that the subclass $Q \subset H$ is algorithmically nonrecognizable in H, i.e., there is no algorithm such that, acting according to a single program, we can determine whether the set of generators and relations given at the input define a group from class Q or not. This theorem is nontrivial and we will skip its proof.

14.2. Fundamental Groups of Four-Dimensional Manifolds

Theorem 14.1. Any finitely generated group is presentable as the fundamental group of a four-dimensional smooth compact connected closed manifold.

Proof. Let G be defined by generators $a_1, \ldots, a_n$ and the relations $W_1, \ldots, W_k$, where $n, k < \infty$, i.e., $G = F_n(a_1, \ldots, a_n)/N(W_1, \ldots, W_k)$. Consider the sphere S^4 and attach to it n handles of index 1, i.e., let us perform n reconstructions of index 1 (see Secs. 8 and 11). We obtain the manifold M^4. Clearly, $F(a_1, \ldots, a_n) = \pi_1(M^4)$, where generators $a_1, \ldots, a_n$ are realized geometrically in M^4 in terms of circles S_i generated by axes of handles of index 1 (see Fig. 169). Following the scheme of Sec. 13.1, we realize each word W_j by a circle S_j smoothly embedded in M^4 so that the consecutive bypass along the circle intersections with handles will constitute the word W_i. We may assume that circles that realize words $W_1, \ldots, W_k$ do not intersect. Consider the circle $\bar{S}_j$ and let U be its sufficiently small tubular neighborhood. We claim that U is the direct product $\bar{S}_j \times D^3$. In fact, since U is a fibration over a circle with the fiber D^3, it can be only either (a) the direct product, or (b) the nonorientable manifold with boundary. But the case (b) is impossible since M^4 is, evidently, orientable. Therefore, we can make a reconstruction of index 2 (see Sec. 11) of M^4 along $\bar{S}_j$. In other words, consider $S^2 \times D^2$ and, ejecting $U = \bar{S}_j \times D^3$, glue $S^2 \times D^2$ instead, according to the identifications $\partial(S^2 \times D^2) = S^2 \times S^1 = \partial U = \partial(\bar{S}_j \times D^3)$. This reconstruction having been made along all circles $\bar{S}_1, \ldots, \bar{S}_k$, we obtain the new manifold Q_4, whose fundamental group is equal to G since handles of index 2 being attached to all circles $\{\bar{S}_j\}$ eliminate elements $W_1, \ldots, W_k$ in the group $F_n(a_1, \ldots, a_n)$, making them equal to zero (unity), thus giving us the group G. The theorem is proved.

Note that these arguments are not valid in dimension 3, since here we would have to attach, after the reconstruction of index 2, the handle $S^1 \times D^2$ (instead of the neighborhood $U = \bar{S}_j \times D^2$ of $\bar{S}_j$) with respect to identification of boundaries $\partial U = \bar{S}_j \times S^1 = \partial(S^1 \times D^2) = T^2$. Clearly, the annihilation of W_j realized by $\bar{S}_j$ adds a new and, generally speaking, nontrivial element instead, realized by another circle which is the generator of the torus, i.e., new generators are introduced which are not presupposed by the presentation of G. As we have already seen, some groups G are not, actually, fundamental groups of three-dimensional manifolds. Clearly, any finitely generated group G is realizable as the fundamental group of a certain manifold M^n for any $n \geq 4$.

14.3. On the Impossibility of Classifying Smooth Manifolds in Dimensions Greater Than 3

Each smooth compact manifold admits a triangulation, i.e., a presentation in the form of a simplicial complex. Hence, by generating a table where all these simplexes and their faces and incidence coefficients are listed, we can define a manifold, considering the table as a code of the manifold. The problem of algorithmic classification of manifolds of a given dimension arises: Is there an algorithm that acts according to a unique program and determines whether two arbitrary codes define diffeomorphic (homeomorphic) manifolds or not? (See [17, 18].)

Theorem 14.2. There is no algorithm defined on the set of codes of all four-dimensional manifolds (defined, for example by its simplicial decompositions) that determines whether two codes that enter its input define diffeomorphic manifolds or not.

Proof. We will refer here to some purely algebraic results whose proof extends far beyond the limits of this book. If the group is given by a table listing generators and relations, we will say that a copresentation of the group is given. First, we construct a finitely generated group G_0, defined by generators and relations such that the identity problem for words of G_0 is algorithmically unsolvable, i.e., there is no algorithm that determines whether two words given as the input to the algorithm define the same element of G_0. The reason is that, since elements of G_0 are given as conjugate classes with respect to the normal subgroup generated by words-relations, the same element is expressed by infinitely many words resulting in the "identity of words" problem. Copresentation of this interesting group has 10 generators a_i and 29 relations R_j. We first list generators of G_0 and then all the relations.

$$G_0 = \{\text{generators: } s_1, s_2, s_3, s_4, s_5, c, d, e, k, t;$$
$$\text{relations: } d^{10}s_i = s_i d; \quad es_i = s_i e^{10}, \quad s_i c = cs_i; \quad i = 1, 2, \ldots, 5;$$
$$ds_1s_3ec = cds_3s_1c; \qquad d^2s_1s_4e^2c = cd^2s_4s_1e^2; \qquad d^3s_2s_3e^3c = cd^3s_3s_2e^3;$$
$$d^4s_2s_4e^4c = cd^4s_4s_2e^4; \; d^5s_5s_3s_1e^5c = cd^5s_3s_5e^5; \; d^6s_5s_4s_2e^6c = cd^6s_4s_5e^6;$$
$$d^7s_3s_4s_3s_1e^7c = cd^7s_3s_4s_3s_1s_5e^7; \quad d^8s_3s_1{}^3e^8c = cd^8s_1{}^3e^8; \quad d^9s_4s_1{}^3e^9c =$$
$$= cd^9s_1{}^3e^9; \; ct = tc; \; dt = td; \; ck = kc; \; ek = ke; \; s_1{}^{-3}ts_1{}^3k = ks_1{}^{-3}ts_1\} =$$
$$= \{a_i, R_j\}.$$

The countable set of copresentations of groups is based on this copresentation of G_0, in which the problem of triviality of the group is algorithmically unsolvable. Let $G_0 = \{a_i, R_j\}$ be the group described above. Let us order words in the alphabet $(a_1, a_1{}^{-1}, \ldots, a_{10}, a_{10}{}^{-1})$ lexicographically, i.e., let us construct an (infinite) list

$$a_1, a_1^{-1}, \ldots, a_{10}^{-1}; a_1 a_1^{-1}, a_1 a_2, \ldots, a_1 a_{10}^{-1}; a_1^{-1} a_1, \ldots .$$

Let w_k be the k-th word of this list. Furthermore, denote by G_k, where $k \geqq 1$, the group defined by generators $a_1, \ldots, a_{11}$, t, c, s, b, d, u and relations

$$R_1 = 0, \; R_2 = 0, \; \ldots, \; R_{29} = 0;$$

$$u = a_{11}\omega_k a_{11}^{-1}\omega_k;\ t^2 u = ut;\ c^2 t = tc;$$
$$s^2 u = us;\ b^2 s = sb;$$
$$a_i b^i c b^{-i} = d^i c d^{-i};\ i = 1, 2, \ldots, 11;$$
$$b^{12} c b c^{-1} b^{-12} = d^{12} c d c^{-1} d^{12}.$$

In this copresentation of G_k we have 17 generators and 46 relations. The following important statement holds: There is no algorithm defined on this countable set of copresentations of groups that determines whether the copresentation of a group given as its input defines the trivial (unit) group or not.

In other words, we cannot algorithmically recognize unit groups in the sequence of groups G_k.

We return to the problem of classification of manifolds. Making use of Theorem 14.1, let us construct the sequence of four-dimensional manifolds M_k^4 such that $\pi_1(M_k^4) = G_k$. Let p be the number of generators and q the number of relations in G_k. Actually, p = 17, q = 46. The numbers p and q do not depend on k. Recall that each manifold M_k^4 is of the form $S^4 + (p, 1) + (q, 2)$, where (p, 1) stands for p reconstructions of index 1 and (q, 2) for q reconstructions of index 2. Let us construct new four-dimensional manifolds $N_k^4 = S^4 + (p, 1) + (q, 2) + (p, 2) = M_k^4 + (p, 2)$, i.e., let us attach handles of index 2 to M_k^4 so as not to affect handles attached earlier. The latter condition means, in particular, that we do not affect the fundamental group, i.e., $\pi_1(N_k^4) = \pi_1(M_k^4) = G_k$.

<u>Lemma 14.1</u>. If G_k is trivial, then N_k^4 is diffeomorphic to the standard smooth simply connected manifold $M_0^4 = S^4 + (q, 2)$, i.e., to S^4 with q handles of index 2 attached in the standard way.

<u>Proof</u>. Let $\pi_1(N_k^4) = 0$. The constructed manifolds N_k^4 and M_k^4 also have the following important property. They are boundaries of certain five-dimensional smooth manifolds. This follows immediately from Sec. 11.4, where we have proved that the Morse reconstruction of index λ is realized as distinguishing the boundary of the manifold whose dimension is greater by 1 and which is situated between two level surfaces of the function f, which has in this layer one critical point of index λ. Making use of the triviality of the fundamental group, we may close off all its p generators by two-dimensional disks. Since generators of π_1 are realized as "axes" of

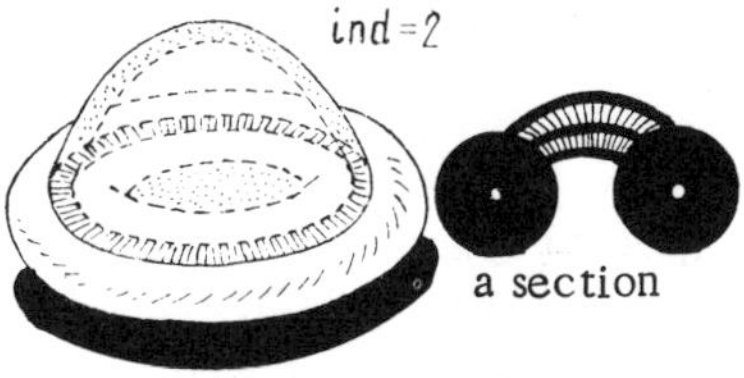

Fig. 170

handles of index 2, then removing handles (p, 2) so as to creep over these generators, we close off (annihilate) all generators of π_1. We skip the technical details here, referring the reader to the original paper [17]. Here we make use of the fact that handles of indices 1 and 2 can annihilate each other if the sole of the handle of index 2 is attached to the axis of the handle of index 1 (it can be done because of triviality of π_1). This mutual annihilation of two handles is depicted in Fig. 170. The lemma is proved.

Returning to the proof of Theorem 14.2, let us assume the contrary: Suppose there is a classifying algorithm that enables one to compare different codes of four-dimensional manifolds, which determines whether the corresponding manifolds are diffeomorphic or not. Let us restrict this algorithm to the infinite series of the above manifolds $M_0{}^4 = S^4 + (q, 2)$, $N_1{}^4$, $N_2{}^4$, $N_3{}^4$, ... Taking an arbitrary group G_k of the above series, we can make use of the above recognition algorithm by comparing the manifold $N_k{}^4$ corresponding to G_k with $M_0{}^4$. If they happen to be diffeomorphic, then G_k is trivial since $\pi_1(M_0{}^4) = 0$. If the algorithm says that $N_k{}^4$ and $M_0{}^4$ are not diffeomorphic, then, by Lemma 14.1, the group $G_k = \pi_1(N_k{}^4)$ is nontrivial since, otherwise, we would have obtained the diffeomorphism $N_k{}^4 = M_0{}^4$. Hence, the existence of a topological algorithm for recognition (classification) yields the existence of an algorithm that enables one to recognize in the series of groups $\{G_k\}$ (more exactly their copresentations) a trivial group, which is impossible by the construction of these copresentations. Theorem 14.2 is proved.

In the three-dimensional case, this theorem still does not have an analog, since the possibility for realization of a series of groups of type G_k as fundamental groups of three-dimensional manifolds is unclear. The following circumstance should also be mentioned. Suppose that the following statement holds: A series of groups G_k (or another series with analogous properties) is realized as fundamental groups of three-dimensional manifolds, i.e., $G_k = \pi_1(N_k^3)$. Which statement corresponds, then, to Lemma 14.1 proved above? Since any three-dimensional compact closed connected and simply connected manifold is homotopically equivalent to the standard sphere (see Sec. 13), then the following lemma might be the three-dimensional analog of Lemma 14.1: If G_k is the trivial group, then the manifold N_k^3 is diffeomorphic to the standard three-dimensional sphere.

Chapter 4

SYMMETRIC SPACES

15. MAIN PROPERTIES OF SYMMETRIC SPACES, THEIR MODELS AND ISOMETRY GROUPS

15.1. Definition of Symmetric Spaces

Definition 15.1. The connected Riemannian manifold V is called a Riemannian symmetric space if for any point $p \in V$ there is an isometry $s_p: V \to V$ that preserves p and "upsets" geodesics passing through p. This means that if γ is a geodesic such that $\gamma(0) = p$, then $s_p\gamma(t) = \gamma(-t)$.

Symmetric spaces can also be defined as follows: For any point $p \in V$ there is an involutive isometry s_p (i.e., its square is an identity) different from the identity such that p is its isolated fixed point.

15.2. Lie Groups as Symmetric Spaces

As our first example, we consider Lie groups. We will mostly consider compact Lie groups $\mathfrak{G}$; corresponding Lie algebras will be denoted by G.* Let us assume that the compact group $\mathfrak{G}$ is a closed subgroup in the orthogonal group SO_N or in the special unitary group SU_N for a sufficiently large $N < \infty$. This will simplify many constructions and will not lead to loss of generality since any compact group admits such a presentation. (We will not prove this fact.)

*Or vice versa – Translator.

On a compact Lie group $\mathfrak{G}$, fix a left- and right-invariant (biinvariant) Riemannian metric. For proof of the existence of such a metric, see, for example, [2]. Recall also explicit formulas that define such a metric. Let $\mathfrak{G}$ be embedded into SO_N. A biinvariant metric on $\mathfrak{G}$ is introduced as a restriction of the biinvariant metric from SO_N onto $\mathfrak{G}$. Consider the standard presentation of SO_N as the set of $(N \times N)$-matrices A such that $A^{-1} = A^T$. Consider the linear space $\mathbf{R}^{N^2}$ of all real $(N \times N)$-matrices and introduce the scalar product on it by the formula $\langle A, B\rangle = \operatorname{Tr} A \cdot B^T$, where $A, B \in \in \mathbf{R}^{N^2}$ are two arbitrary matrices. Clearly, if $A = (A_{ij})$, $B = = (B_{ij})$, then $\langle A,B\rangle = \sum_{i,j} A_{ij}B_{ij}$ is the Euclidean scalar product. The obvious property of this product is its invariance with respect to transformations $X \to CXC^{-1}$, where $X \in \mathbf{R}^{N^2}$ and $\det C \neq 0$, i.e., $\operatorname{Tr}(CXC^{-1}CYC^{-1}) = \operatorname{Tr} XY$. The transformation $X \to \to CXC^{-1}$ is sometimes denoted by Ad_C. The group SO_N is realized as the closed compact submanifold in $\mathbf{R}^{N^2}$. Since for any $g \in SO_N$ the identity $gg^T = E$ holds, then $|g|^2 = \langle g, g\rangle = N = = \operatorname{Tr} gg^T$, i.e., the submanifold SO_N is embedded into the sphere S^{N^2-1} of radius $\sqrt{N}$ with center at 0.

On SO_N, define the metric by restriction of the Euclidean metric of the enveloping space onto the submanifold SO_N (see Fig. 171).

<u>Lemma 15.1</u>. The obtained Riemannian metric on the space of matrices $\mathbf{R}^{N^2}$ is invariant both with respect to left and right shifts by elements of SO_N.

<u>Proof</u>. Let $X = gA$, $Y = gB$. Then $\langle X, Y\rangle = \langle gA, gB\rangle = = \operatorname{Tr} gAB^Tg^T = \operatorname{Tr} AB^T$, since $g^T = g^{-1}$. The lemma is proved.

Biinvariance of the obtained metric on subgroups $\mathfrak{G} \subset SO_N$ is equivalent to the invariance of the scalar product $\langle , \rangle$ at the unit $E \in \mathfrak{G}$ with respect to inner automorphisms $X \to gXg^{-1}$, where $X \in G = T_e\mathfrak{G}$, $g \in \mathfrak{G}$.

<u>Statement 15.1</u>. Let $\mathfrak{G}$ be a compact Lie group with a biinvariant metric. Then $\mathfrak{G}$ is a symmetric space. Involution $S_g\colon \mathfrak{G} \to \mathfrak{G}$ is defined by the formula $s_g(x) = gx^{-1}g$; evidently, $s_g(g) = g$.

<u>Proof</u>. Consider the smooth mapping $\nu\colon \mathfrak{G} \to \mathfrak{G}$, $\nu(x) = = x^{-1}$. Then $d\nu\colon T_E\mathfrak{G} \to T_E\mathfrak{G}$ inverts tangent vectors to $\mathfrak{G}$ at

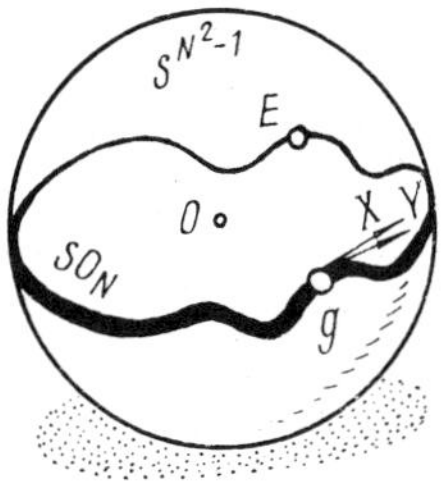

Fig. 171

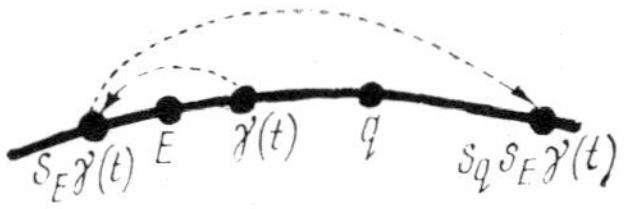

Fig. 172

E; in particular, $d\nu$ is an isometry of G. Since $\nu = R_g^{-1} \times \nu L_g^{-1}$, where R_g and L_g stand for right and left shifts, respectively, then $(d\nu)_x : T_x \mathfrak{G} \to T_{x^{-1}} \mathfrak{G}$ is an isometry, too. Since $s_g = R_g s_E R_g^{-1}$, then s_g is an isometry that inverts geodesics at g. The statement is proved.

It is easy to verify that one-parameter subgroups in $\mathfrak{G}$ are exactly geodesics that pass through E. In fact, if γ is a geodesic in $\mathfrak{G}$ such that $E = \gamma(0)$, $q = \gamma(c)$, then $s_q s_E \gamma(t) = \gamma(t + 2c)$ (see Fig. 172). Since $s_g(x) = gx^{-1}g$, then $s_q s_E(x) = \gamma(c)x\gamma(c) = qxq$, whence $\gamma(c)\gamma(t)\gamma(c) = \gamma(t + 2c)$, where $x = \gamma(t)$. Setting $t = 0$, we get $\gamma(n \cdot c) = \gamma(c)^n$ for any natural n. If c_1/c_2 is rational, i.e., $c_1 = n_1 c$, $c_2 = n_2 c$ for some c and $n_1, n_2 \in \mathbf{Z}$, then

$$\gamma(c_1 + c_2) = \gamma((n_1 + n_2)c) = \gamma(c)^{n_1+n_2} = \gamma(c_1) \cdot \gamma(c_2) \cdot$$

From continuity considerations, we get $\gamma(c_1) \cdot \gamma(c_2) = \gamma(c_1 + c_2)$ for arbitrary c_1 and c_2, i.e., γ is a homomorphism and hence a one-parameter subgroup. Since geodesics and a one-parameter subgroup are defined by the tangent vector to the unit, the statement is proved.

15.3. Properties of the Curvature Tensor

The existence of geodesic symmetries on symmetric spaces places rigid restrictions on differential-geometrical structures of the space. Let us explain this by the example of Lie groups.

Proposition 15.1. Let $\mathfrak{G}$ be a compact Lie group with a biinvariant metric (put $\langle , \rangle$ for the corresponding scalar product); let X, Y, Z, W be left-invariant vector fields on $\mathfrak{G}$. Then the following identities hold, where R stands for the Riemannian curvature tensor:

1) $\langle [X,Y],Z\rangle = \langle X,[Y,Z]\rangle$,

2) $R(X,Y)Z = {}^1/_4 [[X,Y],Z]$,

3) $\langle R(X,Y)Z, W\rangle = {}^1/_4 \langle [X,Y],[Z,W]\rangle$.

Proof. On $\mathfrak{G}$, consider the Riemannian symmetric connection ∇ generated by the biinvariant metric. Let $\nabla_X Y$ be the covariant derivative of the vector field Y along the vector field X (see [1]). Then for any left-invariant field X on $\mathfrak{G}$, the identity $\nabla_X X \equiv 0$ holds. In fact, integral trajectories of a left-invariant field are left shifts of one-parameter subgroups, hence geodesics, i.e., $\nabla_X X = 0$. Therefore, $0 = \nabla_{X+Y}(X + Y) = \nabla_X X + \nabla_Y Y + \nabla_X Y + \nabla_Y X$. Hence, $\nabla_X Y + \nabla_Y X = 0$. On the other hand, the definition of a covariant derivative easily implies that $\nabla_X Y - \nabla_Y X = [X, Y]$. Comparing the latter two equalities, we get $2\nabla_X Y = [X, Y]$. The properties of the covariant derivative imply $Y\langle X, Z\rangle = \nabla_Y\langle X, Z\rangle = \langle\nabla_Y X, Z\rangle - \langle X, \nabla_Y Z\rangle$. Since X and Z are left-invariant, then $\langle X, Z\rangle = \text{const}$ on $\mathfrak{G}$, i.e., $Y\langle X, Z\rangle = 0$. Hence, $\langle [Y, X], Z\rangle + \langle X, [Y, Z]\rangle = 0$, i.e., $\langle X, [Y, Z]\rangle = \langle [X, Y], Z\rangle$. Relation 1 is proved. By definition of R(X, Y)Z, we have $R(X, Y)Z = -\nabla_X \times \nabla_Y Z + \nabla_Y\nabla_X Z + \nabla_{[X,Y]}Z$. Since $\nabla_X Y = \frac{1}{2}[X, Y]$, then $R(X, Y)Z = -\frac{1}{4}[X, [Y, Z]] + \frac{1}{4}[Y, [X, Z]] + \frac{1}{2}[[X, Y], Z]]$. Furthermore, using the Jacobi identity we get $R(X, Y)Z = \frac{1}{2}[[X, Y], Z]$. The proposition is proved.

Recall that if M^n is a Riemannian manifold, then the sectional curvature with respect to X, Y is $\langle R(X, Y)X, Y\rangle$. If $M^n = \mathfrak{G}$ is a Lie group, then the above proposition implies $\langle R(X, Y)X, Y\rangle = \frac{1}{4}\langle [X, Y], [X, Y]\rangle = \frac{1}{4}\|[X, Y]\|^2 \geqq 0$. Thus, for compact Lie groups, sectional curvature is always nonnegative and vanishes if this section "commutes," i.e., $\|[X, Y]\| = 0$.

15.4. Involutive Automorphisms and the Corresponding Symmetric Spaces

Lie groups do not exhaust the list of symmetric spaces. For example, the standard sphere $S^n \subset \mathbf{R}^{n+1}$ is a symmetric space, although S^n is not a group for $n \neq 1, 3$. Let $\mathfrak{G}$ be a connected, simply connected compact Lie group and $\sigma: \mathfrak{G} \to \mathfrak{G}$ an involutive automorphism of the group. Let $\mathfrak{H}$ be the set of fixed points of σ in the group $\mathfrak{H}$. Clearly, $\mathfrak{H}$ is a closed compact subgroup in $\mathfrak{G}$. The Lie algebra G splits into the direct sum of two spaces $G = H + B$, where $H = \{X \in G \mid d\sigma(X) = X\}$, $B = \{X \in G \mid d\sigma(X) = -X\}$. Since $(d\sigma)^2 = 1_G$ (the identity mapping of G onto itself), then all eigenvalues of $d\sigma$ are ± 1. Since σ is an automorphism of the group, then $d\sigma$ is an automorphism of its Lie algebra, i.e., $d\sigma[X, Y] = [d\sigma X, d\sigma Y]$. The definition of H implies that H is a Lie subalgebra in G and it is the Lie algebra of $\mathfrak{H}$.

It is possible to prove that $\mathfrak{H}$ (the set of fixed points of the involution on the connected, simply connected compact group) is connected. Moreover, the set of fixed points of any automorphism of a simply connected, connected compact group is connected. We will not prove these statements in the general case since, in all subsequent examples, the connectedness of $\mathfrak{H}$ will be obvious. Note here the useful corollary of the connectedness of $\mathfrak{H}$: the homogeneous space $\mathfrak{G}/\mathfrak{H}$ is simply connected. This follows from the exact homotopic sequence of the bundle

$$\begin{array}{ccccc} \pi_1(\mathfrak{G}) & \to & \pi_1(\mathfrak{G}/\mathfrak{H}) & \to & \pi_0\mathfrak{H}. \\ \| & & & & \| \\ 0 & & & & 0 \end{array}$$

If on the group $\mathfrak{G}$ a biinvariant metric is given, then on any homogeneous space $\mathfrak{G}/\mathfrak{H}$ where $\mathfrak{H}$ is a closed subgroup of $\mathfrak{G}$, a Riemannian metric invariant with respect to the $\mathfrak{G}$-action on $\mathfrak{G}/\mathfrak{H}$ automatically arises. To define this metric, it suffices to define it on spaces orthogonal to the conjugate classes with respect to $\mathfrak{H}$ in $\mathfrak{G}$ (see Fig. 173). Since actions are developed in $\mathfrak{G}$, then for such a metric it suffices to take the restriction of a biinvariant metric of the group onto the planes indicated. Clearly, the metric resulting from $\mathfrak{G}/\mathfrak{H}$ is invariant with respect to the left $\mathfrak{G}$-action on $\mathfrak{G}/\mathfrak{H}$: the plane orthogonal to one of the conjugate classes passes, after a left shift, into the plane also orthogonal to the new conjugate class. Thus, $\mathfrak{G}$ acts on $\mathfrak{G}/\mathfrak{H}$ as a subgroup in the group of isometries of the introduced metric. In what follows, we will always assume that on the homogeneous space $\mathfrak{G}/\mathfrak{H}$ is introduced the metric generated by a biinvariant metric of $\mathfrak{G}$.

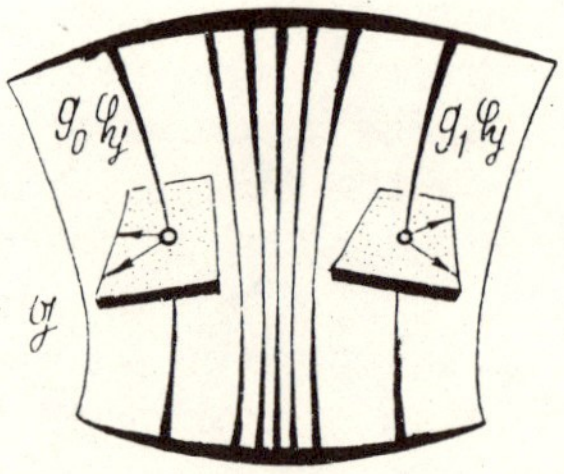

Fig. 173

Proposition 15.2. Let $\mathfrak{G}$ be a connected compact Lie group and $\sigma: \mathfrak{G} \to \mathfrak{G}$ an arbitrary involutive automorphism of the group. Then the homogeneous space $\mathfrak{G}/\mathfrak{H}$ with the Riemannian metric generated by a biinvariant metric of $\mathfrak{G}$ is symmetric.

Proof. Since on $\mathfrak{G}/\mathfrak{H}$ is introduced the metric generated by a biinvariant metric of the group, then the action of this group on $\mathfrak{G}/\mathfrak{H}$ is an isometry. For a point $v \in \mathfrak{G}/\mathfrak{H}$, we must construct a geodesic symmetry in it that inverts geodesics. Since $\mathfrak{G}$ acts transitively on $V = \mathfrak{G}/\mathfrak{H}$, it suffices to construct such a symmetry only at one point $v_0 \in V$. Take, as an example of such a point, a conjugate class that coincides with $\mathfrak{H}$. Then the involution in $\mathfrak{G}$, preserving $\mathfrak{H}$, evidently defines a geodesic involution on the quotient space $V = \mathfrak{G}/\mathfrak{H}$ too. The proposition is proved.

15.5. The Cartan Model of Symmetric Spaces

Consider the decomposition $G = H + B$. Since H is a subalgebra of fixed points of the automorphism $\theta = (d\sigma)_E$ and B is the invariant space corresponding to the eigenvalue −1 of θ, then elements of H and B commute, as follows: $[H, H] \subset H$, $[B, H] \subset B$, $[B, B] \subset H$. In particular, the triple commutator $[[B, B], B]$ (i.e., the set of all elements of the form $[[X, Y], Z]$, where $X, Y, Z \in B$) is contained in B. It turns out that the decomposition of the Lie algebra generates a "decomposition" of the corresponding Lie group. In $\mathfrak{G}$ is contained the subgroup $\mathfrak{H}$ which is the stationary subgroup of V. Question: Is it possible to embed the symmetric space V in $\mathfrak{G}$? If V were just a homogeneous space without any symmetries, its embedding into $\mathfrak{G}$ would be impossible in the general case (such examples are easy to construct). But for a symmetric space, the situation is more favorable. It turns

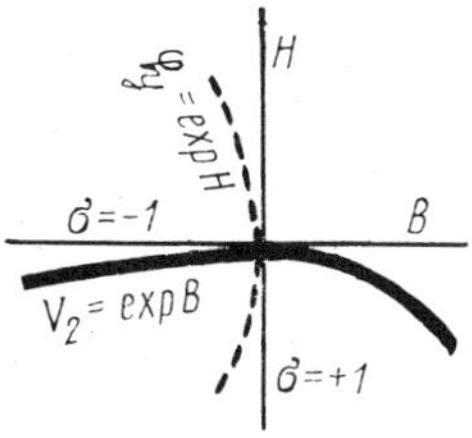

Fig. 174

out that in this case V can be realized as a "homogeneous surface" in $\mathfrak{G}$.

In $\mathfrak{G}$, consider the connected component V_1 of the subset of elements g such that $\sigma(g) = g^{-1}$ and also the subset V_2 that consists of all geodesics γ of $\mathfrak{G}$ that pass through the unit and such that $\dot{\gamma}(0) \in B$, $\gamma(0) = E$. Since $\mathfrak{G}$ is realized as a subgroup in the group of orthogonal matrices, then V_2 is the set of all matrices of the form exp X, where $X \in B$, i.e., $V_2 = \exp B$. Finally, in $\mathfrak{G}$, consider the subset V_3 which is the image of $\mathfrak{G}$ with respect to the following mapping $p: g \to g \times \sigma(g^{-1})$ (see Fig. 174).

<u>Theorem 15.1</u>. Subsets V_1, V_2, V_3 in $\mathfrak{G}$ coincide. This subset is a smooth submanifold in $\mathfrak{G}$, diffeomorphic to the symmetric space $V = \mathfrak{G}/\mathfrak{H}$. Moreover, this submanifold is totally geodesic, i.e., any geodesic of $\mathfrak{G}$ tangent to V belongs to this submanifold. The continuous mapping p defines the locally trivial bundle (the so-called principal bundle) $p: \mathfrak{G} \xrightarrow{\mathfrak{H}} V$ with the fiber $\mathfrak{H}$.

<u>Proof</u>. Let us prove that V_1 and V_2 coincide. Since σ is a geodesic symmetry at E, then the coincidence of V_1 and V_2 follows from the identity $(\exp X)^{-1} = \exp(-X)$. Let us prove the coincidence of V_2 and V_3. Let $v \in V_3$, i.e., there exists a g such that $v = g\sigma(g^{-1})$. Then $\sigma(v) = \sigma(g)g^{-1} = (\sigma(g^{-1}))^{-1}g^{-1} = v^{-1}$, i.e., $v \in V_1 = V_2$. Conversely, let $v \in V_2$, i.e., $\sigma(v) = v^{-1}$ or (see above) $v = \exp X$, where $X \in B$. In $\mathfrak{G}$, consider the element $v_1 = \exp(\frac{1}{2}X)$; then $p(v_1) = v_1^2 = \exp X = v$, i.e., $v \in \operatorname{Im} p$, as required. We have made use of the fact that v_1 also belongs to V_2. Let us prove that p defines the principal bundle with $\mathfrak{H}$ as a fiber. In $\mathfrak{G}$, consider conjugate classes $g\mathfrak{H}$ with respect to $\mathfrak{H}$. If elements g_1 and g_2 belong to one conjugate class, then $p(g_1) =$

$= p(g_2)$. Conversely, if $p(g_1) = p(g_2)$, then $g_1\sigma(g_1^{-1}) = g_2 \times \times \sigma(g_2^{-1})$, or $\sigma(k^{-1}) = k^{-1}$, where $k = g_2^{-1}g_1$; i.e., $k \in \mathfrak{H}$ and $g_1 = g_2h$, since the set of fixed points of an involution is connected. This implies that the submanifold V in $\mathfrak{G}$ is diffeomorphic to the symmetric space $\mathfrak{G}/\mathfrak{H}$. It remains to prove that this submanifold is totally geodesic, i.e., that any geodesic submanifold V is at the same time a geodesic in the whole $\mathfrak{G}$. Consider the curvature tensor $R_{\mathfrak{G}}$ on $\mathfrak{G}$ and the curvature tensor R_V on V. We claim that R_V is obtained from $R_{\mathfrak{G}}$ by restriction onto the submanifold $V \subset \mathfrak{G}$. Above we have computed $R_{\mathfrak{G}}$ explicitly. It turned out that $R\mathfrak{G}(X, Y)Z = = \frac{1}{4}[[X, Y], Z]$. Let $X, Y, Z \in T_*V$, where $V = \exp B$. The triple commutator of elements of B belongs again to B; hence $R_{\mathfrak{G}}(X, Y):B \to B$ if $X, Y \in B$. But this means that R_V is obtained from $R\mathfrak{G}$ by restriction onto V. Let $\gamma \subset V$ be a geodesic in the induced Riemannian metric. Then the equation that this geodesic satisfies is of the form (see [1]) $\nabla_{\dot{\gamma}}\dot{\gamma} = = 0$, and an arbitrary Jacobi field I along γ (on V) satisfies $(\nabla_{\dot{\gamma}})^2 I + R(\dot{\gamma}, I)\dot{\gamma} = 0$. Hence, any geodesic variation on V is the same on $\mathfrak{G}$. Since $V = \exp B$, then V is covered by the bunch of one-parameter subgroups. The theorem is proved.

The constructed embedding $V \to \mathfrak{G}$ is called the Cartan model of the symmetric space $\mathfrak{G}/\mathfrak{H}$. We caution the reader not to believe that this embedding is a section of the bundle $p: \mathfrak{G} \to V$. Let us study in detail how conjugate classes with respect to $\mathfrak{H}$ intersect with the submanifold $V \subset \mathfrak{G}$.

Proposition 15.3. Each conjugate class $g_0\mathfrak{H}$ has a nonempty intersection with V.

Proof. Assume the contrary: Suppose there is a class $g_0\mathfrak{H}$ such that $(g_0\mathfrak{H}) \cap V = \emptyset$. On V, consider the point $m_0 = = p(g_0\mathfrak{H})$ and let $m' = \sqrt{m_0}$ be the notation for a point of V such that $(m')^2 = m_0$. If there are several such points, then take any of them. If $v \in V$, then $v = g\sigma g^{-1}$ for some $g \in \mathfrak{G}$; hence $\sigma(v) = (g\sigma(g^{-1}))^{-1}$, i.e., $\sigma(v) = v^{-1}$ and the choice of g does not matter. The mapping p transforms $\mathfrak{G}$ into V. We may consider the through mapping $pi:V \to V$, where $i:V \to \mathfrak{G}$, $pi(v) = = v\sigma(v^{-1})$, i.e., $pi(v) = v^2$ and pi acts on V by "raising it to the second power." Since $m_0 = (m')^2$, then $m_0 = pi(m')$. Consider the class $m'\mathfrak{H}$; then $m' \in m'\mathfrak{H} \cap V$ and $m_0 = p(m'\mathfrak{H})$, i.e., the pre-image of m_0 with respect to p contains points of two conjugate classes: $m'\mathfrak{H}$ and $g_0\mathfrak{H}$. Since p defines the principal bundle, then $m'\mathfrak{H} = g_0\mathfrak{H}$, i.e., $g_0\mathfrak{H} \cap V \neq \emptyset$, and contains at least one point m'. This contradiction proves our proposition.

Thus, an arbitrary class $g\mathfrak{H}$ is presentable in the form $m\mathfrak{H}$, where $m \in V$. If $g \in \mathfrak{G}$, let $\sqrt{g}$ be an element g' such that $(g')^2 = g$ and $\{\sqrt{g}\}$ be the set of all such elements.

<u>Proposition 15.4</u>. Let $m\mathfrak{H}$ be an arbitrary conjugate class, where $m \in V$. Then $m\mathfrak{H} \cap V = \{\sqrt{m^2}\} \cap V$ (see Fig. 175).

<u>Proof</u>. Let us prove that $\{\sqrt{m^2}\} \cap V \subset m\mathfrak{H} \cap V$. Let m_1, $m_2 \in V$ and $m_1{}^2 = m_2{}^2$. Then $m_1\sigma(m_1{}^{-1}) = m\sigma(m^{-1})$ and $m^{-1}m_1 \times \times\sigma(m_1{}^{-1}m) = E$, i.e., $k\sigma(k^{-1}) = E$, where $k = m^{-1}m_1$. Since $p(k) = E$, then $k \in \mathfrak{H}$ and $m_1 = mh$, where $h \in \mathfrak{H}$. Let us show that the converse inclusion holds. Let $v \in m\mathfrak{H}$, $v \in V$; then $v = mh$, $h \in \mathfrak{H}$ and $v^2 = v\sigma(v^{-1}) = m^2$, i.e., $v \in \{\sqrt{m^2}\}$, as required $\times\sigma(m_1{}^{-1}m) = E$, i.e.,

The intersection of the "zero-conjugate class," i.e., $\mathfrak{H} \cap V$, consists of the set of v such that $v^2 = E$.

The action of $\mathfrak{G}$ on V may be described as follows: $g(a) = g a\sigma(g^{-1})$, where $g \in \mathfrak{G}$, $a \in V$. If, in particular, $g \in V$, then $g(a) = gag$; if $g \in \mathfrak{H}$, then $g(a) = gag^{-1}$, i.e., $\mathfrak{H}$ acts on V by rotations and V acts on itself by shifts. The action of the projection $p: \mathfrak{G} \to V$ is shown in Fig. 176.

15.6. Geometry of Cartan Models

Consider the simplest examples of realization of a symmetric space as a submanifold in the group $\mathfrak{G}$. Let $\mathfrak{G} = SU_2 = \left\{\begin{pmatrix} x & y \\ -\bar{y} & \bar{x} \end{pmatrix}\right\}$ where $x, y \in \mathbf{C}$, $|x|^2 + |y|^2 = 1$. As an involution σ, take the homomorphism $\sigma(g) = \bar{g}$ (the complex conjuga-

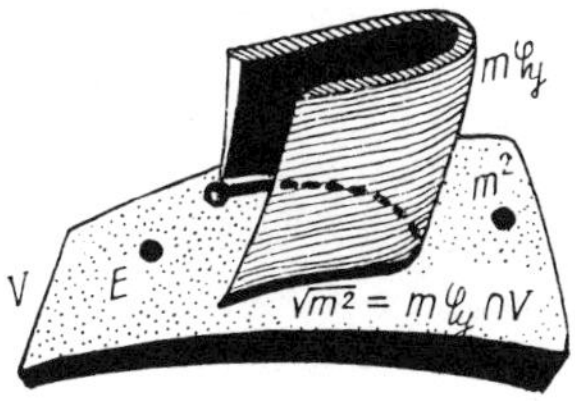

Fig. 175

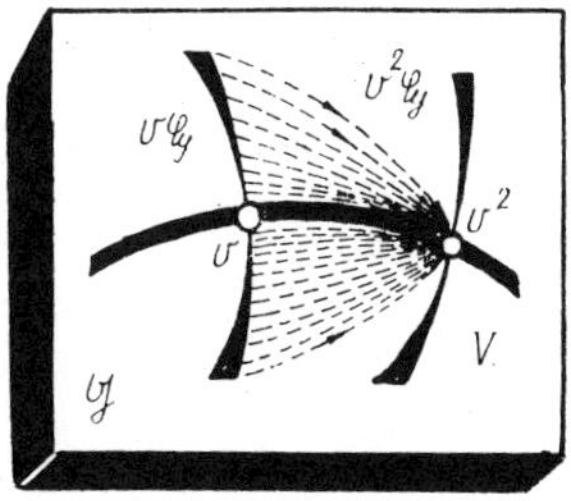

Fig. 176

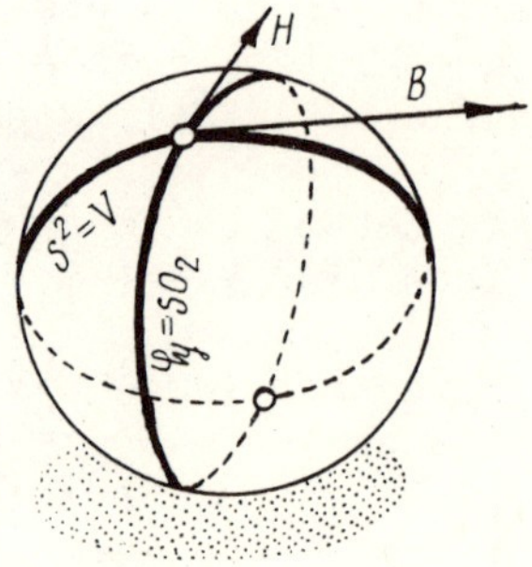

Fig. 177

tion in the group of matrices). Then $\mathfrak{H} = SO_2 = \{g \in SU_2 \mid g = g\}$; the subgroup of real unitary matrices coincides with SO_2. Consider the decomposition $G = H + B$. Then

$$H = \begin{pmatrix} 0 & \varphi \\ -\varphi & 0 \end{pmatrix}, \quad B = i \begin{pmatrix} a & b \\ b & -a \end{pmatrix}; \; a, b \in \mathbf{R}.$$

The symmetric space $V = SU_2/SO_2$ is homeomorphic to the sphere. Let us present the Cartan model of the sphere S^2 in SU_2. Let $i \begin{pmatrix} a & b \\ b & -a \end{pmatrix} = Q, \; a^2 + b^2 = 1$; then

$$\exp(tQ) = E\left(1 - \frac{t^2}{2!} + \dots\right) + Q\left(t - \frac{t^3}{3!} + \dots\right) = E\cos t + Q\sin t.$$

Since the matrix Q runs through the circle, then matrices exp (tQ) fill in the two-dimensional sphere $S^2 \subset S^3$, where S^2 is embedded as an equator in S^3 (see Fig. 177). The "zeroth" conjugate class $\mathfrak{H}$ intersects S^2 at two points: E and $-E$.

As was proved earlier, Lie groups are symmetric spaces. Let V be a Lie group. Let $\mathfrak{G} = V \times V$, and consider the involution $\sigma: \mathfrak{G} \to \mathfrak{G}$ such that $\sigma(v_1, v_2) = (v_2, v_1)$. The set of fixed points of σ is the subgroup $\mathfrak{H} = (v, v)$, i.e., the diagonal in the direct product $V \times V$. Let G and K be Lie algebras of the groups $\mathfrak{G}$ and V, respectively. Then $G = K \oplus \oplus K$. On the other hand, G splits into the direct sum of two invariant subspaces of the automorphism Θ, where $\Theta(X, Y) =$

$= (Y, X)$, $(X, Y) \in K \oplus K$, $G = B + H$, $\Theta H = H$. Clearly, $H = (X, X)$, $B = (X, -X)$. Since $\sigma(v, v^{-1}) = (v, v^{-1})^{-1}$, the space V is embedded into $\mathfrak{G}$ as a totally geodesic submanifold consisting of points of the form (v, v^{-1}). The mapping $p: \mathfrak{G} \to V$ is of the form $p(g) = (v_1 v_2^{-1}, v_2 v_1^{-1}) \in iV$, where $g = (v_1, v_2)$; hence $pi(g) = g^2$, and p is a diffeomorphism of V and $(V \times V)/\mathfrak{H}$.

Consider Riemannian metrics on symmetric spaces $\mathfrak{G}/\mathfrak{H}$ invariant with respect to the $\mathfrak{G}$-action.

Proposition 15.5. Consider the Cartan model $V \subset \mathfrak{G}$ of the symmetric space $\mathfrak{G}/\mathfrak{H}$ and let $\langle , \rangle$ be a biinvariant metric on $\mathfrak{G}$. On V, consider the Riemannian metric $\langle , \rangle_V$ induced by the embedding $i: V \to \mathfrak{G}$. Then $\langle , \rangle_V$ is invariant with respect to the $\mathfrak{G}$-action on V.

Proof. Recall that we have realized $\mathfrak{G}$ as a smooth submanifold in S^{N^2-1} in the space $\mathbf{R}^{N^2}$ of all $(N \times N)$-matrices. If $X, Y \in G$, then $\langle X, Y\rangle = \mathrm{Tr}\, XY^T$, where $X, Y = \mathbf{R}^{N^2}$. Therefore, the totally geodesic manifold $V \subset \mathfrak{G}$ is embedded into a Euclidean space and the metric $\langle , \rangle_V$ is induced on V by this embedding. If $g \in \mathfrak{G}$, $v \in V$, then $g(v) = gv\sigma(g^{-1})$. Let $X, Y \in B$; then, under the action of $\mathfrak{G}$, the space B is mapped onto the space $T_{g(E)}V$, where $g(E) = g\sigma(g^{-1})$. The mapping $g_*: B \to T_{g(E)}V$ is of the form $X \to gX\sigma(g^{-1})$. It remains to verify the identity

$$\langle g_*X, g_*Y\rangle_V |_{g(E)} = \langle X, Y\rangle_V |_E,$$

i.e.,

$$\mathrm{Tr}\, gX\sigma(g^{-1})(gY\sigma(g^{-1}))^T = \mathrm{Tr}\, XY^T.$$

We have proved above that each conjugate class $g_0\mathfrak{H}$ with respect to $\mathfrak{H}$ has a nonzero intersection with V; hence, any element $g \in \mathfrak{G}$ is presentable (nonuniquely) in the form of the product $g = v'h$, where $v' \in V$, $h \in \mathfrak{H}$. Substituting this decomposition into the previous formula, we obtain

$$\mathrm{Tr}\, v'hXh^{-1}v'v'^T h^{-1^T} Y^T h^T v'^T = \mathrm{Tr}\, XY^T,$$

since matrices v', h are orthogonal ($\mathfrak{G} \subset SO_N$). The proposition is proved.

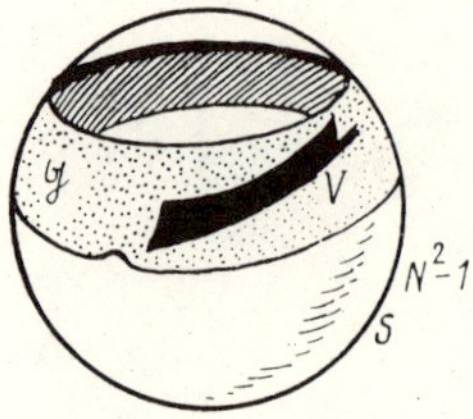

Fig. 178

This theorem is useful in various concrete computations on symmetric spaces. In fact, we have described explicitly the invariant Riemannian metric on the symmetric space realized in the Euclidean space, and this metric turns out to be induced by the Euclidean metric of the enveloping space (see Fig. 178). Thus, if we must solve a problem related to the invariant metric on a symmetric space, it is useful to realize the latter as the Cartan model in the group embedded into the standard sphere. Below we will list several concrete matrix realizations of symmetric spaces.

15.7. Several Important Examples of Symmetric Spaces

1) The space SU_n/SO_n. The isometric group $\mathfrak{G}$ of the space V is SU_n. The involutive automorphism Θ in the Lie algebra $G = su_n$ is of the form $\Theta(X) = \bar{X}$, where the bar stands for the complex conjugation of matrices. The automorphism $\Theta: G \to G$ is extendable up to the involutive automorphism $\sigma: SU_n \to SU_n$, $\sigma(g) = \bar{g}$. In G, consider the space B, where Θ is equal to -1. This space consists of all symmetric, purely imaginary matrices of order n with trace 0. The space H on which θ equals +1 consists of real skew symmetric matrices (it is the Lie algebra of SO_n). The space B is the tangent space to the Cartan model $V = \{g\sigma(g^{-1})\} = \{gg^T\}$, and since $v^T = v$ for $v \in V$, the model of the symmetric space consists only of all unitary symmetric matrices.

2) The space SU_{2m}/Sp_m. The isometric group of the space V is SU_{2m}. The involutive automorphism Θ in the Lie algebra su_{2m} is of the form $\Theta(X) = J\bar{X}J^{-1}$, where $J = \begin{pmatrix} 0 & E \\ -E & 0 \end{pmatrix}$; the bar stands for the complex conjugation. The automorphism Θ

is extendable up to the involution $\sigma: SU_{2m} \to SU_{2m}$, $\sigma(g) = J\bar{g}J^{-1}$. The set of fixed points of σ coincides with Sp_m, i.e., with the set of elements such that $gJ = J\bar{g}$. In fact, unitary operators g satisfying this extra condition preserve the skew symmetric form in $\mathbf{C}^{2m}$ of the form $\sum_{k=1}^{m} z_k \wedge z_{m+k}$. This form is defined by the skew symmetric matrix J. The totally geodesic submanifold V (the model of the symmetric space) consists of all elements of the form $g\sigma(g^{-1})$, i.e., gJg^TJ^{-1}, where g runs over the entire group SU_{2m}. For a more useful description of V, consider its isometric shift by the element $J \in SU_{2m}$. We obtain the model $V' = V \cdot J = \{gJg^T\}$. If $v' = gJg^T$, then $v'^T = -v'$ and, clearly, V' (which is isometric to V) is contained in the set of all skew symmetric unitary matrices of SU_{2m}. The tangent space B is of the form

$$\begin{pmatrix} Z_1 & Z_2 \\ \bar{Z}_2 & -\bar{Z}_1 \end{pmatrix}, \quad \text{where } Z_1 \in su_m,\ Z_2 \in so(m, \mathbf{C}).$$

3) The space SO_{2n}/U_n. The isometric group of the space V is SO_{2n}. The involutive automorphism Θ in the Lie algebra so_{2n} is of the form $\Theta(X) = JXJ^{-1}$, where $J = \begin{pmatrix} 0 & E \\ -E & 0 \end{pmatrix}$. The automorphism Θ is extendable up to the involution $\sigma: SO_{2n} \to SO_{2n}$, $\sigma(g) = JgJ^{-1}$. The set of fixed points of σ is the subgroup $\mathfrak{H} = \{g \in SO_{2n}, Jg = gJ\}$. Since J defines the complex structure on $\mathbf{R}^{2n}$, identifying it with $\mathbf{C}^n$, then $\mathfrak{H}$ is isomorphic to U_n. The embedding $U_n \to SO_{2n}$ is given by the standard formula $C + iD \to \begin{pmatrix} C & D \\ -D & C \end{pmatrix}$. The totally geodesic submanifold V (the Cartan model) in SO_{2n} consists of all elements of the form $g\sigma(g^{-1})$, i.e., $gJg^{-1}J^{-1}$, where g runs over SO_{2n}. For a more useful description of V, consider its isometric shift by the element J. We obtain the model $V' = VJ = \{gJg^T\}$. If $v' = gJg^T$, then $(v')^T = -v'$, i.e., V' is contained in the set of skew symmetric orthogonal matrices in SO_{2n} (but does not coincide with it). The tangent space B is of the form

$$\begin{pmatrix} Z_1 & Z_2 \\ Z_2 & -Z_1 \end{pmatrix}, \text{ where } Z_1, Z_2 \in so_{2n}.$$

4) The space Sp_n/U_n. The isometric group $\mathfrak{G}$ of the space V is the group Sp_n. We assume that Sp_n is realized as the standard subgroup in SU_{2n}. The involution Θ in the Lie al-

gebra sp_n is of the form $\Theta(X) = \bar{X}$ (the complex conjugation on su_{2n}), i.e., it coincides with the automorphism $\Theta(X) = JXJ^{-1}$, where $J = \begin{pmatrix} 0 & E \\ -E & 0 \end{pmatrix}$. Recall that an element g of SU_{2n} belongs to Sp_n iff $\bar{g} = JgJ^{-1}$, i.e., $\bar{g}J = Jg$. The automorphism Θ is extendable up to the involution $\sigma: Sp_n \to Sp_n$, $\sigma(g) = \bar{g}$. The set of fixed points of σ is the subgroup $\{g \in Sp_n, g = \bar{g}\}$, i.e., $Sp_n \subset SO_{2n}$, which coincides with the subgroup $U_n \cap SO_{2n}$ with respect to the standard identification $C + iD \to \begin{pmatrix} C & D \\ -D & C \end{pmatrix}$. The totally geodesic manifold V (the Cartan model) in Sp_n consists of matrices of the form $g\sigma(g^{-1}) = gg^T$. The tangent space B is of the form $\begin{pmatrix} Z_1 & Z_2 \\ Z_2 & -Z_1 \end{pmatrix}$, where $Z_1 \in U_n$ is a purely imaginary matrix and Z_2 is a purely imaginary symmetric matrix.

5) The space $SU_{p+q}/S(U_p \times U_q)$ – the complex Grassmanian manifold. The isometric group $\mathfrak{G}$ of V is SU_{p+q}. The involution Θ in the Lie algebra su_{p+q} is of the form $\Theta(X) = J_{p,q}XJ_{p,q}$, where $J_{p,q} = \begin{pmatrix} -E_p & 0 \\ 0 & E_q \end{pmatrix}$, and E_α is the unit matrix of order α. The automorphism Θ is extendable up to the involution $\sigma: SU_{p+q} \to SU_{p+q}$, $\sigma(g) = J_{p,q} \cdot g \cdot J_{p,q}$. The set of fixed points of σ is $\{g \in SU_{p+q}, J_{p,q} \cdot g = gJ_{p,q}\} = S(U_p \times U_q)$. The Lie algebra H is of the form $\begin{pmatrix} C & 0 \\ 0 & D \end{pmatrix}$, where $C \in u_p$, $D \in u_q$, $\mathrm{Tr}\,(C + D) = 0$. The totally geodesic submanifold V (the Cartan model) is generated in SU_{p+q} by matrices $gJ_{p,q}g^{-1}J_{p,q}$. The tangent space V is of the form $\begin{pmatrix} 0 & Z \\ -\bar{Z}^T & 0 \end{pmatrix}$, where Z is the complex matrix with p rows and q columns. For p = 1, the space V is the complex projective space.

6) The space $SO_{p+q}/SO_p \times SO_q$ – the real Grassmanian manifold. The isometry group $\mathfrak{G}$ is SO_{p+q}. The involution Θ is of the form $\Theta(X) = J_{p,q}XJ_{p,q}$. It is extendable up to the involution $\sigma(g) = J_{p,q} \cdot g \cdot J_{p,q}$. The set of fixed points of σ is the subgroup $\{g \in SO_{p+q}, J_{p,q} \cdot g = g \cdot J_{p,q}\} = SO_p \times SO_q$.

The Lie subalgebra H is of the form $\begin{pmatrix} C & 0 \\ 0 & D \end{pmatrix}$, where $C \in so_p$, $D \in so_q$. The Cartan model V is constructed in SO_{p+q} by matrices $gJ_{p,q}J^{-1}J_{p,q}$. Consider the shift $V' = V \cdot J_{p,q}$; then $v = gJ_{p,q}g^{-1}$, $v^T = v$, i.e., V' is contained in the set of symmetric orthogonal matrices. The condition $g^T = g$ is equivalent to the equality $g^2 = E$. The set $\{g^2 = E\}$ in SO_{p+q} is identified with the union of Grassmanian manifolds $P = \bigcup_\alpha SO_{p+q}/S(O_\alpha \times O_{p+q-\alpha})$ and V is one of the connected components of P. Each element $gJ_{p,q}g^{-1} \in V$ can be interpreted as a p-dimensional plane in $\mathbf{R}^{p+q}$ if we assign to $gJ_{p,q}g^{-1}$ its invariant subspace corresponding to the eigenvalue −1. Thus, the invariant metric on V is induced by the Euclidean metric of $\mathbf{R}^{N^2}$ (where N = p + q) when V is embedded as one of the connected components of the set $\{g^2 = E\} \cap SO_{p+q}$. The case of the complex Grassmanian manifold is similar.

The tangent space B is of the form $\begin{pmatrix} 0 & Z \\ -Z^T & 0 \end{pmatrix}$, where Z is a real matrix with p rows and q columns. For p = 1, the space V coincides with S^q.

7) The space $Sp_{p+q}/Sp_p \times Sp_q$ — the quaternionic Grassmanian manifold. Here $\mathfrak{G} = Sp_{p+q}$, $\theta(X) = K_{p,q}XK_{p,q}$, where $K_{p,q} = \mathrm{diag}\,(-E_p, E_q, -E_p, E_q)$. The subspace B is of the form

$$\begin{pmatrix} 0 & C & 0 & D \\ -\overline{C}^T & 0 & D^T & 0 \\ 0 & -\overline{D} & 0 & \overline{C} \\ -\overline{D}^T & 0 & -C^T & 0 \end{pmatrix},$$

where C, D are arbitrary complex matrices with p rows and q columns.

We have listed all compact symmetric spaces that cannot be decomposed into a direct product of other symmetric spaces and such that the group $\mathfrak{G}$ is not an "exceptional" Lie group. For "exceptional" Lie groups, see below.

A connected finite-dimensional compact noncommutative Lie group is called simple if it does not contain nontrivial connected subgroups that are normal (of positive dimension) and different from $\mathfrak{G}$. In terms of the Lie algebra G of $\mathfrak{G}$, this means that in G there are no nonzero ideals different from G. It turns out that simple Lie groups are those "elementary bricks" from which any compact Lie group is constructed. The following nontrivial theorem holds (we skip its proof):

Any connected compact simply connected Lie group splits into the direct product (as a manifold and as a group) of simple Lie groups.

It is possible to list all simple Lie groups. The so-called classical series comes first: 1) orthogonal groups SO_n, 2) special unitary groups SU_n, 3) simplectic groups Sp_n. Besides, there are five exceptional groups, usually denoted G_2, F_4, E_6, E_7, E_8. These groups do not enter into a finite series, unlike classical ones, and they keep aloof. The very existence of these special groups is based on very specific algebraic principles. To complete the picture, we will give here the description of three of the five exceptional groups, namely of G_2, F_4, E_6. The description of E_7 and E_8 is quite complicated and therefore is omitted.

a) Description of G_2. Let $\mathbf{O}$ be the octonion algebra (the algebra of Caley numbers), i.e., $\mathbf{O}$ is the eight-dimensional real algebra with a unit over $\mathbf{R}$. The multiplication in O may be defined, for example, as follows. Fix in O the orthobasis whose vectors will be denoted by 1, e_2, e_3, ..., e_8. Let 1 be the unit, i.e., $1 \cdot x = x \cdot 1 = x$ for $x \in O$ and the multiplication of other generators is anticommutative: $e_ie_j + e_je_i = -2\delta_{ij}$. The multiplication table is depicted by the scheme shown in Fig. 179. For example, $e_2e_3 = e_5$, $e_4e_2 = -e_6$, and so on, i.e., the direction of arrows in Fig. 179 shows the sign of the product. Each segment of the triangle is considered here as a closed circle oriented with respect to the direction of the arrow. Consider the group of linear transformations of $\mathbf{R}^8$ that are automorphisms of the octonion algebra $\mathbf{O}$. By definition, G_2 is this group. Since automorphisms preserve the unit, then G_2 is a subgroup in SO_7. In fact, in O, we may define the scalar product $\langle a, b\rangle = a \cdot \bar{b}$, where $\bar{b} = b_1 \cdot 1 - b_2e_2 - \dots - b_8e_8$ (conjugation). Clearly, this scalar product coincides with the Euclidean one $\langle a, b\rangle = \sum_{i=1}^{8} a_ib_i$. The group G_2 is the 14th dimensional smooth manifold.

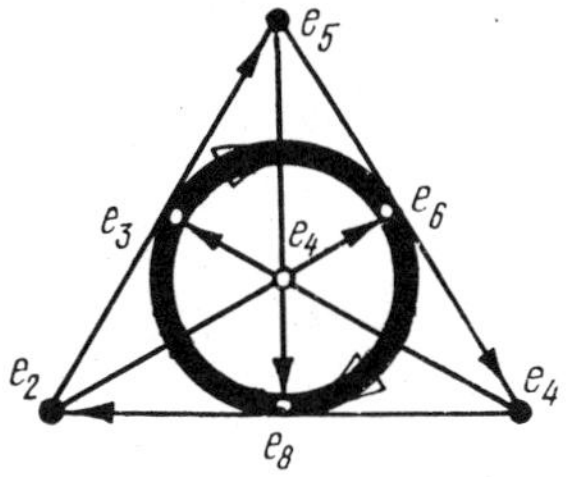

Fig. 179

b) Description of F_4. Consider the linear space L formed by all (3 × 3)-matrices of the form $\begin{pmatrix} a_1 & x_3 & \bar{x}_2 \\ \bar{x}_3 & a_2 & x_1 \\ x_2 & \bar{x}_1 & a_3 \end{pmatrix}$, where $x_i \in \mathbf{O}$ and a_i are complex (real) numbers, $1 \leq i \leq 3$, and $\bar{x}_i$ is the conjugation in $\mathbf{O}$. The addition of matrices and their multiplication of elements of **C** (or **R**) are defined as usual, making L a 27-dimensional linear space. The algebraic structure in L is introduced by the following operation $X \circ Y = \frac{1}{2}(XY + YX)$, where XY and YX are ordinary products of matrices. The operation $X \circ Y$ makes L a nonassociative algebra, so that 1) $X \circ Y = Y \circ X$, and 2) $(X^2 \circ Y) \circ X = X^2 \circ (Y \circ X)$. The Lie group F_4 can be described as the group of automorphisms of L.

c) Description of E_6. Let $A \in L$ be an arbitrary element. Denote by $R_A: L \to L$ the right shift $R_A(X) = X \circ A$. Now, consider the Lie algebra f_4 of F_4. This Lie algebra is the algebra of differentiations of the algebra L (see above). Now, let us extend the algebra f_4, embedding it into a wider Lie algebra which is isomorphic to the Lie algebra e_6 of the group E_6. Denote by e_6 the linear space of all linear transformations of L that are of the form $H = R_A + D$, where $A \in L$, $\mathrm{Tr}\, A = 0$, $D \in f_4$. Commutation in e_6 is introduced by the formulas $[H_1, H_2] = H_1H_2 - H_2H_1$, where H_iH_j stands for the composition of transformations H_i and H_j. It is easy to verify that this operation transforms e_6 into a Lie algebra. By definition, the Lie group whose algebra coincides with e_6 is called E_6. The definition implies that $E_6 \subset SU_{27}$.

16. GEOMETRY OF LIE GROUPS

16.1. Semisimple Lie Groups and Lie Algebras

In this section, we will briefly describe the algebraic structure of semisimple Lie algebras. For this, we introduce the so-called roots of Lie algebras. This root theory, although it is geometrically quite clear, is, at the same time, technically quite nontrivial and we are unable to describe it here fully and generally. On the other hand, a knowledge of roots is absolutely necessary to effectively use the Lie group and Lie algebra machinery in various problems of applied nature (examples of such applications will be given below). Therefore, we decided to arrange our exposition as follows. We formulate in complete form all the main necessary results and definitions of root theory; however, the proof will be given mainly for a unique simple algebra, namely for $sl(n, \mathbf{C})$. This Lie algebra is chosen as a model example due to the following considerations. First, all main effects connected with the root structure of semisimple Lie algebras can be illustrated completely by the example of $sl(n, \mathbf{C})$. Secondly, having studied the structure of this algebra, the interested reader can master, without great difficulty, the finer details of the theory.

Let $\mathfrak{G}$ be a Lie group and G its Lie algebra, i.e., the tangent space to the unit of the group. Since we consider only matrix groups, i.e., groups realized as subgroups in GL_N, then commutation in G may be defined as $[X, Y] = XY - YX$, where XY and YX are the products of the matrices $X, Y \in G$. The group $\mathfrak{G}$ naturally acts on its Lie algebra G as the group of linear transformations denoted by $Ad_g: G \to G$ and defined as follows: $Ad_g X = gXg^{-1}$, where $X \in G$, $g \in \mathfrak{G}$, and gXg^{-1} is the product of matrices. The differential of Ad_g at $E \in \mathfrak{G}$ is of the form $ad_X Y = [X, Y] = XY - YX$ (verify!). The correspondence $g \to Ad_g$ is called the adjoint representation of $\mathfrak{G}$ and the correspondence $X \to ad_X$ is called the adjoint representation of G. Our coefficient field is either $\mathbf{C}$ or $\mathbf{R}$. In this section, we will consider mainly $sl(n, \mathbf{C})$ or $sl(n, \mathbf{R})$.

<u>Definition 16.1</u>. The bilinear complex-valued form $(X, Y) = \mathrm{Tr}\, ad_X \cdot ad_Y$ is called the Killing form. Here ad_X and ad_Y are linear transformations of G. This form defines the real-valued scalar product $\mathrm{Re}\,(X, Y) = \langle X, Y \rangle$. In our model

example G = sl(n, **C**) ; this form may be defined as follows (up to a constant multiple): (X, Y) = Tr XY.

Lemma 16.1. The complex-valued form (,) on sl(n, **C**) is nondegenerate and satisfies 1) (X, Y) = (Y, X), and 2) ([X, Y], Z) = (X, [Y, Z]).

The proof follows immediately from the formula Tr [X, Y] = = 0 and the explicit formula for (,). Property 2 means that the equality $\langle \mathrm{ad}_X Y, Z\rangle = -\langle Y, \mathrm{ad}_X Z\rangle$ holds, i.e., linear operators ad_X are defined by skew symmetric matrices, or ad_X is skew symmetric with respect to the scalar product $\langle\ ,\ \rangle$.

Definition 16.2. The Lie algebra G and the corresponding group $\mathfrak{G}$ are called semisimple if the Killing form is nondegenerate on G.

The following important algebraic fact holds. Any semisimple Lie algebra splits into the direct sum (as a Lie algebra) of its subalgebras, each of which is a simple Lie algebra (in the sense discussed in Sec. 15.7). In particular, as follows from Lemma 16.1, sl(n, **C**) is simple.

Lemma 16.2. The Killing form is invariant with respect to transformations Ad_g, i.e., $(\mathrm{Ad}_g X, \mathrm{Ad}_g Y) = (X, Y)$ for $g \in \mathfrak{G}$, X, $Y \in G$.

For sl(n, **C**), the proof follows immediately from the explicit formula for (,).

We have already seen that the form Tr $\mathrm{ad}_X \cdot \mathrm{ad}_Y$ can be expressed in a simpler way if we make use of the concrete matrix representation of sl(n, **C**) namely (X, Y) = Tr XY (up to a constant multiple). This situation reflects a more general fact: for any simple Lie algebra G (over **C**), the form (,) is uniquely defined up to a scalar multiple in the sense that if G is a simple algebra, then for any of its nontrivial linear representations ρ, the identity $(X, Y) = a_\rho \operatorname{Tr} \rho X \cdot \rho Y$ holds, where the coefficient a_ρ does not depend on X, Y, but is defined only by ρ. Here ρX·ρY stands for the usual matrix product of matrices ρX and ρY.

16.2. Cartan Subalgebras

Let G be a semisimple finite-dimensional Lie algebra over the field of complex numbers [e.g., sl(n, **C**)] and X its arbitrary element. Let Ann X be a linear subspace in G consisting of elements Y such that $\mathrm{ad}_Y X = 0$, i.e., $[Y, X] = 0$. Thus, Ann X consists of all elements of G that commute with X.

Lemma 16.3. Ann X is a subalgebra in G.

Proof. Let $Z, Y \in$ Ann X; then $[X, Z] = [Y, X] = 0$, implying $[[Z, Y], X] = -[[X, Z], Y] - [[Y, X], Z] = 0$, i.e., $[Z, Y] \in$ Ann X. Here we make use of the Jacobi identity $[[X, Y], Z[+ [[Z, X], Y] + [[Y, Z], X] = 0$, valid in any Lie algebra.

The subalgebra Ann X is sometimes called the annihilator of X.

Definition 16.2. Element $X \in G$ is called regular if the dimension of its annihilator is the minimum possible. Other elements of the Lie algebra are called singular. The annihilator of a regular element is called a Cartan subalgebra of G.

Clearly, $X \in$ Ann X. Regular elements in G constitute the open dense set Reg G and singular elements fill in the closed set of zero measure. This follows from the fact that the singularity condition is described as a system of algebraic equations on elements of the matrix X. Hence, any small perturbance of a general kind kills the singularity. In this sense, regular elements are elements "of general position."

Proposition 16.1. If G is a semisimple Lie algebra, then any of its Cartan subalgebras are commutative. Any two Cartan subalgebras T_1 and T_2 are conjugate in G, i.e., there exists an element $g \in \mathfrak{G}$ such that $gT_2g^{-1} = T_1$. Moreover, a Cartan subalgebra is the maximum commutative subalgebra in G.

Proof. See, for example, [19]. Consider the example of $sl(n, \mathbf{C})$. As a Cartan subalgebra, we may take, for example, the set of diagonal matrices of the form $h = \begin{pmatrix} a_1 & & 0 \\ & \ddots & \\ 0 & & a_n \end{pmatrix}$, where $\{a_i\}$ are eigenvalues of the operator (matrix) h and $\mathrm{Tr}\, h = \sum_i a_i = 0$. This evidently implies that $h \in T$ is regular iff all eigenvalues of h are different, i.e., $a_i \neq a_j$ for

$i \neq j$. Thus, regular elements in T are really elements of "general position," i.e., "typical elements" of sl(n, **C**). Let us find the annihilator of the regular element $h \in T$. By definition, it is the set of matrices commuting with h. But since all eigenvalues α_i are different, then any matrix commuting with h must be diagonal (verify!). Thus, we have proved that the annihilator of a regular element is a subalgebra of diagonal matrices. Evidently, this subalgebra is commutative. Moreover, it is the maximum commutative subalgebra.

By Proposition 16.1, we may now speak of one fixed Cartan subalgebra, since all others are obtained from it by automorphisms of the Lie algebra. For example, for sl(n, **C**) we fix the Cartan subalgebra T consisting of diagonal matrices. In this example, singular elements are characterized by the fact that some eigenvalues of matrices that preserve them coincide. Clearly, this means that the dimension of the annihilator of such an element is larger than in the general case: there are new matrices that commute with the fixed one. Note that gl(n, **C**) containing sl(n, **C**) as a Lie subalgebra is not semisimple. In fact, gl(n, **C**) contains a nonzero ideal λE, where $\lambda \neq 0$ and E is the unit matrix. This ideal is the center; hence, the Killing form is not nondegenerate on gl(n, **C**).

16.3. Roots of a Semisimple Lie Algebra and Its Root Decomposition

Let G be a semisimple Lie algebra over **C** and T a fixed Cartan subalgebra. It turns out that this subalgebra defines a set of linear functions that can be depicted by vectors of the algebra and are called roots of the algebra. The geometrical meaning of roots is very simple. First, let us give an informal explanation. Suppose an exact linear finite-dimensional representation of G is given, i.e., there is given a monomorphism of G onto the Lie algebra of linear transformations of the linear space V, where $\dim V = N < \infty$. Hence, all elements of G are described by linear operators acting in V. In particular, this concerns elements of the Cartan subalgebra T. The set of $(N \times N)$-matrices that describes elements of T possesses an important property: matrices from this set commute.

Suppose that in V it is possible to choose a basis where all matrices of T are described (simultaneously) by diagonal operators, i.e., a basis consisting of vectors that are eigenvectors for all operators of T. The fact that simultaneous diagonalizability of all elements of T exists in this basis implies that eigenvectors $\lambda_i(h)$ of the transformation $h \in T$, which stand on the diagonal of the matrix, are linear functions on T, since adding two diagonal matrices is equivalent to the summation of eigenvalues that stand on the same basis. Thus, each eigenvalue of the set $\lambda_1(h), \ldots, \lambda_N(h)$ can be considered as a linear function on the subalgebra T in G. These linear functions are called weights of this representation. It turns out that properties of a representation are defined to a great extent by properties of the set of linear functionals on T. It is also clear that, for another representation of G, this set of linear functions (eigenvalues) will be, generally speaking, another one.

Among the set of all linear representations of G a remarkable representation that we already know is fixed. It is the adjoint representation. It is given by the mapping $X \to \mathrm{ad}_X$, where $\mathrm{ad}_X Y = [X, Y]$, $\mathrm{ad}_X: G \to G$. Here the linear space V coincides with the space of the Lie algebra G itself, in particular $\dim V = N = \dim G$. Each element $X \in G$ is depicted by an $(N \times N)$-matrix. Weights of this representation are called roots of G. It turns out that roots completely define the structure of semisimple Lie algebras; in particular, the classification of all such Lie algebras can be given in terms of roots. These roots are linear functions defined on the Cartan subalgebra T. Now let us give the formal definition.

Definition 16.3. The linear function $\alpha(h)$ on the Cartan subalgebra T in the semisimple Lie algebra G is called a root if there is an element $E_\alpha \in G$, $E_\alpha \neq 0$ such that $[h, E_\alpha] = \alpha(h)E_\alpha$ for any $h \in T$. In other words, a vector $E_\alpha \in G$ must exist that is an eigenvector for all operators of the form $\mathrm{ad}_h: G \to G$, $\mathrm{ad}_h E_\alpha = [h, E_\alpha]$.

Theorem 16.1 (the root decomposition of a semisimple Lie algebra). Let G be a semisimple Lie algebra, $T \in G$ a Cartan subalgebra, and $\Delta = \{\alpha\}$ the set of roots of G. Let G_α be the eigensubspace corresponding to the root α. Then G splits into the direct sum of linear subspaces T and $\{G_\alpha\}$, i.e., $G = T \oplus \sum_{\alpha \neq 0} G_\alpha$. All subspaces G_α corresponding to $\alpha \in \Delta$ are one-dimensional over **C**.

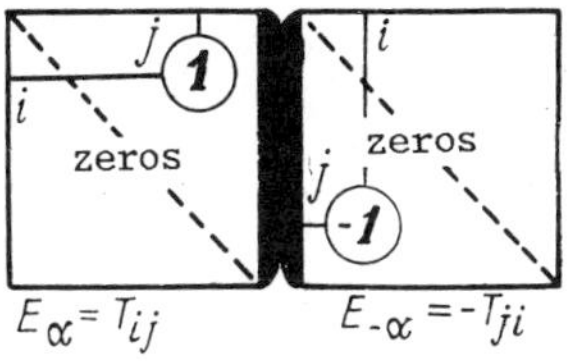

Fig. 180

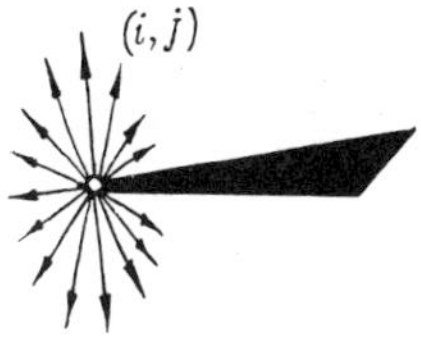

Fig. 181

Subspaces G_α are called root subspaces.

Proof. As earlier, we will give the proof only for sl(n, **C**) (the model example). The Lie algebra sl(n, **C**) is presented as the algebra of complex-valued (n × n)-matrices. Denote by T_{ij} an elementary (n × n)-matrix that has only one nonzero entry in the (i, j)-th position equal to unity, where i is the number of a row and j is the number of a column. The Cartan subalgebra T consists of diagonal matrices $h = \begin{pmatrix} a_1 & & 0 \\ & \ddots & \\ 0 & & a_n \end{pmatrix}$, where $a_i \in \mathbf{C}$ and $\sum_{i=1}^{n} a_i = 0$. Vectors E_α that are eigenvectors for all transformations $ad_h: G \to G$ are of the form T_{ij} if i < j, and $-T_{ij}$ if i > j (see Fig. 180). Computing $ad_h\, T_{ij}$, we obtain $[h, T_{ij}] = a_i - a_j$, i.e., $\alpha(h) = a_i - a_j$. Thus, roots α of sl(n, **C**) are numbered by a pair of indices. We will now write $\alpha = \alpha_{ij}$, $\alpha_{ij}(h) = a_i - a_j$ (see Fig. 181). Since $(-\alpha_{ij})(h) = a_j - a_i = \alpha_{ji}(h)$, then $-\alpha_{ij} = \alpha_{ji}$ (see Fig. 182). Thus, sl(n, **C**) is presentable in the form $T \oplus \sum_{i,j} \mathbf{C} T_{ij}$. Hence, as the root decomposition of sl(n, **C**) we may take its standard decomposition into the direct sum of one-dimensional subspaces $\mathbf{C}T_{ij}$ and the space T. The theorem is proved.

Note that the Cartan subalgebra T can also be considered as the eigenspace of the representations ad_T corresponding to the eigenvalue zero. The multiplicity of the zero eigenvalue equals r, i.e., the rank of G, the dimension of a Cartan subalgebra (see Fig. 181).

16.4. Several Properties of a Root System

Consider the Killing form on G. In what follows, it is important to know the properties of the basis $\{E_\alpha\}$ in the space complementary to T in G.

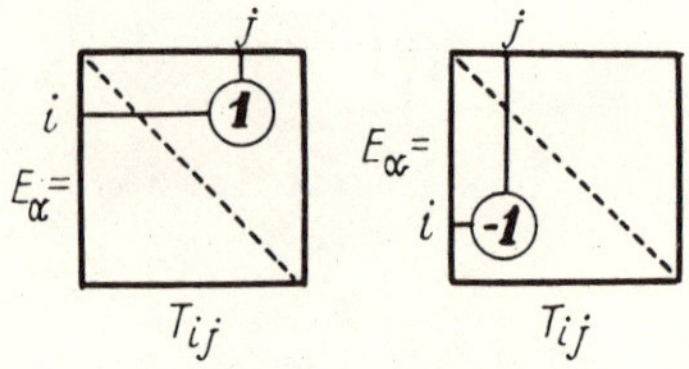

Fig. 182

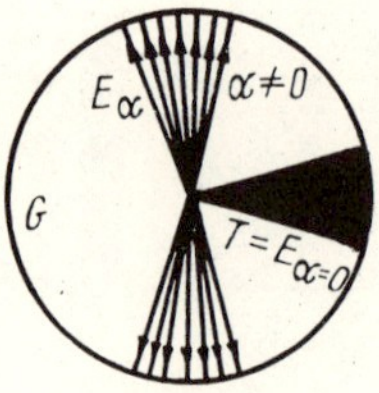

Fig. 183

Proposition 16.2. Let G_α be the one-dimensional subspace spanned by E_α, where $\alpha \neq 0$. Then $[G_\alpha, G_\beta] \subset G_{\alpha+\beta}$, i.e., $[E_\alpha, E_\beta] = N_{\alpha\beta}E_{\alpha+\beta}$, where $N_{\alpha\beta}$ is a number. If $\alpha + \beta \neq$ $\neq 0$, then the corresponding vectors are orthogonal to each other with respect to the Killing form. Conversely, vectors E_α and $E_{-\alpha}$, where $\alpha \neq 0$, are not orthogonal. If α, β, $\alpha + \beta$ are nonzero roots of G, then $[G_\alpha, G_\beta] = G_{\alpha+\beta}$, i.e., $N_{\alpha\beta} \neq 0$. The only roots proportional to $\alpha \neq 0$ are the following: 0, $\pm\alpha$.

Proof. Let $G = sl(n, \mathbf{C})$. Since $E_\alpha = T_{ij}$, then by commuting matrices $E_\alpha = T_{ij}$ and $E_\beta = T_{pq}$, we evidently obtain the statement: $[T_{ij}, T_{pq}] = N_{\alpha\beta}E_{\alpha+\beta}$, since $N_{\alpha\beta} = 0$ if all indices i, j, p, q are different and $[T_{ij}, T_{pq}] = T_{iq}$ if $j =$ $= p$. Clearly, $N_{\alpha\beta} = N_{-\alpha,-\beta}$, since if $E_\alpha = T_{ij}$, then $E_{-\alpha} =$ $= -T_{ji}$. The explicit form of the Killing form on $sl(n, \mathbf{C})$ is as follows: $(X, Y) = \mathrm{Tr}\, XY$; therefore, if $\alpha + \beta \neq 0$ (i.e., the corresponding eigenvectors E_α and E_β are expressed by matrices T_{ij} and T_{pq}, whose nonzero entries are in different positions, namely $(j, i) \neq (p, q)$, then $(E_\alpha, E_\beta) = \mathrm{Tr}\, T_{ij}T_{pq} =$ $= 0$. Thus, E_α and E_β are orthogonal when $\alpha + \beta \neq 0$. If $\alpha =$ $+ \beta = 0$, then $E_\alpha = T_{ij}$ and $E_{-\alpha} = -T_{ji}$ are not orthogonal, since $(E_\alpha, E_{-\alpha}) = \mathrm{Tr}\, T_{ij}(-T_{ji}) = -1$. Finally, the equality $[G_\alpha, G_\beta] = G_{\alpha+\beta}$, where, say, $\alpha = \alpha_{ij}$ for $i < j$ and $\beta = \alpha_{pq}$ for $p < q$, may hold if some of indices i, j, p, q coincide: either $j = p$ or $i = q$. It is equivalent to the fact that some of the roots $\alpha + \beta$ are also roots. In fact, $\alpha_{ij}(h) +$ $+ \alpha_{pq}(h) = (\alpha + \beta)(h) = a_i - a_j + a_p - a_q$ equals $a_i - a_q$ if $j = p$, and equals $a_p - a_j$ if $i = q$. This is depicted in Fig. 184, where the root $\alpha_{ij}(h) = a_i - a_j$ for $i < j$ is depicted by the arrow that shows that a_j is subtracted from a_i. The statement is proved.

The explicit form of the Killing form implies that the restriction of the form onto the Cartan subalgebra T is nondegenerate. Hence, this restriction is defined by the canon-

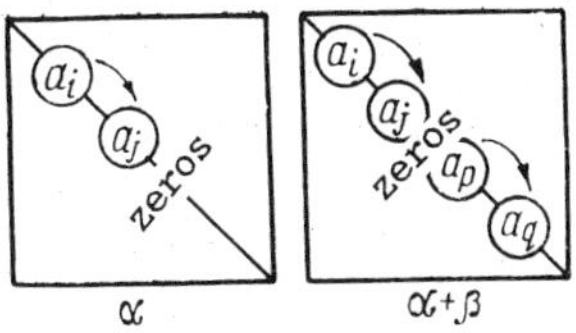

Fig. 184

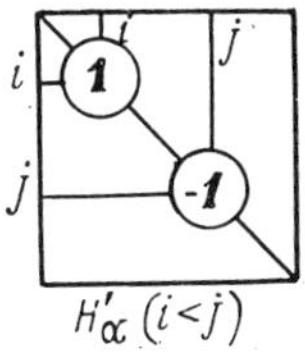

Fig. 185

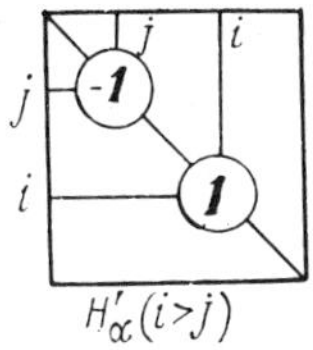

Fig. 186

ical identification of T with the dual space T* which is the space of linear functions on T. Thus, each root α that is a linear function on T can be uniquely expressed as a vector on T, i.e., there is the unique vector H_α' such that $\alpha(h) = (h, H_\alpha')$ for all $h \in T$. This vector H_α', corresponding to α, will also be called a root. Thus, if $\alpha \neq 0$, then $[E_\alpha, X] = (E_\alpha, X)H_\alpha'$, where $X \in G_\alpha$ and $(\alpha, \alpha) \neq 0$. For $sl(n, \mathbf{C})$, these statements are subject to a straightforward verification (verify!).

Lemma 16.4. The linear span of all roots $\{H_\alpha'\}$, $\alpha \in \Delta$, $\alpha \neq 0$ coincides with the Cartan subalgebra T.

The proof for $sl(n, \mathbf{C})$ follows from the definition of H_α'. In this case, the root H_α', where $\alpha = \alpha_{ij}$, $i < j$, is expressed by the matrix depicted in Fig. 185. If $\alpha = \alpha_{ij}$, $i > j$, then H_α' is shown in Fig. 186. Thus, $H_\alpha' = \alpha_{ij} = T_{ii} - T_{jj}$. This implies that the linear span of matrices $\{T_{ii} - T_{jj}\}$ coincides with the subalgebra $T = \sum_{i=1}^{n} \mathbf{C}T_{ii}$. The lemma is proved.

At the same time, we see that the number of roots is larger than the dimension of T. In other words, roots form a superfluous basis in T. Consider in T the subspace T_0, generated by all vectors H_α' over the field of real numbers, i.e., T_0 is the linear span of $\{H_\alpha'\}$ with real coefficients.

Lemma 16.5. The restriction of the Killing form onto T_0 is nondegenerate. Moreover, this restriction is a positive-definite form on T_0 with real values. Moreover, $\dim_{\mathbf{R}} T_0 = \dim_{\mathbf{C}} T = \frac{1}{2}\dim_{\mathbf{R}} T$, i.e., T_0 is the "real part" of the Cartan subalgebra T. The Cartan subalgebra T is presentable as the direct sum $T = T_0 + iT_0$, where i is the imaginary unit, i.e., T is the complexification of $T_0 \subset T$.

Proof. Since (X, Y) = Tr X·Y, then for $X = \begin{pmatrix} a_1 & & 0 \\ & \ddots & \\ 0 & & a_n \end{pmatrix}$ and $Y = \begin{pmatrix} b_1 & & 0 \\ & \ddots & \\ 0 & & b_n \end{pmatrix}$, we have $(X, Y) = \sum_{i=1}^{n} a_i b_i$. The lemma is proved [for G = sl(n, **C**)].

In particular, all values $\alpha_{ij}(h) = a_i - a_j$ are real on $T_0 \subset T$. It turns out that in the set of all roots a natural ordering arises that is useful in various computations. Above we have denoted by Δ the set of all nonzero roots of G. Let $H_1, \ldots, H_r$ be a fixed basis in $T_0 \subset T$ and r = rank G. If λ and μ are linear functions on T_0, then we say that $\lambda > \mu$ if $\lambda(H_i) = \mu(H_i)$ for i = 1, 2, ..., k and $\lambda(H_{k+1}) > \mu(H_{k+1})$. Recall that if λ and μ are roots of the algebra, then their values λ(h) and μ(h) are real numbers for any $h \in T_0$; hence, the inequality $\lambda(H_{k+1}) > \mu(H_{k+1})$ is well defined. Thus, in the set of all roots a linear ordering arises (when a basis is fixed).

Definition 16.4. The root $\alpha \in \Delta$ is called positive if $\alpha > 0$, i.e., if $\alpha(H_i) = 0$ for i = 1, 2, ..., k and $\alpha(H_{k+1}) > 0$. In other words, positivity of α is equivalent to the positivity of its first nonzero coordinate (with respect to the basis $\{H_i\}$).

It is clear that the described ordering depends on the choice of the basis $H_1, \ldots, H_r$ in T_0; when the basis is changed, so is the ordering. To remove this ambiguity, let us assume that the basis $H_1, \ldots, H_r$ in T_0 is fixed. Thus, in the root system Δ, two subsystems are distinguished: the positive roots that constitute the set $\Delta^+ \subset \Delta$ and the negative roots that constitute the set $\Delta^- \subset \Delta$. Evidently, $\Delta = \Delta^+ \cup \Delta^-$ and $\Delta^+ \cap \Delta^- = \emptyset$. Furthermore, there is a one-to-one correspondence between Δ^+ and Δ^- established by the involution $\alpha \to -\alpha$. Clearly, if $\alpha \in \Delta^+$, then $-\alpha \in \Delta^-$.

Definition 16.5. A positive root $\alpha \in \Delta^+$ is called simple if it is impossible to present it as the sum of other positive roots.

Proposition 16.3. If G is a semisimple Lie algebra over the field of complex numbers, then in the subspace T_0 there always exists a basis of simple roots $\alpha_1, \ldots, \alpha_r$, where r = = rank G. These vectors also constitute a basis of T over **C**. Moreover, each root $\alpha \in \Delta$ is presentable as the sum

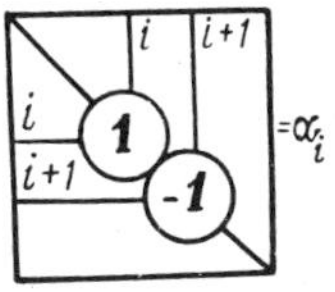

Fig. 187

$\sum_{i=1}^{r} m_i\alpha_i$, where m_i are integers of the same sign (or zero), and $m_i \geqq 0$ if $\alpha \in \Delta^+$, while $m_i \leqq 0$ if $\alpha \in \Delta^-$.

The system of simple roots is usually denoted by Π.

Proof. Let $G = sl(n, \mathbf{C})$; then for a system of simple roots, we may take roots $\alpha_{12}, \alpha_{23}, \ldots, \alpha_{n-1,n}$, i.e., $\alpha_{i,i+1}(h) = a_i - a_{i+1}$, where $1 \leqq i \leqq n-1$. Making use of the Killing form, we describe these functionals as the following vectors of T_0 ("real part" of the Cartan subalgebra): $\alpha_1 = T_{11} - T_{22}$, $\alpha_2 = T_{22} - T_{33}$, ..., $\alpha_{n-1} = T_{n-1,n-1} - T_{nn}$; here $r = n - 1$ (see Fig. 187). Clearly, these roots are positive and simple. Furthermore, it is evident that any root $\alpha(h) = a_i - a_j$ decomposes with respect to this basis with integer coefficients of the same sign. The positive roots $\alpha \in \Delta^+$ are of the form α_{ij}, $i < j$ and are depicted in T_0 by vectors $T_{ii} - T_{jj}$, $i < j$, while negative roots $\alpha \in \Delta^-$ are depicted by vectors $T_{ii} - T_{jj}$, $i > j$. The statement is proved.

Thus, for the initial basis $H_1, \ldots, H_r$, with respect to the ordering of roots in Δ defined above, we may now take the basis $H_{\alpha_1}', \ldots, H_{\alpha_r}'$ consisting of simple roots. Furthermore, we will assume that in the subalgebra T_0 (over $\mathbf{R}$) or in T (over $\mathbf{C}$), this basis of simple roots is fixed. Note that this basis is not orthonormal. Exactly this fact forms the basis for the classification of simple Lie algebras. Thus, for instance, angles between simple roots and lengths of roots are important features of a Lie algebra. For our future purposes, fix the following remarkable basis in G.

Proposition 16.4. In the semisimple Lie algebra G there exists a basis consisting of vectors $H_{\alpha_1}', \ldots, H_{\alpha_r}'$ that are simple roots of the algebra and constitute the basis of the Cartan subalgebra T (over $\mathbf{C}$) and of the root vectors $\{E_\alpha\}$, where $E_\alpha \in G_\alpha$ for $\alpha \neq 0$, $\alpha \in \Delta$. Vectors E_α can be chosen so that the following conditions will hold:

1) $\mathrm{ad}_h E_\alpha = [h, E_\alpha] = \alpha(h)\cdot E_\alpha$, where $h \in T$ and $\alpha(h) \in \mathbf{R}$ if $h \in T_0 \subset T$;

2) $[E_\alpha, E_{-\alpha}] = -H_\alpha' \in T$;

3) $$[E_\alpha, E_\beta] = \begin{cases} N_{\alpha\beta}\cdot E_{\alpha+\beta}, & \text{if } \alpha + \beta \neq 0 \text{ is a root} \\ 0, & \text{if } \alpha + \beta \neq 0 \text{ is not a root;} \end{cases}$$

4) $(h, H_\alpha') = \alpha(h)$ for $h \in T$.

We may assume that $N_{\alpha\beta} = 0$ if $\alpha + \beta \neq 0$ is not a root. If $\alpha + \beta$ is a nonzero root, then $N_{\alpha\beta} \neq 0$ and $N_{\alpha\beta} = N_{-\alpha,-\beta}$. The constants $N_{\alpha\beta}$ can be chosen to be real.

The basis $\{E_\alpha, H_{\alpha_1}', \ldots, H_{\alpha_r}'\}$ is called the Weyl basis. The structure of T is therefore completely defined by numbers $N_{\alpha\beta}$, i.e., by the geometry of the roots $\{H_\alpha'\}$, realized by vectors of T.

The proof of this proposition for sl(n, **C**) was in fact obtained earlier, since we have already proved Proposition 16.2. Now it is possible to write the root decomposition of G in the following form: $G = T \oplus V^+ \oplus V^-$, where $V^+ = \sum_{\alpha>0} G_\alpha$ and $V^- = \sum_{\alpha<0} G_\alpha$. In the case sl(n, **C**), the subspace V^+ evidently coincides with the subspace of all upper-triangular matrices with zeros on the main diagonal, while the subspace V^- is identified with the subspace of all lower-triangular matrices with zeros on the main diagonal. This decomposition is depicted in Fig. 188. Note that the scalar product of vectors of the basis (generated by the Killing form) satisfies $\langle E_\alpha, E_{-\alpha}\rangle = -1$, $\langle E_\alpha, E_\alpha\rangle = 0$. This implies in particular, that vectors E_α are isotropic with respect to the nondegenerate Killing form. This is due to the fact that the Killing form is indefinite on G. In particular, restrictions of this form onto V^+ and V^- vanish identically, as follows from the explicit form of the form.

Lemma 16.6. Subspaces V^+ and V^- are subalgebras in the Lie algebra G. These subalgebras are nilpotent.

Proof. Let $E_\alpha, E_\beta \in V^+$; then by Proposition 16.4 we have $[E_\alpha, E_\beta] = N_{\alpha\beta}E_{\alpha+\beta}$, and since $\alpha > 0$, $\beta > 0$, then $\alpha + \beta \neq 0$ and $\alpha + \beta > 0$, i.e., $E_{\alpha+\beta} \in V^+$, as required. The case

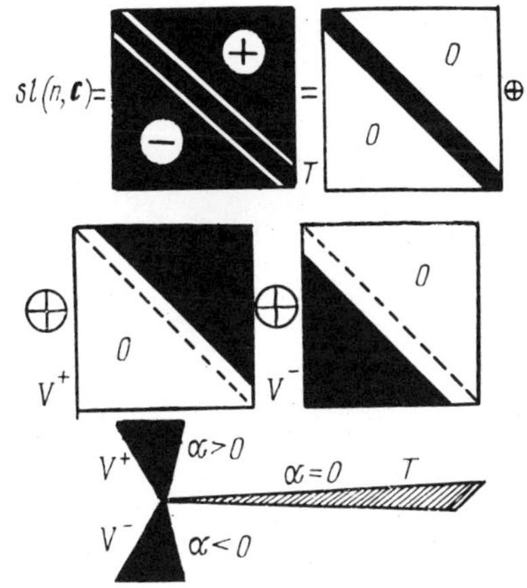

Fig. 188

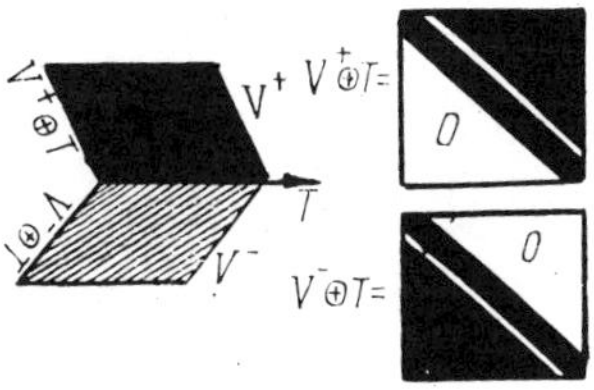

Fig. 189

V^- is similar. Nilpotency of both subalgebras follows immediately from the explicit form of the matrices which constitute them.

Lemma 16.7. Subspaces $V^+ \oplus T$ and $V^- \oplus T$ are subalgebras in G. These subalgebras are solvable.

The proof also follows immediately from Proposition 16.4. The difference between it and Lemma 16.6 is the extra commutators of the form $[H_\alpha', E_\beta]$ which coincide with $\beta(H_\alpha')E_\beta$ (see Fig. 189).

Problem. Prove that the Killing form vanishes identically on $V^+ \oplus T$ and $V^- \oplus T$. This Killing form does not coincide with the restriction of the Killing form defined on the enveloping semisimple Lie algebra G onto $V^\pm \oplus T$.

Problem. Prove that if the Killing form of a Lie algebra vanishes identically, then this Lie algebra is solvable.

We gave this detailed study of the roots of a semisimple Lie algebra because the following important theorem holds (its proof surpasses the limits of our course and is contained for example, in [19]).

Theorem 16.2. A simple Lie algebra over **C** is defined up to an isomorphism by its root system in the Cartan subalgebra.

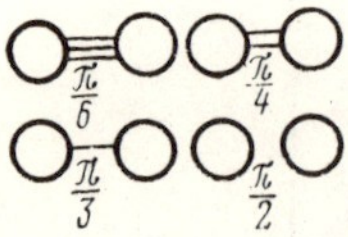

Fig. 190

It turns out that all such root systems can be listed explicitly giving the classification of all simple Lie algebras.

16.5. Root Systems of Simple Lie Algebras

The preceding section implies that to define a simple Lie algebra it suffices to define its root system Δ. Moreover, since by Proposition 16.3 all systems Δ are recovered from the subsystem of simple roots Π, it suffices to define Π. This system of vectors H_{α_1}', ..., H_{α_r}', where r = rank G = = dim T, is defined by the lengths of vectors and angles that these vectors form with each other. We will give here the main result of a classification producing the explicit form of systems of simple roots of all simple Lie algebras. Let $\mathbb{R}^{n+1}$ be a Euclidean space and $\{e_i\}$ its orthobasis. Each system Π will be depicted as the following convenient scheme—graph. Each vector of Π is a point of a two-dimensional plane labeled by the squared length of this vector (root) in the Killing metric on T. Since simple roots constitute a basis in T, the number of roots is equal to rank G. It is now necessary to depict angles formed by these vectors with one another. It turns out that these angles cannot be arbitrary. Moreover, there are only a few variants, namely π/6, π/4, π/3, π/2. Let us make the following convenient assumption that enables us to depict all this information graphically. If vectors form the angle π/6, then join the corresponding points by three parallel segments (see Fig. 190). If the angle between vectors is π/4, then join the points of two segments, and if the angle equals π/3, then let us draw only one segment; finally, if vectors are orthogonal, then leave them as they are. The obtained graph will be called the diagram of roots.* Clearly, this planar graph makes it possible to completely recover the system of simple roots.

*It is also called the Dyukin scheme — Translator.

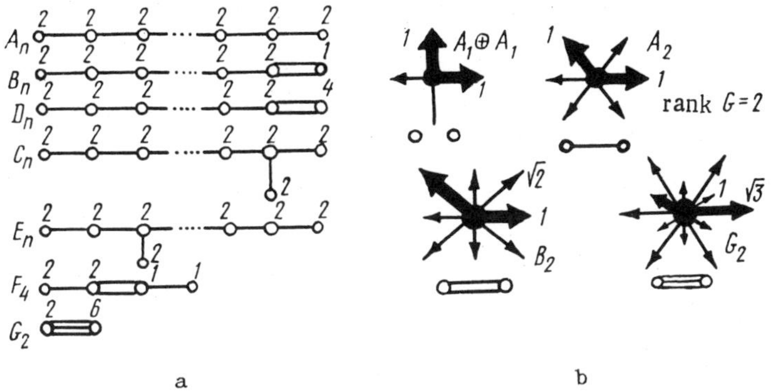

Fig. 191

<u>Theorem 16.3</u>. Let $e_1, \ldots, e_{n+1}$ be an orthobasis in R^{n+1}. Let G be a simple Lie algebra over the field of complex numbers. Then the system Π of simple roots of this algebra coincides with one of the systems of vectors of R^n, where n = rank G, listed below:

1) $\Pi(A_n) = (e_1 - e_2, \ldots, e_n - e_{n+1}),\ n \geqslant 1;$

2) $\Pi(B_n) = (e_1 - e_2, \ldots, e_{n-1} - e_n, e_n),\ n \geqslant 2;$

3) $\Pi(C_n) = (e_1 - e_2, \ldots, e_{n-1} - e_n, 2e_n),\ n \geqslant 3;$

4) $\Pi(D_n) = (e_1 - e_2, \ldots, e_{n-1} - e_n, e_{n-1} + e_n),\ n \geqslant 4;$

5) $\Pi(G_2) = (e_1 - e_2,\ 2e_2 - e_1 - e_3);$

6) $\Pi(F_4) = (e_1 - e_2,\ e_2 - e_3,\ e_3,\ {}^1/_2(e_4 - e_1 - e_2 - e_2));$

7, 8, 9) $\Pi(E_n) = (e_1 - e_2, \ldots, e_{n-1} - e_n,\ -(\bar{e}_1 + \bar{e}_2 + \bar{e}_3) +$

$+ e_{n+1}(9/n - 1)^{1/2}),\ n = 6, 7, 8,$

where $\bar{e}_i = e_i - \frac{1}{n}\sum_{k=1}^{n} e_k$, i.e., $\bar{e}_i$ is the orthogonal projection of e_i onto the hypersurface R^{n-1} in $R^n = R^n(e_1, \ldots, e_n)$, orthogonal to the vector

Vectors from each of the listed systems are linearly independent and the dimension of their linear span is equal to n, i.e., the rank of the Lie algebra. Conversely, each of these systems defines a simple Lie algebra (over **C**) and different systems define nonisomorphic Lie algebras. The system $\Pi(A_n)$ defines $sl(n + 1, \mathbf{C})$ and the system $\Pi(B_n)$ defines $so(2n + 1, \mathbf{C})$. The system $\Pi(C_n)$ defines $sp(n, \mathbf{C})$. The system $\Pi(D_n)$ defines $so(2n, \mathbf{C})$. These are Lie algebras of classical series. Other root systems define exceptional Lie algebras: $\Pi(G_2)$ defines the Lie algebra $g_2(\mathbf{C})$ of the Lie group $G_2(\mathbf{C})$, the system $\Pi(F_4)$ defines the Lie algebra $f_4(\mathbf{C})$ of $F_4(\mathbf{C})$, and the system $\Pi(E_n)$ defines the Lie algebra $e_n(\mathbf{C})$ of the Lie group $E_n(\mathbf{C})$, where n = 6, 7, 8. Diagrams of all these systems are seen in Fig. 191.

We will also list root systems corresponding to the above simple Lie algebras. The reason is that in various applied problems simple roots are not sufficient and we must know the system of all roots of the algebra. Although the system of simple roots suffices to recover all root systems, this recovery procedure involves several technical difficulties; therefore, we will give here the final result.

1) Series A_n. As we have already seen in the study of the model example, all roots of A_n are labeled by two indices (i, j), where $1 \leqq i < j \leqq n + 1$, $i \neq j$. Denote these roots by α_{ij}. Then $\alpha_{ij} = e_i - e_j$. Simple roots are vectors $\alpha_{i,i+1} = e_i - e_{i+1}$, $1 \leqq i \leqq n$, where n is rank A_n. Any positive root $\alpha_{ij} \in \Delta^+$, where $i < j$, is uniquely presentable in the form of a sum (decomposition with respect to simple roots):

$$\alpha_{ij} = \alpha_{i,i+1} + \alpha_{i+1,i+2} + \ldots + \alpha_{j-1,j} =$$
$$= e_i - e_{i+1} + e_{i+1} - e_{i+2} + \ldots - e_j = e_i - e_j.$$

2) Series B_n. Simple roots are $\alpha_i = e_i - e_{i+1}$ for $1 \leqq i \leqq n - 1$ and $\alpha_n = e_n$. The root system is: $\pm e_i$, $\pm e_i \pm e_j$, where $i \neq j$. The system of positive roots consists of roots $a_{ij} = e_i - e_j = \alpha_i + \ldots + \alpha_{j-1}$, $1 \leqq i < j \leqq n$; $b_{ij} = \alpha_i + \ldots + \alpha_{j-1} + 2\alpha_j + \ldots + 2\alpha_{n-1} + 2\alpha_n$, $1 \leqq i < j \leqq n$; $c_i = \alpha_i + \ldots + \alpha_{n-1} + \alpha_n$.

3) Series C_n. Simple roots are vectors $\alpha_i = e_i - e_{i+1}$ for $1 \leqq i \leqq n - 1$ and $\alpha_n = 2e_n$. Then positive roots in their decomposition with respect to simple roots $\alpha_1, \ldots, \alpha_{n-1}, \alpha_n$

are of the form $a_{ij} = \alpha_i + \ldots + \alpha_{j-1}$ for $1 \leqq i < j \leqq n$ and $b_{ij} = \alpha_i + \ldots + \alpha_{j-1} + 2\alpha_j + \ldots + 2\alpha_{n-1} + \alpha_n$. The root $b_{in} = \alpha_i + \ldots + \alpha_{n-1} + \alpha_n$ can be formally included into the first series $\{a_{ij}\}$ by setting $b_{in} = a_{i,n+1}$.

4) Series D_n. Simple roots are $\alpha_i = e_i - e_{i+1}$ for $1 \leqq i \leqq n - 1$ and $\alpha_n = e_{n-1} + e_n$. Positive roots in their decomposition with respect to simple roots $\alpha_1, \ldots, \alpha_{n-1}, \alpha_n$ are of the form $a_{ij} = \alpha_i + \ldots + \alpha_{j-1}$, where $1 \leqq i < j \leqq n$, and $b_{ij} = \alpha_i + \ldots + \alpha_{j-1} + 2\alpha_j + \ldots + 2\alpha_{n-2} + \alpha_{n-1} + \alpha_n$, where $1 \leqq i < j \leqq n$. Clearly, b_{ij} can be expressed as $b_{ij} = e_i + e_j = (e_i - e_{n-1}) + (e_j - e_n) + (e_{n-1} + e_n)$. For $j = n$ we have $b_{in} = \alpha_i + \ldots + \alpha_{n-2} + \alpha_n = e_i + e_n$.

5) The exceptional Lie algebra g_2. Simple roots: α_1, α_2. Positive roots: $\alpha_1, \alpha_2, \alpha_1 + \alpha_2, \alpha_1 + 3\alpha_2, 2\alpha_1 + 3\alpha_2, \alpha_1 + 2\alpha_2$.

6) The exceptional Lie algebra f_4. Roots of this algebra are the following linear functions:

$$\pm\omega_i \pm \omega_j,\ (i < j) = 1, 2, 3, 4;\ \pm\omega_i,\ \pm\Lambda_i,\ \pm M_i;$$

where

$$\Lambda_i = {}^1/_2(\omega_1 + \omega_2 + \omega_3 + \omega_4) - \omega_i;$$

$$M_1 = {}^1/_2(\omega_1 + \omega_2 + \omega_3 + \omega_4);\quad M_2 = 1/2(\omega_1 + \omega_2 - \omega_3 - \omega_4);$$

$$M_3 = {}^1/_2(\omega_1 - \omega_2 + \omega_3 - \omega_4);\ M_4 = {}^1/_2(\omega_1 - \omega_2 - \omega_3 + \omega_4).$$

Simple roots are:

$$\alpha_1 = {}^1/_2\omega_1 - {}^1/_2\omega_2 - {}^1/_2\omega_3 - {}^1/_2\omega_4;\ \alpha_2 = \omega_4;$$
$$\alpha_3 = \omega_3 - \omega_4;\ \alpha_4 = \omega_2 - \omega_3.$$

7) The exceptional Lie algebra e_6. Roots of this algebra are linear functions:

$$\pm\omega_i \pm \omega_j,\ (i < j) = 1, 2, 3, 4;\ \pm(\omega_i \pm {}^1/_2(\omega_6 - \omega_5)),$$
$$\pm(\Lambda_i \pm {}^1/_2(\omega_7 - \omega_5)),\ \pm(M_i \pm {}^1/_2(\omega_7 - \omega_6)),$$

where Λ_i, M_i are defined above for f_4. Simple roots are:

$$\alpha_1 = -\omega_1 + {}^1/_2(\omega_6 - \omega_5),\ \alpha_2 = \omega_1 - \omega_2,\ \alpha_3 = \omega_2 - \omega_3,$$
$$\alpha_4 = \omega_3 + \omega_4,\ \alpha_5 = -M_1 + {}^1/_2(\omega_7 - \omega_6),\ \alpha_6 = \omega_3 - \omega_4.$$

8) The exceptional Lie algebra e_8. Simple roots are α_1, α_2, ..., α_8. Consider the following set of vectors:

$$\begin{aligned}
\lambda_1 &= 3(\alpha_1 + \alpha_2 + \alpha_3 + \alpha_4 + \alpha_5) + 2\alpha_6 + \alpha_7 + \alpha_8,\\
\lambda_2 &= 3(\alpha_2 + \alpha_3 + \alpha_4 + \alpha_5) + 2\alpha_6 + \alpha_7 + \alpha_8,\\
\lambda_3 &= 3(\alpha_3 + \alpha_4 + \alpha_5) + 2\alpha_6 + \alpha_7 + \alpha_8,\\
\lambda_4 &= 3(\alpha_4 + \alpha_5) + 2\alpha_6 + \alpha_7 + \alpha_8,\\
\lambda_5 &= 3\alpha_5 + 2\alpha_6 + \alpha_7 + \alpha_8,\\
\lambda_6 &= 2\alpha_6 + \alpha_7 + \alpha_8,\\
\lambda_7 &= -\alpha_6 + \alpha_7 + \alpha_8,\\
\lambda_8 &= -\alpha_6 - 2\alpha_7 + \alpha_8.
\end{aligned}$$

Then the system of all roots of e_8 is expressed as follows:

$$\lambda_i - \lambda_j,\ \pm(\lambda_i + \lambda_j + \lambda_k),\ \pm(\lambda_i + \lambda_j + \lambda_k + \lambda_l + \lambda_m + \lambda_n),$$
$$\pm(2\lambda_i + \lambda_j + \lambda_k + \lambda_l + \lambda_m + \lambda_n + \lambda_p + \lambda_q),$$

where all indices are different and run over the set (1, 2, ..., 8).

9) The exceptional lie algebra e_7. Simple roots α_2, α_3, ..., α_8 are obtained from roots of e_8 by ejecting α_1. Furthermore, the root system of e_7 is obtained from the root system of e_8 by ejecting all sums listed above that contain λ_1.

Returning to the formulation of Theorem 16.3, note that the restriction on the rank r (e.g., $n \geqq 4$ for D_n) is mentioned in order to exclude isomorphic Lie algebras in smaller dimensions. For example, we already know that Lie algebras so_3 and su_2 are isomorphic. We leave to the reader, as a useful exercise, the proof of the following isomorphisms: $A_1 = B_1 = C_1$, $B_2 = C_2$, $A_3 = D_3$, $D_2 = A_1 \oplus A_1$. All other Lie algebras listed in the theorem are mutually nonisomorphic.

17. COMPACT LIE GROUPS

17.1. Real Forms

Until now we have studied complex semisimple Lie algebras. However, a great role is also played by different real subalgebras contained in these algebras. One of these is especially remarkable because the corresponding Lie group is compact.

Definition 17.1. Let G be a semisimple Lie algebra over the field of complex numbers. The real subalgebra G_0 of G (considered as a Lie algebra over $\mathbf{R}$) is called a real form of G if the canonical mapping of the complexification $G_0^{\mathbf{C}} = G_0 \otimes_{\mathbf{R}} C$ of G_0 onto G is an isomorphism. In this case, we have $\dim_{\mathbf{R}} G_0 = \dim_{\mathbf{C}} G$.

This means that complexifying the real algebra G_0, i.e., considering linear combinations of its elements with complex coefficients, we obtain the whole enveloping algebra G.

Let G_0 be a real form of the complex semisimple Lie algebra G. Then any element of G is uniquely presentable in the form $X + iY$, where $X, Y \in G_0$. This decomposition of G generates the natural involution σ that maps G onto itself, namely, $\sigma(X + iY) = X - iY$. This involution depends on G_0 and has the following evident properties: $\sigma^2 = 1_G$, $\sigma X = X$ if $X \in G_0$; $\sigma(A + B) = \sigma A + \sigma B$, $A, B, \in G$; $\sigma(\lambda A) = \lambda\sigma A$, $\sigma[A, B] = [\sigma A, \sigma B]$.

Lemma 17.1. Conversely, let an involution σ with the above properties be defined on G. Then this involution defines a Lie subalgebra G_0 in G that is a real form.

Proof. Denote by G_0 a set of fixed points of σ in G. Properties of σ imply that G_0 is a real subalgebra in G. On the other hand, any element A of G is presentable in the form $X + iY$, where $X, Y \in G_0$. In fact, $A = \frac{1}{2}(A + \sigma A) + i((A - \sigma A)/2i)$, where $X = \frac{1}{2}(A+\sigma A) \in G_0$, $Y = 1/2i(A - \sigma A) \in G_0$, since $\sigma X = \sigma(A + \sigma A) = X$, $\sigma Y = \sigma((A - \sigma A)/2i) = Y$. The lemma is proved.

In G, consider the Killing form $(\, , \,)_G$. We may restrict this form onto the real subalgebra G_0; denote this restriction by $(\, , \,)_{G_0}'$. On the other hand, on G_0 its own Killing form $(\, , \,)_{G_0}$ is defined. The natural question arises: Do these two forms coincide (up to a scalar multiple)? Above (see Sec. 16) we have seen that, generally speaking, the restriction of the Killing form of an enveloping algebra onto an arbitrary subalgebra does not coincide with the Killing form of this subalgebra. However, for real forms of simple algebras the situation is more favorable.

Lemma 17.2. If G_0 is a real form of the semisimple Lie algebra G, then $(\, , \,)_{G_0}' = (\, , \,)_{G_0}$ (up to a constant multiple).

Proof. By definition, the Killing form is of the form $\mathrm{Tr}\, \mathrm{ad}_X \mathrm{ad}_Y$. Since G_0 is the "real part" of G, then for $X, Y \in G_0$ the endomorphism $\mathrm{ad}_X \mathrm{ad}_Y$ preserves G_0; therefore, the trace of its restriction onto G_0 coincides with the Killing form on G_0. In particular, the Killing form assumes real values on G_0. The lemma is proved.

17.2. The Compact Form

Definition 17.2. A real Lie algebra is called compact if its Killing form is negative definite, i.e., the corresponding quadratic form satisfies $\langle X, X\rangle < 0$ if $X \neq 0$.

Definition 17.3. The real form G_0 of the complex Lie algebra G is called the compact real form of G if G_0 is a compact real algebra.

The notion "compact algebra" is due to the fact that it so happens that a Lie group with a compact Lie algebra is compact as a topological space (see the examples below). The compactness of the real form G_0 may be established from considerations of properties of the Killing form $\langle \, , \, \rangle_G$ on G.

Lemma 17.3. The real form G_0 of G is compact iff the Hermitian form $\langle A, \alpha A\rangle$ on G is negative definite.

Proof. Let G_0 be compact and $A = X + iY \in G$; then

$$\langle A, \sigma A\rangle = \langle X + iY, X - iY\rangle = \langle X, X\rangle + \langle Y, Y\rangle < 0.$$

Conversely, if the form $\langle A, \sigma B\rangle$ is negative definite, then for $A \in G_0$, $A \neq 0$, we have $\sigma A = A$ and $\langle A, A\rangle < 0$. The lemma is proved.

In what follows, the compact form will be denoted by G_u. The classification of all real forms of G is reduced to the description of all nonequivalent involutions of semisimple Lie algebras. It turns out that the compact form is defined by a special involution that exists on each semisimple Lie algebra.

First, consider the model example sl(n, **C**). On G, consider the involution $\sigma A = \bar{A}$, i.e., the complex conjugation of the matrix A. The set of fixed points of this involution evidently coincides with the subalgebra of real matrices, which is sl(n, **R**). Clearly, the Killing form on sl(n, **R**) is the real form presentable in the form $\langle X, X\rangle = \mathrm{Tr}\, X^2 = \sum_{i,j} x_i^j x_j^i$. This form is evidently not negative definite (it is indefinite). Therefore, sl(n, **R**) is real but the compact form of sl(n, **C**) is not. The compact form G_u is constructed in this example as follows. Consider the involution $\tau: G \to G$, $\tau A = -\bar{A}^T$. Fixed points of this involution are matrices A, such that $\bar{A}^T = -A$, i.e., the set of fixed points coincides with the space of all skew Hermitian matrices. As we already know, this space is the Lie algebra of the Lie group SU_n, which is compact. In fact, computing for this real form G_0 the Killing form, we get $\langle X, Y\rangle = \mathrm{Tr}\, XY = -\mathrm{Tr}\, X\bar{Y}^T$, i.e.,

$$\langle X, X\rangle = -\mathrm{Tr}\, X\bar{X}^T = -\sum_{i,j} x_i^j \bar{x}_i^j < 0, \text{ if } X \neq 0.$$

In this case, the Killing form coincides (up to multiplication by −1) with the standard Hermitian scalar product. Thus, the Lie algebra su_n of SU_n is compact in the sense of Definition 17.2, also.

It turns out that having considered the case sl(n, **C**), we have constructed the general model for all complex simple Lie algebras.

Theorem 17.1. Each semisimple complex Lie algebra G possesses a compact real form G_u.

Proof. We will just produce the explicit embedding of the compact subalgebra G_u into G. This embedding will be used subsequently since it is connected with various geometric properties of Lie groups. Consider the Weyl basis in G (see Proposition 16.4). In G, consider the involution σ defined on the Weyl basis by the formulas $\sigma E_\alpha = E_{-\alpha}$ if $\alpha \neq 0$ and $\sigma h = -h$ for any $h \in T_0$, where $T_0 \subset T$ is the "real part" of the Cartan subalgebra T. We will assume that $\sigma(\lambda X) = \bar{\lambda}\sigma X$. Thus, σ acts as is shown in Fig. 192, i.e., $\sigma: V^+ \to V^-$, $\sigma: V^- \to V^+$, $\sigma T_0 \to -T_0$, $\sigma: iT_0 \to iT_0$. Properties of the Weyl basis (see Proposition 16.4) immediately imply that σ is an automorphism of the Lie algebra G. Thus, for instance, if $\alpha > 0$, $\beta > 0$, then

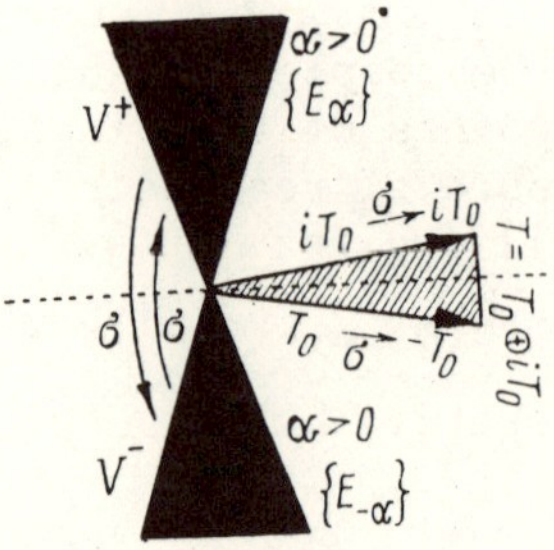

Fig. 192

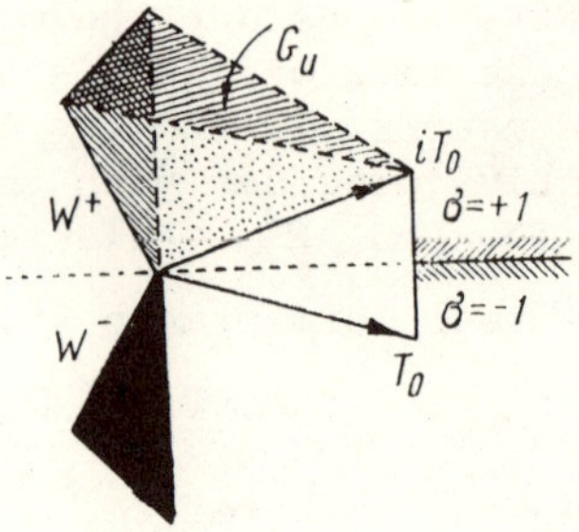

Fig. 193

$$\sigma[E_\alpha, E_\beta] = \sigma(N_{\alpha\beta}E_{\alpha+\beta}) = N_{\alpha\beta}E_{-\alpha-\beta} = N_{-\alpha,-\beta}E_{-\alpha-\beta} =$$
$$= [E_{-\alpha}, E_{-\beta}] = [\sigma E_\alpha, \sigma E_\beta].$$

The fact that other commutative relations are preserved is similarly verified. Now let us find the real form corresponding to this involution. The explicit form of σ implies that for a basis (over $\mathbb{R}$) in the subalgebra of fixed points of σ, we may take the following vectors: $\{E_\alpha + E_{-\alpha};\ i(E_\alpha - E_{-\alpha});\ iH_\alpha'\}$. We claim that this is a compact subalgebra in G. In fact, since $\langle E_\alpha, E_\alpha\rangle = 0$ and $\langle E_\alpha, E_{-\alpha}\rangle = -1$, it suffices to compute the following scalar products:

$$\langle E_\alpha + E_{-\alpha}, E_\alpha + E_{-\alpha}\rangle = -2,\ \langle i(E_\alpha - E_{-\alpha}), i(E_\alpha - E_{-\alpha})\rangle =$$
$$= 2\langle E_\alpha, E_{-\alpha}\rangle = -2,\ \langle E_\alpha + E_{-\alpha}, i(E_\alpha - E_{-\alpha})\rangle = 0,$$
$$\langle iH_\alpha', iH_\alpha'\rangle = -\alpha(H_\alpha') < 0,$$

since $\alpha(H_\alpha') = 0$ and H_α' is dual to the linear form α. Hence, the Killing form is negative definite on the whole subalgebra of fixed points, which proves the compactness of this subalgebra. The theorem is proved.

Now consider in detail the embedding of G_u into G. In G we may evidently choose the following basis over $\mathbb{R}$:

$$\{E_\alpha + E_{-\alpha},\ i(E_\alpha - E_{-\alpha}),\ E_\alpha - E_{-\alpha},\ i(E_\alpha + E_{-\alpha}),\ H_\alpha',\ iH_\alpha'\}.$$

This means that, besides the root decomposition of G, i.e., $G = V^+ \oplus V^- \oplus T$ (over $\mathbb{C}$), there is another natural decomposition (over $\mathbb{R}$):

$$G = W^{+} \oplus W^{-} \oplus T_0 \oplus i\,T_0,$$

where

$$W^{+} = \{E_\alpha + E_{-\alpha},\ i\,(E_\alpha - E_{-\alpha})\},$$

$$W^{-} = \{E_\alpha - E_{-\alpha},\ i\,(E_\alpha + E_{-\alpha})\},\quad T_0 = \{H'_\alpha\},\ iT_0 = \{iH'_\alpha\}.$$

In parentheses we put vectors that form the basis of the corresponding subspace (see Fig. 193). Clearly, $\sigma = +1$ on $W^+ \oplus \oplus\, iT_0$ and $\sigma = -1$ on $W^- \oplus T_0$. Hence, the space $W^+ \oplus iT_0$ is a subalgebra (unlike the space $W^- \oplus T_0$) and the space of fixed points coincides with the compact subalgebra G_u in G, i.e., $G_u = W^+ \oplus iT_0$. Thus we have established that the compact Lie algebra G_u in G is spanned by vectors of the form

$$G_u = \{E_\alpha + E_{-\alpha},\ i\,(E_\alpha - E_{-\alpha}), iH'_\alpha\} = W^+ \oplus iT_0.$$

Consider the adjoint action of G_u on itself, i.e., let us study the action of transformations of the form $ad_h: G_u \to G_u$, where $h \in iT_0$. Since h belongs to the Cartan subalgebra, then ad_h transforms the space W^+, orthogonal to the space iT_0, into itself. Here we make use of the fact that operators ad_h are skew symmetric with respect to the Killing form and therefore preserve the orthogonal complement, transforming it into the circle. Let $h = iq$, where $q \in T_0$. Clearly, $ad_h(E_\alpha + E_{-\alpha}) = i\, ad_q\,(E_\alpha + E_{-\alpha}) = i\alpha(q)(E_\alpha - E_{-\alpha})$, where $\alpha(q)$ is a real number (see the defintion of T_0 – "the real part" of the Cartan subalgebra). Hence, $ad_h\,(E_\alpha + E_{-\alpha}) = \alpha(q)(i(E_\alpha - E_{-\alpha}))$. Similarly, $ad_h\,(i(E_\alpha - E_{-\alpha})) = -\alpha(q)(E_\alpha + E_{-\alpha})$. Hence, ad_h transforms the two-dimensional real plane spanned by vectors $E_\alpha + E_{-\alpha}$ and $i(E_\alpha - E_{-\alpha})$ into itself and on this plane is defined by the following skew symmetric (2×2)-matrix: $ad_h = \begin{pmatrix} 0 & \alpha(q) \\ -\alpha(q) & 0 \end{pmatrix}$, where $h = iq$. Thus, we see the difference between the action of ad_h on the compact algebra and the similar action on the complex Lie algebra. If, in the complex case, this operator is reduced to the diagonal form in the basis that consists of root vectors, then in the real case this operator has no real eigenvectors in the orthogonal complement to the Cartan subalgebra iT_0 and is reduced to the block-diagonal form, i.e., can be written as a matrix with (2×2)-matrices on its main diagonal. Each of these blocks corresponds to one root (eigenvalue over $\mathbb{C}$) and is described by the above matrix. In our model example, the compact Lie algebra $G_u = su_n$, where $G_u \subset G = sl(n, \mathbf{C})$, splits into the direct sum of the following subspaces: $su_n = W^+ \oplus iT_0$, where

$$iT_0 = \begin{pmatrix} i\varphi_1 & & 0 \\ & \ddots & \\ 0 & & i\varphi_n \end{pmatrix}, \quad \varphi_1 + \ldots + \varphi_n = 0;$$

and φ_i are real numbers, $W^+ = \operatorname{Re} W^+ \oplus \operatorname{Im} W^+$, where

$$\operatorname{Re} W^+ = \{E_\alpha + E_{-\alpha}\} = \left\{\begin{pmatrix} & 0 & 1 \\ & & \ddots \\ -1 & & 0 \end{pmatrix}\right\},$$

$$\operatorname{Im} W^+ = \{i(E_\alpha - E_{-\alpha})\} = \left\{\begin{pmatrix} 0 & & i \\ & \ddots & \\ i & & 0 \end{pmatrix}\right\}.$$

The two-dimensional invariant subspace spanned by vectors $E_\alpha + E_{-\alpha}$ and $i(E_\alpha - E_{-\alpha})$ is, in this case, of the form

$$\begin{pmatrix} & 0 & a + ib \\ & \ddots & \\ -a + ib & & 0 \end{pmatrix}, \quad \text{where } a, b \in \mathbf{R}.$$

The role of the constructed canonical embedding of the compact subalgebra G_u into the complex semisimple Lie algebra G especially increases in view of the following fact: the compact form of G is unique up to automorphisms of G.

Proposition 17.1. Let G_u and G_u' be any two compact real forms of the semisimple Lie algebra G over $\mathbf{C}$. Then there exists an automorphism ψ of G such that $\psi G_u = G_u'$; and this automorphism is included into a one-parameter subgroup of automorphisms ψ_t, where $\psi = \psi_1$, $\psi_0 = 1_G$, $0 \leqq t \leqq 1$.

Although this fact is useful in order to understand the general picture, it will not be used and, therefore, we skip its proof.

Now consider our model example $sl(n, \mathbf{C})$. Above, we have completely studied the structure of the root decomposition of this algebra; therefore, now it is easy to write explicitly the canonical embedding of the compact form G_u into

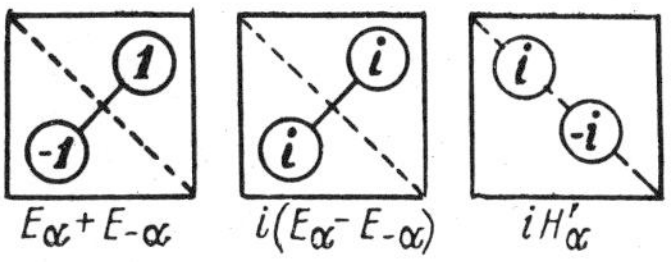

Fig. 194

sl(n, C). In fact, root vectors E_α coincide with elementary matrices T_{pq}, $p < q$ if $\alpha > 0$ and $-T_{pq}$, $p > q$ if $\alpha < 0$. Hence, $E_\alpha + E_{-\alpha} = T_{pq} - T_{qp}$ and $i(E_\alpha - E_{-\alpha}) = iT_{pq} + iT_{qp}$ and, finally, $iH_\alpha' = i(T_{pp} - T_{qq})$ if $\alpha(h) = a_p - a_q$, $p < q$ (see Fig. 194). Therefore, the subspace iT_0 coincides with the subspace of all diagonal purely imaginary matrices with trace zero:

$$iT_0 = \begin{pmatrix} i\varphi_1 & & 0 \\ & \ddots & \\ 0 & & i\varphi_n \end{pmatrix}, \quad \varphi_1 + \ldots + \varphi_n = 0, \quad \varphi_i \in \mathbf{R};$$

the subspace $\{E_\alpha + E_{-\alpha}\}$ coincides with the subspace of all real skew symmetric matrices, and the subspace $\{i(E_\alpha - E_{-\alpha})\}$ coincides with the subspace of all symmetric purely imaginary matrices with zeros on the main diagonal. Finally, the subalgebra $G_u = W^+ \oplus iT_0$ coincides with the subspace of all skew Hermitian matrices of trace zero, i.e., with the Lie algebra of su_n. Thus, we have proved the following lemma.

Lemma 17.4. The standard embedding of su_n into sl(n, C) coincides with the canonical embedding of the compact form G_u in G.

Besides the canonical compact form, each semisimple complex Lie algebra possesses the canonical noncompact form, which is sometimes called the normal noncompact form. Consider again the Weyl basis in G and construct the following involution: $\tau: G \to G$, where $\tau E_\alpha = E_\alpha$, $\tau H_\alpha' = H_\alpha'$, i.e., τ is the identity on the real part of V^+ and V^- and also on the real part T_0 of the Cartan subalgebra T. But this involution is not an identity on the whole algebra since $\tau(\lambda X) = \bar{\lambda}\tau X$. Hence, $\tau(iT_0) = -iT_0$, $\tau(iE_\alpha) = -iE_\alpha$. The set of fixed points of τ coincides with Re $V^+ \oplus$ Re $V^- \oplus T_0$, i.e., with the linear span (over $\mathbf{R}$) of vectors $\{E_\alpha, E_{-\alpha}, H_\alpha'\}$. Proposition 16.4 implies that this span is a Lie subalgebra.

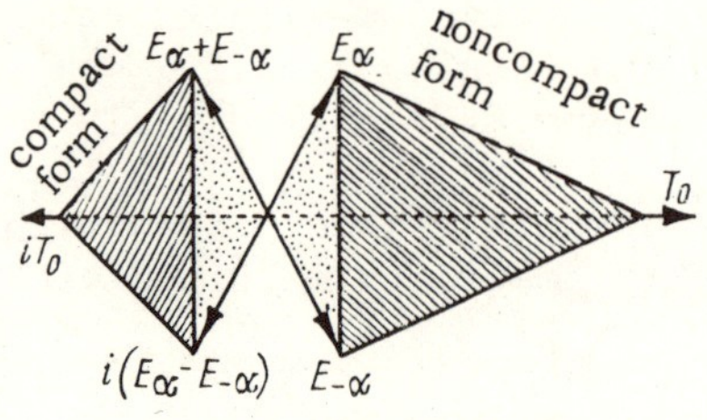

Fig. 195

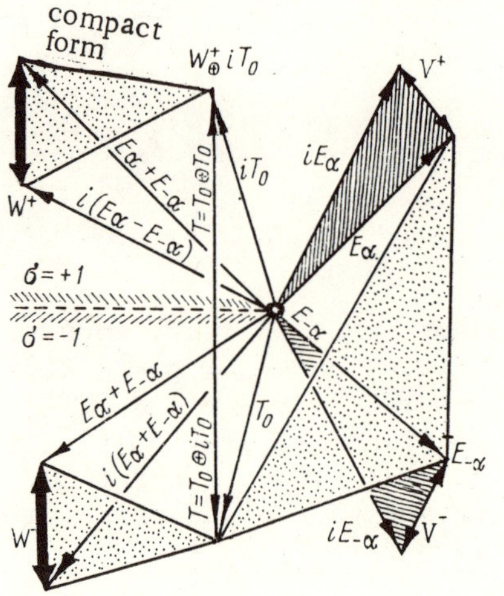

Fig. 196

Definition 17.4. The real subalgebra $\{E_\alpha, E_{-\alpha}, H_\alpha'\}$ is called the normal noncompact form of G.

In our model example, this subalgebra evidently coincides with the subalgebra of real traceless matrices, i.e., with $sl(n, \mathbf{R})$, and τ coincides with $\tau A = \bar{A}$, where the bar stands for the complex conjugation. Like the complex form, the noncompact normal form is defined uniquely up to an automorphism of the enveloping complex Lie algebra. We will not prove this fact here since it will not be used subsequently. Thus, to each complex semisimple Lie algebra G, two of its subalgebras, i.e., the compact and noncompact (normal) forms, are uniquely assigned (up to an automorphism of G). The mutual disposition of compact and noncompact forms in G is depicted in Fig. 195. A more explicit scheme is shown in Fig. 196. This decomposition is shown over $\mathbf{R}$; that is why "real" and "imaginary" subspaces are expressed separately. Note that the compact real form G_u admits one more invariant characterization. Namely, this subalgebra is the maximum compact subalgebra in G. Now let us give the complete list of simple Lie algebras G over $\mathbf{C}$ and their compact real forms G_u (see p. 221). Let us mention one more compact subalgebra of the complex semisimple Lie algebra G that often appears in concrete problems. For this, consider two involutions introduced above: σ (that defines the compact form) and τ (that defines the noncompact normal form). Consider the set G_n of points fixed with respect to both involutions. Since

Cartan's notation	G	G_u	$\dim G_u$
$A_n,\ n \geqslant 1$	$sl(n+1, \mathbf{C})$	su_{n+1}	$n(n+2)$
$B_n,\ n \geqslant 2$	$so(2n+1, \mathbf{C})$	so_{2n+1}	$n(2n+1)$
$C_n,\ n \geqslant 3$	$sp(n, \mathbf{C})$	sp_n	$n(2n+1)$
$D_n,\ n \geqslant 4$	$so(2n, \mathbf{C})$	so_{2n}	$n(2n+1)$
G_2	$g_2(\mathbf{C})$	g_2	14
F_4	$f_4(\mathbf{C})$	f_4	52
E_6	$e_6(\mathbf{C})$	e_6	78
E_7	$e_7(\mathbf{C})$	e_7	133
E_8	$e_8(\mathbf{C})$	e_8	248

$\sigma = 1$ on $W^+ \oplus iT_0$ and $\tau = 1$ on $\{E\alpha\} \oplus \{E\text{-}\alpha\} \oplus T_0$, then G_n is spanned by vectors $\{E_\alpha + E_{-\alpha}\}$, since they, and precisely they, are preserved by both involutions. Hence, G_n is a compact subalgebra. It coincides with the intersection in G of two of its real forms: the compact G_u and the noncompact normal one. Note that G_n is by no means a real form of G (!) since its complexification does not coincide with G. Subalgebra G_n is sometimes called the normal compact subalgebra (but not the normal form!). In our model example sl(n, **C**), we get $G_n = G_u \cap$ (normal noncompact form) $= su_n \cap sl(n, \mathbf{R}) = so_n$, since skew Hermitian real matrices are skew symmetric.

18. ORBITS OF THE COADJOINT REPRESENTATION

18.1. Generic and Singular Orbits

It turns out that with each semisimple Lie algebra there is connected a set of smooth manifolds that are homogeneous spaces and play, as was established especially in recent years, an important role in different problems of classical mechanics and their modern generalizations and analogs. These manifolds are called orbits, and they are naturally embedded into the Lie algebra G and "fibrate" it in the sense that the Lie algebra is the union of these nonintersecting orbits. Moreover, orbits are closely connected with the algebraic structure in G and carry the stable imprint of those relations that distinguish this algebra in the set of all Lie algebras. In the preceding section, studying the Cartan subalgebra T, we were forced to present it as the direct sum of subalgebras $T = T_0 \oplus \oplus\, iT_0$, where iT_0 — the "imaginary part" of T — is the Cartan subalgebra in the compact form G_u. In this section we will deal with the subalgebra iT_0; therefore, to simplify the notation, denote it by H. Thus, $H = iT_0$ in G_u. The Lie subalgebra G_u

will be denoted by G. In this section, we will mostly consider only compact Lie algebras and the corresponding Lie groups. If earlier we illustrated all main ideas connected with complex semisimple Lie algebras by the model example sl(n, **C**), we will now illustrate our constructions by the example of the compact form of this Lie algebra, i.e., by the example of the Lie algebra of the special unitary group SU_n. Let G be a compact algebra and $Ad_{\mathfrak{G}} : G \to G$ the adjoint representation of the group $\mathfrak{G}$ by linear transformations of its Lie algebra. Let $X \in G$. Since we consider matrix Lie algebras, then the action Ad_g is expressed as $Ad_g X = gXg^{-1}$. The set of elements of the form gXg^{-1}, where g runs over the whole group $\mathfrak{G}$, is called the orbit of X (with respect to the adjoint representation of the group) and is denoted by O(X). Clearly, the orbit is a homogeneous space, i.e., is presentable in the form $\mathfrak{G}/C(X)$, where C(X) is the centralizer of X, i.e., the set of elements g of $\mathfrak{G}$ such that $gX = Xg$. The centralizer is a subgroup; therefore, O(X) is identified with the set of cosets with respect to C(X).

Lemma 18.1. The element $X \in G$ is a regular element of the semisimple Lie algebra G iff $\dim C(X) = \dim H$, where H is the Cartan subalgebra in G. In this case, the connected component of the unit in C(X) coincides with the connected subgroup $\tilde{T}$ that corresponds to H and is the maximum connected commutative subgroup in C(X).

Proof. In the compact Lie algebra G, any element may be included into a Cartan subalgebra. In the case su_n, this statement follows immediately from the presentation of X in the form $g^{-1}hg$, where h is a diagonal matrix. Then for a Cartan subalgebra that contains X, we may take the subalgebra $g^{-1}Hg$, where H is the subalgebra of diagonal matrices. Hence, C(X) is the Lie group corresponding to the annihilator of the regular element X. But, this annihilator coincides with the Cartan subalgebra $g^{-1}Hg$, completing the proof.

Corollary 18.1. Let H be a fixed Cartan subalgebra of the compact Lie algebra G. Then each orbit O(X) intersects with H. This means that G is presentable as the union of orbits that "grow" from points of T (see Fig. 197).

The proof (for su_n) follows immediately from the fact that any element X is diagonalizable.

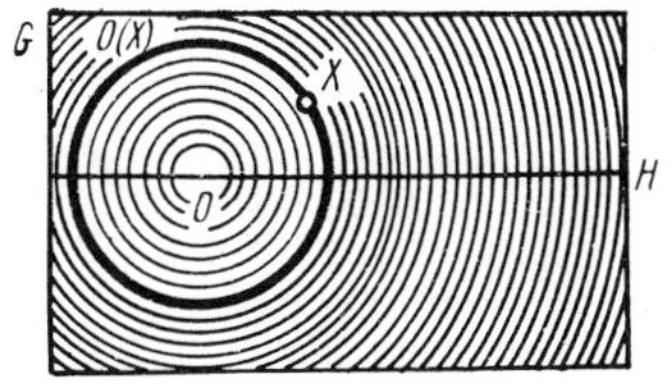

Fig. 197

Note that if X is regular, then all of its orbits consist of regular elements. Hence, singular orbits consist only of singular elements. Let us call orbits that consist of regular elements generic orbits ("typical orbits"). Clearly, the union of all generic orbits is an open dense set in G. Conversely, singular orbits constitute the set of zero measure.

Definition 18.1. The dimension r of the Cartan subalgebra H in the semisimple Lie algebra G is called the rank of G.

Lemma 18.1 implies that the dimension of a generic orbit in G equals N − r, where N is the dimension of G. In other words, codim O(X) = r. If the orbit is singular, then its dimension is strictly less than that of a generic orbit.

Lemma 18.2. Let X be an arbitrary point of O(X) (the orbit may be singular). Let $T_X O(X)$ be the tangent space to O(X) at X. Then $T_X O(X)$ consists of vectors of the form [X, Y], where Y runs over the entire Lie algebra G.

Proof. Since $O(X) \subset G$, then $T_X O(X)$ is contained in G. By definition of an orbit, it consists of points of the form gXg^{-1}, where g runs over $\mathfrak{G}$. Hence, any vector a tangent to O(X) admits a presentation of the form $a = \dot{X}(t)|_{t=0}$, where $X(t) \in O(X)$, $X(t) = g(t)Xg^{-1}(t)$, and g(t) is a smooth curve in $\mathfrak{G}$ which passes through the unit, i.e., g(0) = E (see Fig. 198). Hence, the tangent space $T_X O(X)$ is obtained from X by application of all transformations of the form ad_Y to it, since ad_Y is the differential of the transformation Ad_g at the unit $E \in \mathfrak{G}$. Since $ad_Y X = [Y, X]$, the lemma is proved.

Corollary 18.2. Let Ann X be the annihilator of H in the compact Lie algebra G (X might be singular). Then the

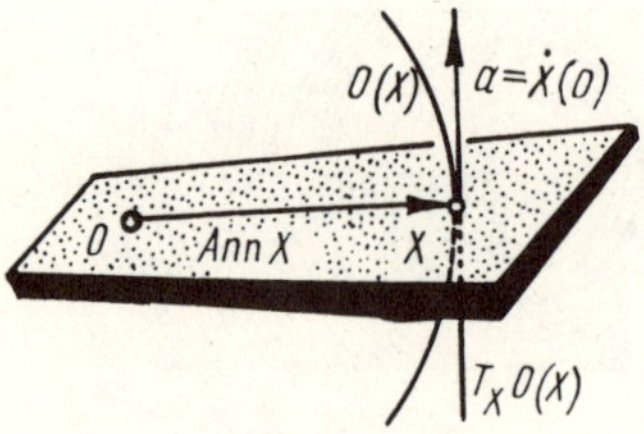

Fig. 198

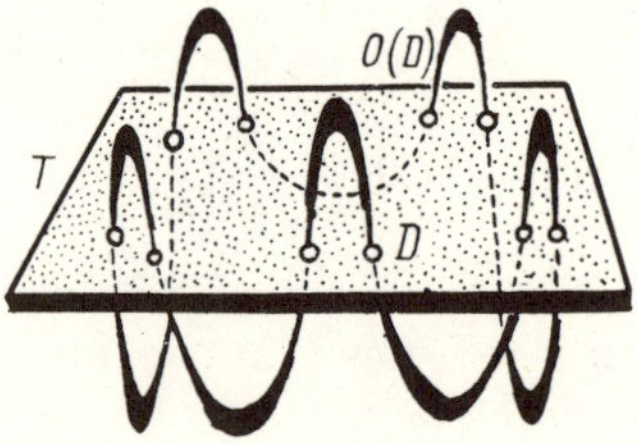

Fig. 199

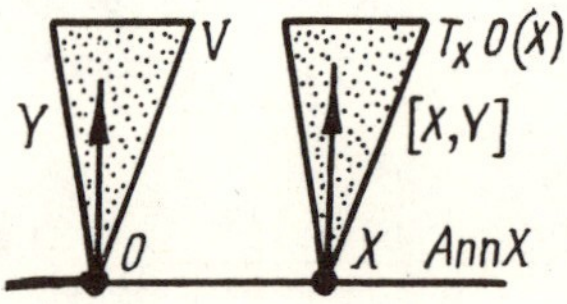

Fig. 200

orbit O(X) of this element is orthogonal to Ann X at X. In particular, spaces Ann X and $T_X O(X)$ (the tangent plane to the orbit) generate, with respect to the sum, all G (see Fig. 199).

Proof. The last statement follows from Lemma 18.2. It remains to prove the orthogonality of Ann X and $T_X O(X)$. By Lemma 16.1, operators ad_Y are skew symmetric with respect to the Killing form; hence, transformations $Ad_g: G \to G$ are orthogonal with respect to this form. Thus, the whole orbit O(X) is contained in the sphere S^{N-1}, whose radius is equal to the length of X, where N = dim G. Let Ann X $\neq$ 0 and X $\in$ Ann X. Then G splits into the direct sum of its two subspaces G = = Ann X $\oplus$ V, where V is the space orthogonal to Ann X with respect to the Killing form. By Lemma 18.2, the plane $T_X O(X)$ consists of vectors of the form [X, Y], where Y $\in$ G (see Fig. 200). Clearly, it suffices to take vectors Y from V, since vectors Y taken from Ann X give zero after commuting with X and, therefore, contribute nothing to the construction of $T_X O(X)$. Thus, $T_X O(X) = [X, V] = \{[X, Y], Y \in V\}$. Each vector of the form [X, Y] = v, where Y $\in$ V, is expressed as follows: $v = ad_X Y$. Since ad_X is skew symmetric with respect to the Killing form (see Lemma 16.1), it transforms the orthogonal complement to Ann X into itself; hence V is invariant with respect to ad_X. This means that $ad_X Y = [X, Y]$ is orthogonal to Ann X if $Y \in V$. The statement is proved.

The proof of this statement enables one to construct the simple geometric picture of orbits in the compact Lie algebra G. Fix the Cartan subalgebra H and generate at each of its points the orbit O of the adjoint action of $\mathfrak{G}$. This orbit, being a smooth manifold, "grows" from the point h orthogonally to H. If the orbit is generic, then the dimension of the tangent space to O at h equals dim G − dim H. If the orbit is singular, then $\dim T_h O < \dim G - \dim H$. One must not think that the orbit O(h) starting from $h \in H$ will never return to H. The intersection of O(h) with the Cartan subalgebra H consists, in the general case, of several points (see Fig. 199). In $\mathfrak{G}$, elements $g_0 \notin \tilde{T}$ may exist (where $\tilde{T}$ stands for the connected commutative subgroup whose tangent space does not coincide with H) such that $g_0 h g_0^{-1} \in H$ and $g_0 h g_0^{-1} \neq h$. This means that O(h) starting from h will return after a while onto H, intersecting it at the point $g_0 h g_0^{-1}$ different from h. Let us illustrate this effect on the simplest example. For $\mathfrak{G}$, take the compact group SO_3 whose Lie algebra so_3 consists of skew symmetric real matrices. Clearly, $\dim so_3 = 3$. The adjoint action $X \to gXg^{-1}$ coincides in this case with the standard action of SO_3 on the Euclidean space R^3. The Killing form coincides here with the usual scalar product (verify!). Hence, orbits of the adjoint action of SO_3 are two-dimensional spheres with center at 0 (if the orbit is generic), and 0 is a unique singular point (in this example, it is a fixed point). The Cartan subalgebra is one-dimensional here and can be identified with a straight line passing through 0. Hence, each generic orbit intersects with the Cartan subalgebra at exactly two points (see Fig. 201). Generic orbits of the compact Lie algebra G are diffeomorphic to the homogeneous space which is the smooth manifold $\mathfrak{G}/\tilde{T}$,

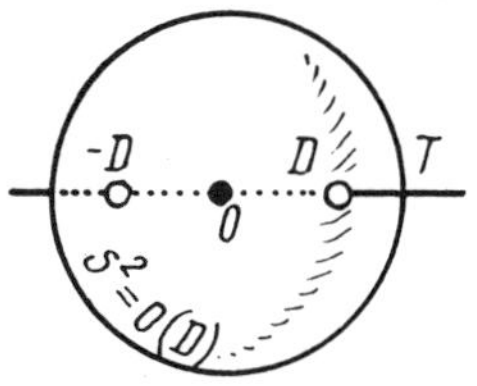

Fig. 201

in particular, all of them are diffeomorphic with respect to each other. This follows immediately from Lemma 18.1.

18.2. Orbits in Lie Groups

Until now we have studied the structure of orbits in the compact Lie algebra. But, in applications, it is often useful to know the structure of orbits in the corresponding connected compact Lie group. The adjoint action $Ad_{\mathfrak{G}}$ of $\mathfrak{G}$ on G is generated by the adjoint action of $\mathfrak{G}$ on itself: $g \to g_0 g g_0^{-1}$, $g, g_0 \in \mathfrak{G}$. Thus, to each element g of $\mathfrak{G}$ corresponds a diffeomorphism of the group onto itself. As in the case of Lie algebras, these smooth $\mathfrak{G}$-actions define a decomposition of the group into the union of nonintersecting orbits, i.e., smooth submanifolds $O(g_0) = \{g g_0 g^{-1}, g \in \mathfrak{G}\}$. Clearly, $O(g_0) = \mathfrak{G}/C(g_0)$, where $C(g_0)$ is the closed (hence compact) subgroup in $\mathfrak{G}$, consisting of elements that preserve g_0. The subgroup $C(g_0)$ is sometimes called the stationary subgroup of $g_0 \in \mathfrak{G}$.

Lemma 18.3. Let g_0 and g_0' be a pair of points that belong to the same orbit $O \subset \mathfrak{G}$. Then stationary subgroups $C(g_0)$ and $C(g_0')$ of these points are conjugate to each other in $\mathfrak{G}$, i.e., there exists an $s \in \mathfrak{G}$ such that $sC(g_0)s^{-1} = C(g_0')$.

The proof follows immediately from the defintion of an orbit.

Let $\tilde{T}$ be a subgroup in $\mathfrak{G}$ corresponding to the Cartan subalgebra H, i.e., $\tilde{T} = \exp H$ (recall that we consider only matrix groups and Lie algebras), where $\exp: G \to \mathfrak{G}$ is the matrix exponent, i.e., $\exp X = \sum_{n=0}^{\infty} \frac{1}{n!} X^n$ (see details in [2]). Our immediate goal is the study of the geometric properties of the splitting of a compact Lie group into a union of orbits.

Lemma 18.4. The subgroup $\tilde{T} = \exp H$, where H is the Cartan subalgebra in the compact Lie algebra G, is diffeomorphic to the torus T^r, where $r = \operatorname{rank} G$ and $\tilde{T}$ is a maximum connected commutative subgroup in $\mathfrak{G}$.

Proof. Consider the subgroup R which is the closure of exp H in $\mathfrak{G}$. Clearly, R is a closed connected commutative subgroup in $\mathfrak{G}$; therefore, its Lie algebra $T_E R$ contains H. If $R \neq \exp H$, then $\dim T_E R > \dim H$, contradicting the property of maximality of H among all commutative subalgebras in G. Therefore, $R = \exp H = \tilde{T}$. The lemma is proved.

This subgroup is sometimes called a maximum torus in $\mathfrak{G}$.

The following theorem is one of the main ones in the theory of compact Lie groups. We could illustrate its validity as earlier only for the model example of the compact group SU_n, i.e., the compact form of $SL(n, \mathbb{C})$, but due to its importance, we will prove it (in Sec. 18.3) for an arbitrary compact group.

Theorem 18.1. Let $\tilde{T}$ be a maximum torus of a connected compact Lie group. Then for any element $g \in \mathfrak{G}$, there is an element $g_0 \in \mathfrak{G}$ such that $g_0 g g_0^{-1} \in \tilde{T}$, i.e., each orbit O of $\mathfrak{G}$ necessarily intersects with at least one point of $\tilde{T}$.

Corollary 18.3. Each element g of a connected compact Lie group $\mathfrak{G}$ belongs to a one-parameter subgroup and, in particular, is of the form $\exp X$ for some $X \in G$.

Proof. By Theorem 18.1, there exists a $g_0 \in \mathfrak{G}$ such that $g_0 g g_0^{-1} \in T$. For elements of a maximum torus, the statement is evident, i.e., $g_0 g g_0^{-1} = \exp Y$ for $Y \in H$, i.e., $g = \exp(g_0^{-1} Y g_0)$, as required.

Corollary 18.4. Any connected commutative subgroup K in the compact Lie group $\mathfrak{G}$ is contained in a maximum torus conjugate to $\tilde{T}$.

Proof. Consider the closure $\bar{K}$ of K in $\mathfrak{G}$. Since K is commutative, then so is $\bar{K}$, i.e., it is a torus of a certain dimension. From algebra it is known that any torus contains an element whose powers constitute a dense subset. Denote this element by t. But, by Theorem 18.1, there is a $g \in \mathfrak{G}$ such that $t \in g\tilde{T}g^{-1}$. Clearly, $t^n \in g\tilde{T}g^{-1}$ for any integer n, implying $\bar{K} = \overline{\{t^n\}} \subset g\tilde{T}g^{-1}$, as required.

Corollary 18.5. Any two maximum tori (i.e., two connected maximum commutative subgroups) in a compact group are conjugate.

Proof. Let $K = \bar{K}$ be a maximum torus. By Corollary 18.4, there is a $g \in \mathfrak{G}$ such that $K = g\tilde{T}g^{-1}$, as required.

Conversely, Corollary 18.5 implies Theorem 18.1.

18.3. Proof of the Theorem on Conjugacy of Maximum Tori in a Compact Lie Group

Let $f:M^n \to M^n$ be a smooth mapping of the compact closed smooth manifold M^n onto itself. Then, to this mapping we may assign an integer called the Lefschetz number $\ell(f)$. One of its possible definitions is related to the homology of M. The mapping f induces homomorphisms $f_q:H_q(M; A) \to H_q(M; A)$, where A is the group of coefficients. For simplicity, take for A the field of real numbers. Then f_q is defined by the $(\beta_q \times \beta_q)$-matrix with real coefficients, where $\beta_q = \dim_{\mathbf{R}} H_q(M^n; \mathbf{R})$. Recall that β_q is called the Betti number and coincides with the rank of $H_q(M; \mathbf{Z})$. Consider the trace of f_q and put $l(f) = \sum_{q=0}^{n} (-1)^q \operatorname{Tr} f_q$. This number is called the Lefschetz number of the mapping f. The number $\ell(f)$ possesses the following properties (we leave the proof to the reader as a useful exercise; see, for example, [4]):

1) $\ell(f)$ is an integer number;

2) if f and g are homotopic, then $\ell(f) = \ell(g)$;

3) if $\ell(g) \neq 0$, then $f:M \to M$ has at least one fixed point;

4) if all the fixed points of f are isolated, then, to any fixed point x_i, a number λ_i, called its index may be assigned, so that $l(f) = \sum_i \lambda_i$, where the sum runs over all fixed points;

5) if x_0 is an isolated fixed point and the differential df of f has no unit eigenvalues at this point, then $\lambda(x_0)$ is equal to the sign of the number $\det(df - E)$, where E is the identity mapping. [Here df is considered as a linear mapping of the tangent space $T_{x_0}M$ onto itself, defined by the matrix $(\partial y_i/\partial x_j)$, where $y_i = f_i(x_1, \ldots, x_n)$ and $\{x_i\}$ are local regular coordinates in a neighborhood of the fixed point.]

For simplicity, let us assume that M is orientable. The Lefschetz number admits one more geometric interpretation.

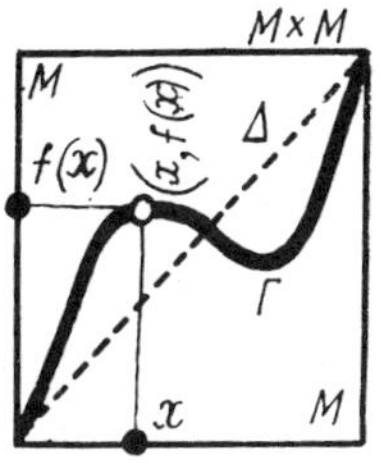

Fig. 202

For this, consider the direct product of M by itself and depict the smooth mapping $f:M \to M$ as the graph of this direct product, i.e., consider in $M \times M$ the subset of points of the form $(x, f(x))$, where $x \in M$ (see Fig. 202). This subset Γ is called the graph of f. For example, if f is the identity mapping, then its graph coincides with the diagonal Δ of the direct product, i.e., with the subset of points of the form (x, x). Thus, we have obtained, in the direct product $M \times M$ of dimension 2n, two compact smooth orientable closed submanifolds Γ and Δ, each of which is of dimension n. Since the sum of their dimensions is equal to the dimension of the enveloping manifold, then, in the "general position," the index of intersection of these manifolds is defined, which we have already considered earlier in Sec. 11. For this we must fix at each point of intersection two n-frames tangent to Γ and Δ, and compare the orientation of the 2n-frame defined by them with the orientation of the enveloping manifold $M \times M$. (We have assumed that n-frames define positive orientations on both submanifolds.) These orientations either coincide or do not and we write +1 or −1 at the point of intersection, accordingly. The sum of these numbers with respect to all points of intersection is the index of intersection of Γ and Δ in $M \times M$. It turns out that this index is equal to the Lefschetz number.

Proposition 18.1. Let Γ be the graph of the smooth mapping $f:M \to M$ and Δ the diagonal (i.e., the graph of the identity mapping $M \to M$). Then, in the general position (i.e., when Γ and Δ intersect transversally in $M \times M$), the Lefschetz number $\ell(f)$ is equal to the index of intersection of Γ and Δ.

Proof. We will make use of the properties of $\ell(g)$ listed above. Since we have assumed the transversality of the intersection of Γ and Δ in $M \times M$, then the number of points

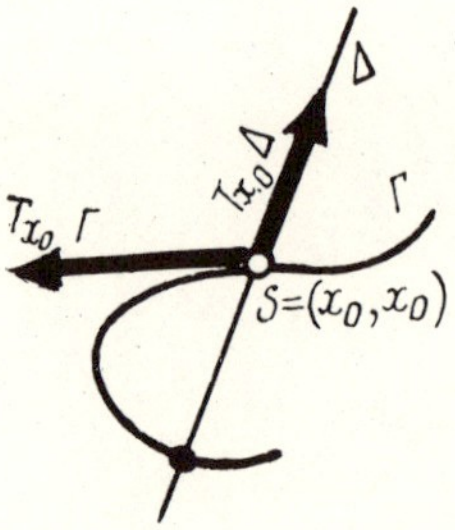

Fig. 203

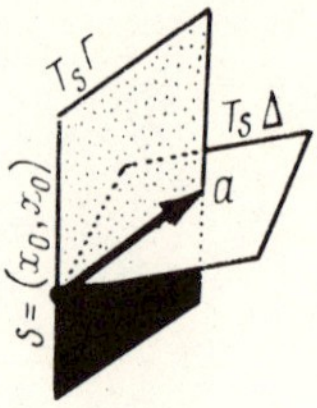

Fig. 204

of intersection is finite and all of them are isolated (due to the compactness of M). At each point of intersection $s = (x_0, x_0)$, tangent spaces $T_{x_0}\Gamma$ and $T_{x_0}\Delta$ intersect only at one point and their sum gives the whole tangent space $T_s(M \times M)$ (see Fig. 203). We claim that the transversality condition for the intersection of Γ and Δ at $s = (x_0, x_0)$ is exactly equivalent to the fact that df has no eigenvalues equal to 1 at x_0. In fact, if $a \in T_{x_0}M$ is an eigenvector of the linear mapping df with eigenvalue 1, then this vector a defines the fixed direction in $T_{x_0}M$, and hence defines the vector (a, a) that belongs to the intersection of $T_s\Gamma$ and $T_s\Delta$, i.e., submanifolds Γ and Δ do not satisfy the transversality condition at s (see Fig. 204). This contradiction proves that df does not have the eigenvalue 1. It remains to be proved that the sign of det (df − E) coincides with the sign defined at the point of intersection of Γ and Δ by two n-frames united into one 2n-frame (see above). Let $x_1, \ldots, x_n$ be regular coordinates in a small neighborhood of the fixed point x_0. Then they define linear nondegenerate coordinates (which will be defined for simplicity by the same letters) on the tangent space $T_{x_0}M$. In the tangent space $T_s(M \times M)$ that splits into the direct sum of two subspaces $T_s\Gamma \oplus T_s\Delta$, the following nondegenerate coordinates arise: $x_1, \ldots, x_n$ in $T_s\Delta$, and $y_1, \ldots, y_n$ in $T_s\Gamma$ (isomorphic to $T_s\Delta$). Thus, the mapping f, being decomposed into a Taylor series in the neighborhood of the fixed point x_0, defines the linear mapping $df: T_s\Delta \to T_s\Gamma$, which is the linearization of f. The action of this mapping is shown in Fig. 205. Here, for convenience, we have established the canonical correspondence between coordinates $x_1, \ldots, x_n$ on $T_s\Delta$ and $y_1, \ldots, y_n$ on $T_s\Gamma$ with respect to the identity mapping E, which is, evidently, the differential of the identity mapping $M \to M$ that defines the diagonal. Thus, the vector df(b) − b is the differ-

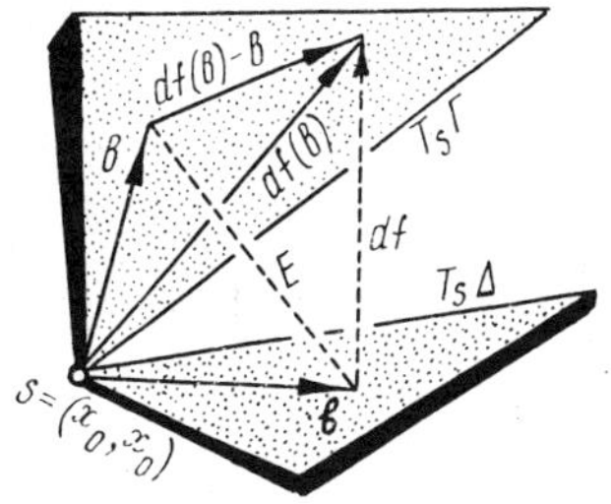

Fig. 205

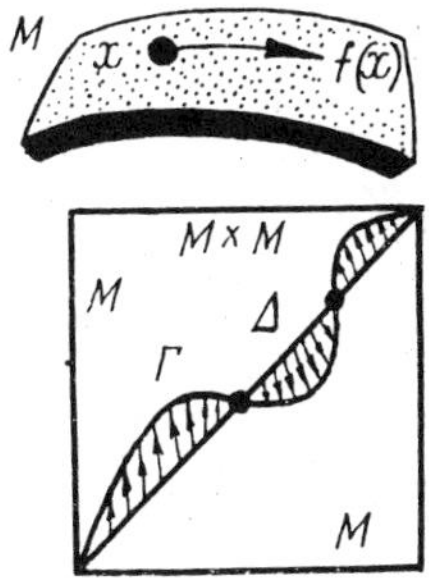

Fig. 206

ference of df(b) and E(b) in $T_s\Gamma$, i.e., it shows the deviation of the linear mapping df from the identity mapping. Thus, we have described the mapping $df: T_s\Delta \to T_s\Gamma$ making use of the endomorphism of $T_s\Gamma$, identifying it with respect to E with the initial space $T_s\Delta$. Now it is evident that the orientation of the 2n-frame is defined by the sign of $\det(df - E)$. We have obtained the definition of the index of a fixed point formulated above in property 5. The statement is proved.

Note an interesting special case. Suppose the smooth mapping f is close to the identity mapping $M \to M$, i.e., it defines a smooth vector field on M (see Fig. 206). Then the graph of this mapping (vector field) is defined by a surface close to Δ. Fixed points of f, i.e., points of intersection of the graph with the diagonal, are evidently zeros of the vector field. Hence, the index of a singular point in the sense of property 5 coincides with the index of a singular point of the vector field. The transversality of the intersection of the graph with the diagonal means here that the singular point of the vector field is nondegenerate (verify!). Thus, Proposition 18.1 implies that the Lefschetz number of such a mapping is equal to the index of the vector field, i.e., to the Euler characteristic of the initial manifold M. If the mapping f is not a "small shift," then surely this interpretation of the Lefschetz number fails, although the fact discovered earlier during the study of the index of a vector field is still preserved: if this index (Lefschetz number) is nonzero, then f has a fixed point.

Now let us return to the study of orbits in a compact Lie group (algebra). Let us proceed directly to the proof of the theorem on conjugacy of maximum tori in a compact group.

As a manifold M mentioned above, take the homogeneous space $\mathfrak{G}/\tilde{T}$ diffeomorphic to a generic orbit in $\mathfrak{G}$ (see above) since the stationary subgroup of a regular element coincides with a maximum torus. Elements of M are conjugate classes $\bar{a} = a\tilde{T}$ with respect to $\tilde{T}$. As a mapping M → M, let us take the mapping f_g generated by the left shift by $g \in \mathfrak{G}$, i.e., $f_g(\bar{a}) = \overline{ga} = ga\tilde{T}$. Here the reader should not confuse the sign "—" with complex conjugation. Thus, for any element $g \in \mathfrak{G}$, we obtain the mapping f_g. Let $\bar{a}_0 \in M$ be a fixed point of f_g. This means that $f_g(\bar{a}_0) = \bar{a}_0$ holds, i.e., $ga\tilde{T} = a_0\tilde{T}$ or $g \in a_0\tilde{T}a_0^{-1}$. Therefore, to prove Theorem 18.1, it suffices to prove that each mapping f_g (for any $g \in \mathfrak{G}$) has at least one fixed point on $\mathfrak{G}/\tilde{T}$. As we already know, this fact may be deduced, for example, from the fact that the Lefschetz number $\ell(f_g)$ of f_g is nonzero [see property 3 of $\ell(f_g)$]. Since $\mathfrak{G}$ is assumed to be arcwise connected, then any two of its points g_1 and g_2 are joined by a continuous path in the group and, hence, any two mappings f_{g_1} and f_{g_2} are homotopic. Hence, by property 2, the Lefschetz numbers coincide. Therefore, it suffices to compute the Lefschetz number for any element $g \in \mathfrak{G}$, which we will choose in a more convenient way. The most appropriate element is the point of $\tilde{T}$ that generates this torus by its powers, i.e., which is, in this sense, a "generator of $\tilde{T}$." During the proof of Corollary 18.4, we already made use of the existence of an element $g \in \tilde{T}$ such that elements g^n for n = 1, 2, ... constitute the dense subset of $\tilde{T}$. Now let us study fixed points of f_g. If $\bar{a}_0$ is a fixed point, i.e., $f_g(\bar{a}_0) = \bar{a}_0$, then $g \in a_0\tilde{T}a_0^{-1}$; hence, $\tilde{T} = a_0\tilde{T}$, i.e., a_0 is contained in the normalizer M of $\tilde{T}$ in $\mathfrak{G}$. A normalizer of a subgroup is the set of elements q of the enveloping group that transforms this subgroup into itself with respect to conjugation $x \to qxq^{-1}$. Elements of N are evidently presentable by automorphisms of $\tilde{T}$, i.e., we may assign to each element $n \in N$ an automorphism $\varphi(n)$, $\varphi(n)h = nhn^{-1}$ of $\tilde{T}$. Thus, we obtain the homomorphism φ of N into the group of automorphisms of T. It is well known that a group of automorphisms of a torus is isomorphic to the group of invertible integer matrices and hence is a discontinuous group. This implies that the homomorphism φ of N into the group of automorphisms of the torus constructed above is constant on the connected component N_0 of the unit of N. Since the normalizer is a closed subgroup in the compact Lie group, then N is the Lie group. Here we will need the following simple lemma.

Lemma 18.5. Let K_0 be the connected component of the unit in the Lie group K. Then K_0 is the normal subgroup in K.

Proof. Let $g \in K$ be an arbitrary element. We must prove that $gsg^{-1} \in K_0$ for any $s \in K_0$. Join the point s with the unit $E \in K_0$ by a continuous path $\gamma(t)$, where $\gamma(1) = s$. Then $g\gamma(0)g^{-1} = gEg^{-1} = E$. Consider the image of γ with respect to the continuous mapping $x \to gxg^{-1}$, where g is fixed. Then γ passes into the new path $g\gamma g^{-1} \subset K_0$ which originates, as before, in the unit of the group and terminates in $g\gamma(1)g^{-1}$, which coincides evidently with gsg^{-1}. Hence, $gsg^{-1} \in K_0$, as required.

Thus, N_0 is the normal subgroup in N.

Lemma 18.6. The group N_0 coincides with the maximum torus $\tilde{T}$.

Proof. Clearly, all elements of N_0 act on $\tilde{T}$ as identity transformations. But this means that $nhn^{-1} \equiv h$, i.e., all elements of N_0 commute with all elements of the maximum torus. Furthermore, $\tilde{T}$ is contained in N_0 (evidently). If $\tilde{T}$ had not coincided with N_0, then the Lie algebra of N_0 would have been bigger than that of $\tilde{T}$ and, hence, we could have found at least one one-parameter subgroup in $\tilde{T}$ that generates together with $\tilde{T}$ the new commutative subgroup greater than $\tilde{T}$. But this is impossible due to maximality of $\tilde{T}$. Hence, $N_0 = \tilde{T}$, as required.

Lemma 18.7. $N/\tilde{T}$ is a finite group.

The proof follows immediately from the two preceding lemmas, since N is compact and $\tilde{T}$ is a normal subgroup in N.

Lemma 18.8. There is a one-to-one correspondence between fixed points of the mapping $f_y: \mathfrak{G}/\tilde{T} \to \mathfrak{G}/\tilde{T}$, where y is a generator of $\tilde{T}$, and elements of $N/\tilde{T}$. In particular, all fixed points of f_y are isolated and there is a finite number of them.

Proof. We have seen above that if $f_y(\bar{a}_0) = \bar{a}_0$, then $a_0 \in N$. Conversely, each element $n \in N$ defines a fixed point of f_y, and it is clear that a change of the element n in the connected component does not affect the fixed point. The lemma is proved.

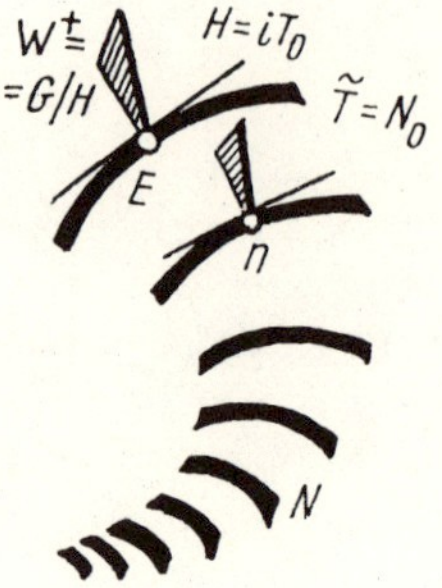

Fig. 207

To compute the Lefschetz number, it suffices to find indices of fixed points of $f_{\bar{y}}$. If $n \in N$, then $\bar{n}$ is a fixed point of f_y, i.e., $f_y(\bar{n}) = \bar{n}$ iff $n\tilde{T}n^{-1} = \tilde{T}$. This means that T is a stationary subgroup of $\bar{n}$ in $\mathfrak{G}/\tilde{T}$. The tangent space to the orbit passing through n (i.e., the tangent space to the manifold $\mathfrak{G}/\tilde{T}$ at $\bar{n}$) is naturally identified with the quotient space G/H, where H is the Cartan subalgebra in G, i.e., the tangent space to $\tilde{T}$ (see Fig. 207). Earlier, while studying embeddings of compact forms into the complex Lie algebra, we established that the compact Lie algebra splits into the sum of subspaces $W^+ \oplus iT_0$, where $iT_0 = H$ (in our notation). Hence, the tangent space to $\mathfrak{G}$ at $n \in N$ also splits into the sum of spaces that are obtained from W^+ and H by the left shift of the unit into n (see Fig. 207). Therefore, the action of df_y on the tangent space to the homogeneous space $\mathfrak{G}/\tilde{T}$ at $\bar{n}$ can be determined from the study of the action induced by df_y on W^+ at E. It remains to find out which linear transformation induces the transformation of df_y on W^+ orthogonal to the Cartan subalgebra H at E. Above we have established that $\tilde{T}$ is realized as the stationary subgroup of $\bar{n}$ in $\mathfrak{G}/\tilde{T}$. Hence, the generator y of $\tilde{T}$ is presentable on the tangent space to $\mathfrak{G}/\tilde{T}$ at $\bar{n}$ by the transformation df_y and, on the other hand, it is presentable on the Lie algebra G by the linear transformation $\mathrm{Ad}\,y: X \to yXy^{-1}$, where $X \in G$, $y \in \tilde{T}$. This transformation Ad y preserves the Killing form and H, since $y \in \tilde{T}$. Hence, Ad y transforms the orthogonal complement W^+ to H in G into itself. It remains to study the action Ad y on W^+. Let us present y in the form exp h, where h is an element of the Cartan subalgebra. Then the differential of Ad y coincides with ad_h. The action of ad_h has already been studied in Sec. 17. In particular, ad_h is presentable by a skew symmetric matrix which is reducible to the block-diagon-

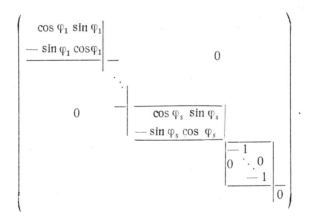

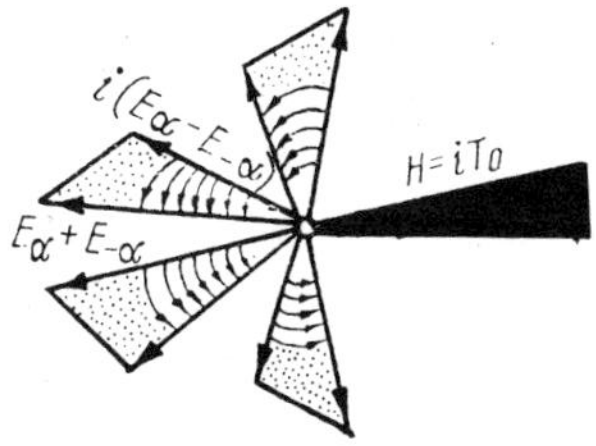

Fig. 208

al form with blocks of size 2 × 2. The space W^+ splits into the direct sum of two-dimensional planes W_α, each of which is spanned by $E_\alpha + E_{-\alpha}$ and $i(E_\alpha - E_{-\alpha})$ (see Sec. 17). In each plane W_α, the transformation Ad_y is therefore an orthogonal transformation (a rotation). We claim that this transformation is not an identity on any of W_α. In fact, if there exists $X \in W^+$ such that $\mathrm{Ad}_y X = X$, then, since y generates all the maximum tori for any $q \in \tilde{T}$, we would have $\mathrm{Ad}_q X = X$, i.e., X would commute with all elements of the torus, hence with all elements of the Cartan subalgebra H, contradicting the maximality of the Cartan subalgebra in G. Hence, Ad_y has on W^+ only the following eigenvalues: −1 (with a certain multiplicity) and a pair of conjugate (complex) eigenvalues φ_α and $i\varphi_\alpha$, where φ_α is real. In terms of an orthogonal rotation of W_α, this means that only rotations by π and by angles different from 0 and 2π are possible (see Fig. 208). If we present Ad_y in the form of a matrix, then it is of the form

$$\begin{pmatrix} \begin{matrix} \cos\varphi_1 & \sin\varphi_1 \\ -\sin\varphi_1 & \cos\varphi_1 \end{matrix} & & & & 0 \\ & \ddots & & & \\ & & \begin{matrix} \cos\varphi_s & \sin\varphi_s \\ -\sin\varphi_s & \cos\varphi_s \end{matrix} & & \\ & & & \begin{matrix} -1 & & 0 \\ & \ddots & \\ 0 & & -1 \end{matrix} & \\ 0 & & & & 0 \end{pmatrix}.$$

Proceeding to the computation of det $(df_y - E)_-$, we get $\det(df_y - E_-) = (-1)^m$, where m is an integer. Here, since y is a generator of a maximum torus, the number m does not depend on the fixed point n, where we compute the determinant. Hence, the Lefschetz number of f_y is nonzero and equals $(-1)^m d$, where d is the number of fixed points of f_y. This implies that the module of the Lefschetz number is equal to the order of $N/\tilde{T}$ (see Lemma 18.8).

18.4. The Weyl Group and Its Relationship to Orbits

Let $\tilde{T}$ be a maximum torus in the compact Lie group $\mathfrak{G}$ and N its normalizer. Then, as was proved in the preceding section, the finite group $\Phi = N/\tilde{T}$ is defined. If $\mathfrak{G}$ is a compact semisimple Lie group, then Φ is called the Weyl group. The number of elements in the Weyl group is equal to the number of points of intersection of a generic orbit with a maximum torus (see Fig. 199). This is a consequence of the following statement.

<u>Statement 18.1</u>. Let $t \in \tilde{T}$ be a regular element of $\mathfrak{G}$. Then its orbits $O(t)$ intersect with the maximum torus $\tilde{T}$ at a finite number of points and this set of points of intersection is in turn the orbit of t under the action of the Weyl group Φ on $\tilde{T}$. The number of points in $\Phi(t)$ is equal to the order of Φ.

<u>Proof</u>. Clearly, the orbit of t under the action of the Weyl group belongs to the intersection of $O(t)$ with $\tilde{T}$. We must prove the converse inclusion $O(t) \cap \tilde{T} \subset \varphi(t)$, where $\varphi \in \Phi$. This is a consequence of the following lemma.

<u>Lemma 18.9</u>. If two elements of a maximum torus are conjugate, then they are transformed into each other by elements of the Weyl group.

<u>Proof</u>. In fact, we will prove a more general statement with other useful consequences besides the one mentioned above. Let $\mathfrak{G}$ be a connected compact group and P a subset in the maximum torus $\tilde{T}$ such that $gPg^{-1} \subset \tilde{T}$ for some $g \in \mathfrak{G}$ (see Fig. 209). Then there is an element $n \in N$ such that $npn^{-1} = gpg^{-1}$ for any $p \in P$. This means that the conjugation that transforms a part of the torus into the torus may be realized by elements of the normalizer (and, finally, by ele-

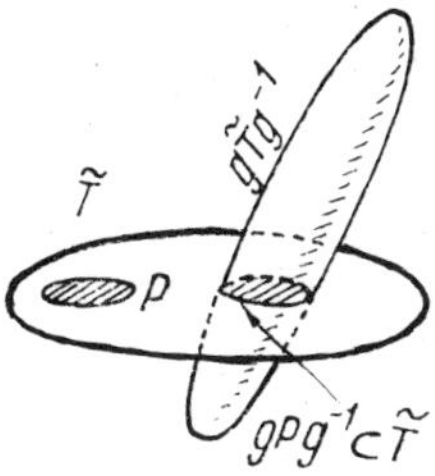

Fig. 209

ments of the Weyl group). Let R be the subgroup in $\mathfrak{G}$ that consists of all elements that commute with elements of P. Let R_0 be the connected component of the unit of R. Clearly, R_0 is the connected compact Lie group that contains a maximum torus. Consider the conjugation of this torus by the element $g \in \mathfrak{G}$ mentioned in the formulation of our statement (see Fig. 209). Since the set gPg^{-1} belongs to $\tilde{T}$, all elements of gPg^{-1} commute with $\tilde{T}$. This is equivalent to the fact that all points of P commute with all points of the "shifted torus," i.e., with points of $g^{-1}\tilde{T}g$ (see Fig. 209). In fact, if $p \in P$, then $(gPg^{-1})t = t(gPg^{-1})$, where $t \in \tilde{T}$; hence $P(g^{-1}tg) = (g^{-1}tg)P$, as required. Therefore, the "shifted torus" $g^{-1}\tilde{T}g$ commutes with all elements of P, i.e., it is contained in R_0. Thus, R_0 contains two commutative subgroups: the initial torus $\tilde{T}$ and the "shifted torus" $g^{-1}\tilde{T}g$. But both these subgroups are maximum tori in the enveloping group $\mathfrak{G}$ and, moreover, they are the maximum commutative subgroups in a smaller group R_0. Hence, by Theorem 18.1, they are conjugate in R_0, i.e., there exists an $r \in R_0$ such that $rg^{-1}\tilde{T}gr^{-1} = \tilde{T}$, i.e., $n = gr^{-1}$. This implies that $npn^{-1} = gmg^{-1}$, since $npn^{-1} = gr^{-1}prg^{-1} = gpg^{-1}$ because $r \in R_0$, i.e., $rp = pr$. The lemma is proved, completing the proof of Statement 18.1.

We leave as an exercise to the reader the proof of the following useful facts:

1) Let S be a connected closed commutative subgroup (i.e., a torus) in the connected compact Lie group $\mathfrak{G}$, and $g \in \mathfrak{G}$ an element commuting with all elements of S. Then, in $\mathfrak{G}$, there is a maximum torus that contains both S and g.

2) Any element of a compact connected group commuting with all elements of the maximum torus belongs to this torus.

3) The center of a compact connected group is contained in each maximum torus of the group, i.e., belongs to the intersection of all maximum tori. Note that the center of a compact connected semisimple group is discontinuous (moreover, it is finite).

4) The Weyl group acts effectively on a maximum torus, i.e., if any of its elements preserves all points of the torus, then this element is the unit in the Weyl group.

5) Any maximum torus in the connected compact group is at the same time a maximum commutative subgroup. This follows from Statement 2 (see above).

6) Let $\mathfrak{G}$ be a simply connected, closed connected simple Lie group, i.e., one of the groups listed above (see the classification of simple Lie groups and Lie algebras and their compact forms). Then the center $Z(\mathfrak{G})$ is an Abelian group:

$$Z(A_n) = \mathbf{Z}_{n+1},\ Z(B_n) = \mathbf{Z}_2,\ Z(C_n) = \mathbf{Z}_2,\ Z(D_{2n}) = \mathbf{Z}_2 \oplus \mathbf{Z}_2,$$
$$Z(D_{2n+1}) = \mathbf{Z}_4,\ Z(G_2) = 0,\ Z(F_4) = 0,\ Z(E_6) = 0,$$
$$Z(E_7) = \mathbf{Z}_2,\ Z(E_8) = 0.$$

Note that if $D \subset Z(\mathfrak{G})$ is a subgroup of the center, then the quotient group $\mathfrak{G}/D$ is also a Lie group, and Lie algebras of $\mathfrak{G}$ and $\mathfrak{G}/D$ are evidently isomorphic, since the centers listed above are discontinuous (even finite) subgroups of $\mathfrak{G}$. Lie groups of this kind are called locally isomorphic. If D is a nontrivial subgroup, then $\mathfrak{G}$ and $\mathfrak{G}/D$ are not homeomorphic (as Lie groups) but have the same Lie algebras. Compact simple Lie groups of series G_2, F_4, E_8 do not have other locally isomorphic groups different from the simply connected ones since their centers are trivial. In the series of locally isomorphic compact Lie groups, two groups are naturally distinguished: the simply connected group $\mathfrak{G}$, i.e., "maximally unfolded" group and the quotient group $\mathfrak{G}/Z(\mathfrak{G})$, i.e., "maximally curled" group. As is known (see, e.g., [2], pp. 551-554), the fundamental group of the base of the regular cover is isomorphic to the monodromy group, i.e., to the fiber of the cover. In our case, the natural projection $\pi: \mathfrak{G} \to \mathfrak{G}/Z(\mathfrak{G})$ is a regular covering whose fiber is $Z(\mathfrak{G})$. Hence, the fundamental group of the base, i.e., of $\mathfrak{G}/Z(\mathfrak{G})$, is isomorphic to $Z(\mathfrak{G})$.

In conclusion, consider the following example: the Weyl group of the unitary group U_n, i.e., of the group of unitary $n \times n$ matrices. As we have established earlier, the subgroup of diagonal matrices $\begin{pmatrix} e^{i\varphi_1} & & 0 \\ & \ddots & \\ 0 & & e^{i\varphi_n} \end{pmatrix}$ is a maximum commutative subgroup (maximum torus) $\tilde{T}$ in U_n. Let N be a normalizer of this maximum torus in U_n. Each matrix $t \in \tilde{T}$ is depicted by the unitary linear transformation in the complex space $\mathbf{C}^n$ endowed with an Hermitian basis $e_1, \ldots, e_n$, invariant with respect to t. Each vector e_s is an eigenvector corresponding to the eigenvalue $e^{i\varphi_s}$. Any matrix q of N transforms any line ℓ invariant with respect to $\tilde{T}$ into the line $\ell' = q(\ell)$ with the same property. In fact, $t(\ell') = q(q^{-1}tq)(\ell) = q(\ell) = \ell'$ for $t \in \tilde{T}$, since $q^{-1}tq \in \tilde{T}$. The unique lines that are invariant with respect to all transformations of the torus are coordinate axes generated by $\{e_i\}$. Thus, any transformation q of N is of the form $e_i \to \lambda_i e_{\sigma(i)}$, where λ_i is a number and σ is a permutation of n elements $(1, 2, \ldots, n)$. Thus, we have explicitly found the Weyl group, since the quotient of N with respect to $\tilde{T}$ yields the permutation group $e_i \to e_{\sigma(i)}$.

Chapter 5

SYMPLECTIC GEOMETRY

19. SYMPLECTIC MANIFOLDS

19.1. The Symplectic Structure and Its Canonical Presentation. The Skew Symmetric Gradient

In this section, we study an important class of smooth manifolds, the so-called symplectic manifolds. They appear in various applied problems, for example in problems of classical mechanics, and, therefore, their investigation is necessary for solving various concrete problems. We are already acquainted with Riemannian manifolds (see [1]), i.e., with manifolds endowed with a nondegenerate symmetric scalar product in tangent spaces (usually we consider positive-definite Riemannian matrices). Another method to introduce an extra structure on a smooth manifold is to define a skew symmetric scalar product that smoothly depends on a point. This leads us to the symplectic manifolds whose geometry is essentially different from the geometry of Riemannian spaces. Since a skew symmetric scalar product (in tangent spaces) is defined by a skew symmetric tensor of rank 2, then to define it it suffices to define an exterior differential form of degree 2.

Definition 19.1. The smooth even-dimensional manifold is called symplectic if an exterior differential form $\omega = \sum_{i<j} \omega_{ij}\, dx_i \wedge dx_j$ of degree 2 is defined on it such that 1) ω is nondegenerate, i.e., the matrix of its coefficients

$(\omega_{ij}(x))$ is nondegenerate at all points; 2) ω is closed, i.e., $d\omega = 0$, where d is the exterior differential (see [1], Chapter 6).

This form ω is sometimes called the symplectic structure on the manifold.

Clearly, ω defines on T_xM^{2n} the nondegenerate skew symmetric scalar product

$$\omega(a,b) = \sum_{i<j} \omega_{ij} a_i b_j, \quad \text{where} \quad a = (a_i),\ b = (b_j),\ a, b \in T_xM.$$

If a point x_0 is fixed, then, as we know from algebra, there is a coordinate transform in the tangent space T_xM generated by a local regular coordinate transform in the neighborhood of x_0 such that the matrix $(\omega_{ij}(x_0))$ is reduced to the canonical form

$$\left\| \begin{array}{ccc} \begin{array}{cc} 0 & 1 \\ -1 & 0 \end{array} & & 0 \\ & \ddots & \\ 0 & & \begin{array}{cc} 0 & 1 \\ -1 & 0 \end{array} \end{array} \right\|.$$

This scalar product evidently defines the canonical identification of the tangent space T_xM and the cotangent one T_x^*M (this identification is considered in greater detail in [1], Chapter 5). Recall that the dual space T_x^*M consists of covectors – linear forms on the tangent space. The presence of the canonical identification of the tangent and the cotangent spaces generated by ω enables one to define the important operation which is an analog of the gradient, i.e., the vector field grad f on a manifold with a symmetric scalar product (Riemannian metric).

Definition 19.2. Let f be a smooth function on M and ω a symplectic structure. The skew symmetric gradient of sgrad f is the smooth vector field sgrad f on M, uniquely recovered from the relation $\omega(v, \text{sgrad } f) = v(f)$, where v runs over the set of all smooth vector fields on M and v(f) is the value of the vector field v at f.

In other words, we must first consider the covector $\left(\frac{\partial f}{\partial x_1}, \ldots, \frac{\partial f}{\partial x_n}\right) \in T_x^* M$ and then, due to the canonical identification of TM and T*M with respect to ω (see above), construct the vector field corresponding to this covector field. This vector field is the skew symmetric gradient. If we had used a Riemannian metric for such an identification, we would have obtained the vector field grad f. The uniqueness of the construction of sgrad f follows from the nondegeneracy of ω. Usually, local coordinates on a symplectic manifold are denoted as follows: $p_1, \ldots, p_n, q_1, \ldots, q_n$, and these coordinates may be chosen so that, at the fixed point x_0, the matrix $(\omega_{ij}(x_0))$ is of the form $\begin{pmatrix} 0 & E \\ -E & 0 \end{pmatrix}$, where E is the unit $(n \times n)$-matrix. If we order the coordinates as $p_1, q_1, p_2, q_2, \ldots, p_n, q_n$, then the matrix $(\omega_{ij}(x_0))$ will be written in the block-diagonal form mentioned above. With respect to such a special choice of local coordinates $p_1, \ldots, p_n, q_1, \ldots, q_n$, the vector field sgrad f at x_0 is expressed especially simply. In fact, since

$$\left(\frac{\partial f}{\partial p_1}, \ldots, \frac{\partial f}{\partial q_n}\right) \in T_{x_0} M,$$

then

$$\operatorname{sgrad} f = \left(\frac{\partial f}{\partial q_1}, \ldots, \frac{\partial f}{\partial q_n}, -\frac{\partial f}{\partial p_1}, \ldots, -\frac{\partial f}{\partial p_n}\right).$$

In coordinates $p_1, q_1, \ldots, p_n, q_n$, we have

$$\operatorname{sgrad} f(x_0) = \left(\frac{\partial f}{\partial q_1}, -\frac{\partial f}{\partial p_1}, \ldots, \frac{\partial f}{\partial q_n}, -\frac{\partial f}{\partial p_n}\right).$$

As we know, a Riemannian metric can also be reduced at one point to the canonical (diagonal) form by an appropriate choice of local coordinates. In this sense, both structures, i.e., the Riemannian and the symplectic, are similar. However, there is a serious difference between them that develops when we begin to consider the whole neighborhood of x_0. As we know (see [1]), a Riemannian metric cannot be, in the general case, transformed by a coordinate transform to the diagonal form in the whole neighborhood because a nonzero tensor of Riemannian curvature may prove to be an obstruction. Unlike a symplectic structure, it is always reducible to the canonical form by a coordinate transform in a sufficiently small neighborhood of a point (the size of a neighborhood is defined by the properties of the form).

Proposition 19.1. Let ω be a symplectic structure on M^{2n}. Then for any point $x_0 \in M$, there is an open neighborhood with local coordinates $p_1, \ldots, p_n, q_1, \ldots, q_n$ such that, in these coordinates, ω is expressed in the simplest canonical way: $\omega = \sum_{i=1}^{n} dp_i \wedge dq_i$.

For the proof, see, for example, [20]. This theorem is useful in various computations related to symplectic structures. Local coordinates, whose existence is presumed in the theorem, that reduce ω to the canonical form are sometimes called symplectic coordinates. Clearly, covering the manifold M by open neighborhoods of the form mentioned in Proposition 19.1, we obtain a special kind of an atlas (see the definition of an atlas in [1], Chapter 3) which is sometimes called symplectic. The simplest example of a symplectic manifold is the Euclidean space $\mathbf{R}^{2n}(p_1, \ldots, q_n)$, endowed with the form $\omega = \sum_{i=1}^{n} dp_i \wedge dq_i$. Another example is a smooth two-dimensional orientable closed Riemannian manifold, i.e., a sphere with handles. Here, as a symplectic structure, we may take the standard form of the two-dimensional Riemannian volume which is a closed nondegenerate exterior 2-form (see [1]). As we already know, in a neighborhood of any point, local coordinates p, q may be chosen so that this form will be written as $\sqrt{(g)}dp \wedge dq$, where $g = \det(g_{ij})$ and g_{ij} is the metric tensor. Now we give one more important example of a symplectic manifold. Let M^n be a smooth manifold and T^*M its cotangent bundle, i.e., the point of T^*M is a pair (x, ξ), where $x \in M$, $\xi \in T_x{}^*M$, i.e., ξ is a covector at x. It is easy to verify that T^*M is a smooth 2n-dimensional manifold. The natural projection $p: T^*M \to M$ is defined as $p(x, \xi) = x$. Clearly, T^*M turns into the fiber bundle with M as the base and the cotangent space $T_x{}^*M$ as the fiber $p^{-1}(x)$ over x. On T^*M we define a symplectic structure. For this, let us first construct on T^*M a smooth 1-form $\omega^{(1)}$. Let $a \in T_y(T^*M)$ be a vector tangent to the cotangent bundle T^*M at $y \in T_x{}^*M$ (see Fig. 210a). The differential of the mapping $p: T^*M \to M$ transforms this vector a into the vector $p_* a$ tangent to M at $x = p(y) = p(x, \xi)$.

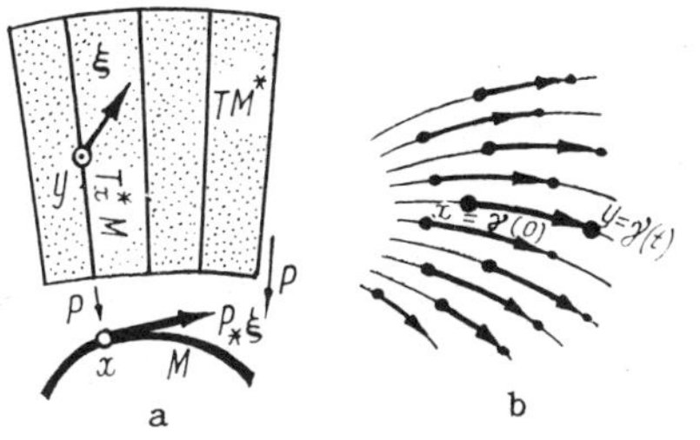

Fig. 210

Now, define the differential 1-form $\omega^{(1)}$ on T*M by the equality $\omega^{(1)}(a) = \langle y, p_*a\rangle$, i.e., the value of the form is equal to the value of y in p_*a. Finally, for the required 2-form ω, let us take the exterior differential of $\omega^{(1)}$, i.e., $\omega = d\omega^{(1)}$. It remains as a useful exercise for the reader to prove that this 2-form is closed and nondegenerate, i.e., that T*M is a symplectic manifold.

19.2. Hamiltonian Vector Fields

Definition 19.3. A smooth vector field v on a symplectic manifold M with the form ω is called Hamiltonian if it is of the form v = sgrad F, where F is a smooth function on M called the Hamiltonian.

In special symplectic coordinates (p_i, q_i) a Hamiltonian vector field is of the form $(\partial F/\partial q_i, -\partial F/\partial p_i)$ (see Sec. 19.1 above). Hamiltonian vector fields (sometimes called Hamiltonian flows) admit another important description in the language of one-parameter groups of diffeomorphisms of M generated by them. Let v be a Hamiltonian field and $\mathfrak{G}^v$ a one-dimensional group of diffeomorphisms of M presented by shifts along integral trajectories of v. This means that $\mathfrak{G}^v$ consists of transformations g_t acting on M as follows: $g_t(x) = y$, where $x = \gamma(0)$, $y = \gamma(t)$, and γ is the integral trajectory of v passing through x at t = 0 (see Fig. 210b). In other words, the diffeomorphism g_t shifts each point x by the time t along γ. Since ω is defined on N, then g_t transforms this form into the new form $(g_t^*\omega)(x) = \omega(g_t(x))$. Hence, a derivative of ω along the vector field v equal to $(d/dt)(g_t^*\omega)$ is defined.

Definition 19.4. The vector field v on the symplectic manifold M is called locally Hamiltonian if this field is preserved by the symplectic structure ω on M, i.e., the derivative of ω with respect to the vector field v vanishes: $(d/dt)(g_t{}^*\omega) = 0$.

In other words, ω is invariant with respect to all transformations of the form g_t generated by v, i.e., it is invariant with respect to the action of the one-parameter group $\mathfrak{G}^v$. The term "locally Hamiltonian field" is due to the following fact.

Proposition 19.2. A smooth vector field v on the symplectic manifold M is locally Hamiltonian iff for any $x \in M$ there exists a neighborhood U(x) of x and a smooth function H_U defined on this neighborhood such that v = sgrad H_U, i.e., in the neighborhood U v is a Hamiltonian field with the Hamiltonian H_U.

For the proof, see [1], pp. 432-433. Clearly, any Hamiltonian field on M is locally Hamiltonian. The converse is false, i.e., a field presentable in the form sgrad H_U on each neighborhood U may admit no global presentation in the form sgrad F, where F is a smooth function defined on the whole manifold. In other words, local Hamiltonians H_U defined on selected neighborhoods U may not "glue" into one smooth function F defined on the whole M. However, in what follows, we will mostly study Hamiltonian fields defined on the whole manifold and of the form sgrad F, where F is a Hamiltonian defined on the whole M.

Problem. Prove that a vector field on a two-dimensional Riemannian manifold is locally Hamiltonian if the divergence of the field vanishes, i.e., this field describes a flow of an incompressible fluid on a two-dimensional surface.

19.3. The Poisson Bracket and Integrals of Hamiltonian Fields

Definition 19.5. The Poisson bracket of two smooth functions f and g on the symplectic manifold M is a smooth function {f, g} defined by the formula

$$\{f, g\} = \omega(\text{sgrad}\, f, \text{sgrad}\, g) = \sum_{i<j} \omega_{ij} (\text{sgrad}\, f)_i (\text{sgrad}\, g)_j.$$

In other words, we must compute the skew symmetric scalar product of skew gradients of f and g. In the explicit form, the Poisson bracket of f and g is expressed as follows:

$$\{f, g\} = \sum_{i<j} \omega^{ij} \frac{\partial f}{\partial x_i} \frac{\partial g}{\partial x_j},$$

where x_i are local coordinates and ω^{ij} are coefficients of the matrix opposite to (ω_{ij}). Here we made use of the fact that $(\mathrm{sgrad}\, f)_i = \sum_j \omega^{ij} \frac{\partial f}{\partial x_j}$. The following statement is easily proved according to the definition (see, for example, [1, 20]).

Proposition 19.3. The Poisson bracket f, g → {f, g} satisfies the following relations: 1) { , } is a bilinear operation; 2) { , } is skew symmetric, i.e., {f, g} = − {g, f}; 3) the Jacobi identity {h, {f, g}} + {g, {h, f}} + {f, {g, h}} = 0 holds for any f, g, h.

Thus, the linear infinite-dimensional space F(M) of smooth functions F on the smooth symplectic manifold M is naturally transformed into the infinite-dimensional Lie algebra over R. Note that the existence of this Lie algebra is due to the existence of the skew gradient since the usual gradient does not generate any Lie algebra. This is another difference between skew symmetric scalar products (symplectic structures) and symmetric ones (Riemannian metrics). In what follows we will often be interested in different finite-dimensional subalgebras in F(M). Let us construct the natural mapping α of F(M) into the Lie algebra V(M) of all smooth vector fields on M. Define this mapping by the formula α(f) = = sgrad f.

Lemma 19.1. The mapping α:F(M) → V(M) is a homeomorphism of Lie algebras, i.e., α{f, g} = [α(f), α(g)]. This means that α transforms the Poisson bracket into the usual commutator of vector fields α(f) and α(g).

The proof follows from Proposition 19.3 and the definition of sgrad (verify!).

The image of the Lie algebra F(M) in V(M) is a subalgebra that will be denoted by H(M). Its elements are vector fields on M presentable in the form sgrad f, i.e., (in our previous terminology) the Hamiltonian fields. Thus, H(M) is a subalgebra consisting of Hamiltonian fields on M. This means, in particular, that the usual commutator of two Hamiltonian fields is again a Hamiltonian field. The Hamiltonian of the commutator is obtained as the Poisson bracket of two initial Hamiltonians of commuted fields. Note that the mapping $\alpha: F(M) \to H(M)$ is an epimorphism, but it is not a monomorphism since α has a nonzero kernel. If the manifold is connected, then the kernel consists of functions constant on the manifold. Hence, Ker α is one-dimensional and $H(M) = F(M)/\text{Ker}\,\alpha = F(M)/\mathbf{R}^1$. Surely, the subalgebra H(M) in V(M) depends on the choice of a symplectic structure in M. If we change the form ω, then the subalgebra H(M) will change; therefore, we should rigorously write $H_\omega(M)$, but we will omit the index ω, assuming it to be implied. The Poisson bracket of f and g has the following simple interpretation: $\{f, g\} = (\text{sgrad}\, f)g$, i.e., it coincides with the derivative of g with respect to the vector field sgrad f. This follows from the definition:

$$\omega(v, \text{sgrad}\, g) = v(g) = \sum_i v_i \frac{\partial f}{\partial x_i}.$$

This simple observation is extremely useful in the study of integrals of Hamiltonian fields.

Definition 19.6. A smooth function f on the manifold M is called an integral of the vector field v if this function is constant along all integral trajectories of v. In other words, the derivative of f along v must vanish (see Fig. 211).

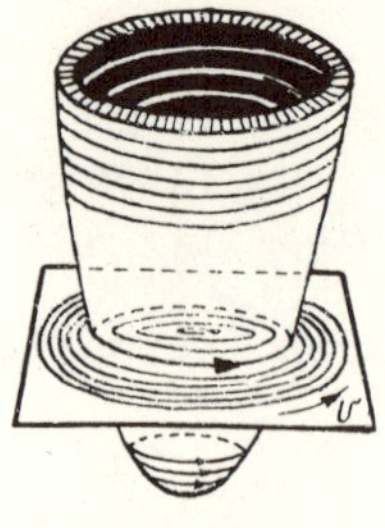

Fig. 211

Proposition 19.4. Let $v = \text{sgrad } F$ be a Hamiltonian field on M and f a smooth function that commutes (with respect to the Poisson bracket) with the Hamiltonian of v, i.e., $\{f, F\} = 0$. Then f is the integral of v.

Proof. Computing the derivative of f along v, we have $v(f) = \omega(v, \text{sgrad } f) = \omega(\text{sgrad } F, \text{sgrad } f) = \{F, f\} = 0$. The proposition is proved.

In particular, F, i.e., the Hamiltonian, is always an integral of v, since $\{F, F\} \equiv 0$ due to the skew symmetry of the Poisson bracket. Thus, any Hamiltonian field has at least one integral, i.e., its Hamiltonian. From here we get the following property of Hamiltonian fields: the integral trajectory of the field cannot be dense on the manifold since it always remains on the hypersurface defined by the equation $F = \text{const}$. Therefore, if a field has an integral trajectory that is dense in an open domain, this field cannot be Hamiltonian. Thus, a Hamiltonian field preserves the foliation of the manifold into hypersurfaces $F = \text{const}$.

Proposition 19.5. If f and g are two integrals of the Hamiltonian vector field $v = \text{sgrad } f$, then so is their Poisson bracket $\{f, g\}$.

Proof. From the Jacobi identity, we get $\{F, \{f, g\} = -\{g, \{F, f\}\} - \{f, \{g, F\}\} = 0$.

This proposition enables one to construct, in principle, new integrals of a Hamiltonian field if two such integrals are known. However, this method of constructing integrals is not always successful, since $\{f, g\}$ can be functionally dependent on the initial functions f and g (i.e., we will obtain nothing new). Here we came close to the problem of finding as many integrals of a Hamiltonian field as possible. Each such field defines a system of ordinary differential equations of order 2n on the manifold M since, in local coordinates $x_1, \ldots, x_{2n}$, the field v is expressed as $\dot{x}_i = v_i(x_1, \ldots, x_{2n})$, where v_i are components of v and $\dot{x}_i$ are derivatives with respect to time t. Integral trajectories of v are solutions of the corresponding system. Thus, if v has an integral f, then we can reduce the order of the system by one, restricting the field onto hypersurfaces along which v "flows." Now suppose that the field v has two integrals f and g so that f and g are functionally independent on the manifold. It is convenient to formulate the condition of in-

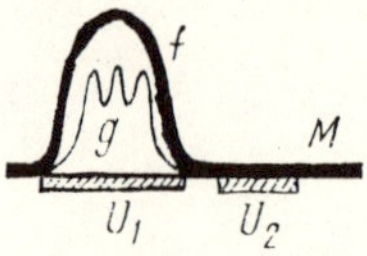

Fig. 212

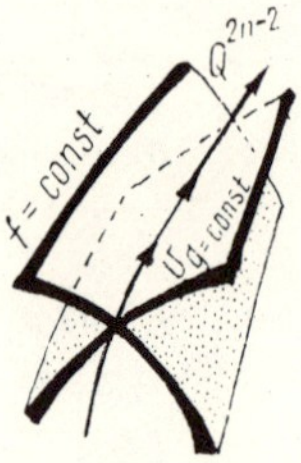

Fig. 213

dependence in terms of the gradients of these functions. In fact, the following simple statement holds.

If two fields grad f and grad g are linearly independent at each point of an open dense subset $U \subset M$, then f and g are functionally independent on the manifold. In what follows, we will deal with polynomial functions on algebraic varieties; therefore, we require that U be dense and open in M. Working with smooth functions, it is necessary to bear in mind that, in this case, a pair of functions can be functionally independent on one domain in M and, on the contrary, dependent on another domain in M (see Fig. 212). For polynomials, these situations are surely impossible.

If f and g are two functionally independent integrals, then they define two different foliations of the manifold M onto hypersurfaces f = const and g = const and, in general position, these hypersurfaces intersect transversely in the submanifold Q of dimension 2n − 2. Since v preserves f and g, then v is tangent to Q^{2n-2} (see Fig. 213). Hence, v can be restricted onto the family of invariant subspaces of dimension 2n − 2 and we have reduced the order of the system by two. In the general case, if we have found r functionally independent integrals of the field, we can reduce the order of the initial system by r and reduce the system to the problem of invariant (with respect to the field v) surfaces of dimension 2n − r. In the ideal case, to solve the system we must find 2n − 1 functionally independent integrals. In this case, the mutual level surfaces of the set $f_1, \ldots, f_{2n-1}$ would be one-dimensional trajectories that coincide with integral trajectories of the initial system. Formally, we need one more function in order to define the points along these trajectories. In other words, we would have integrated the system completely in the sense that we would have presented all of its solutions as mutual level surfaces of the known functions $f_1, \ldots, f_{2n-1}$.

However, in real mechanical systems, such a situation is so rare that we must not rely on the existence of such a set of independent functions in practice. Therefore, we must sometimes be content with a "partial integrability" which, in different problems, assumes various forms. What must be required of such "partial integrability" is clear. We would like to find a set of independent functions $f_1, \ldots, f_r$ on the manifold such that their common level surfaces (along which v flows) are quite simple, for example, all of them (or almost all) would be diffeomorphic to a well-investigated manifold. Then the problem of description of solutions of the system splits into two steps:

1) First we produce r integrals which enable us to describe their common level surfaces Q^{2n-r}.

2) Furthermore, we try to describe the motion of the system (i.e., the motion of points along integral trajectories) on these level surfaces.

A remarkable fact is the existence, for various concrete systems, of sets of integrals that make it possible to realize the theoretical program described above. One of the most famous results of this kind is the Liouville theorem.

19.4. The Liouville Theorem (Commutative Integration of Hamiltonian Systems)

Definition 19.7. We say that two smooth functions f and g on a symplectic manifold are in involution if their Poisson bracket vanishes.

As we have seen for the complete integration of a system, we must know $2n - 1$ integrals of the system. It turns out that for Hamiltonian systems it suffices to know only n functionally independent integrals (where 2n is the dimension of M) that are in involution. In this case, each integral "is counted as 2," i.e., it enables one to reduce the order of the system each time not by one but by two immediately. Moreover, in this case, the initial system is integrated in "quadratures."

Theorem 19.1. Let, on the symplectic manifold M^{2n}, the set of n smooth functions $f_1, \ldots, f_n$ in involution be given, i.e., $\{f_i, f_j\} \equiv 0$ for $1 \leqq i, j \leqq n$. Let M_ξ be a mutual level surface of functions (f_i), i.e., $M_\xi = \{x \in M, f_i(x) = \xi_i, 1 \leqq i \leqq n\}$. Suppose that on this level surface all n functions $f_1, \ldots, f_n$ are functionally independent (i.e., gradients grad f_i, where $1 \leqq i \leqq n$, are linearly independent at all points of M_ξ). Then:

1) M_ξ is a smooth n-dimensional submanifold invariant with respect to each vector field v_i = sgrad f_i whose Hamiltonian is f_i.

2) If M_ξ is connected and compact, it is diffeomorphic to an n-dimensional torus T^n. In the general case, the connected nonsingular variety M_ξ (now not necessarily compact) is the quotient space of the Euclidean space $\mathbf{R}^n$ with respect to a lattice of rank no greater than n.

3) If M_ξ is compact and connected (i.e., is a torus), then in one of its open neighborhoods regular curved coordinates $s_1, \ldots, s_n, \varphi_1, \ldots, \varphi_n$, where $0 \leqq \varphi_i < 2\pi$, may be introduced (called an "action-angle") such that:

a) the symplectic structure ω in these coordinates is expressed in the simplest way, i.e., $\omega = \sum_{i=1}^{n} ds_i \wedge d\varphi_i$ or equivalently, functions $s_1, \ldots, s_n, \varphi_1, \ldots, \varphi_n$ satisfy the following relations: $\{s_i, s_j\} = \{\varphi_i, \varphi_j\} = 0$, $\{s_i, \varphi_j\} = \delta_{ij}$;

b) functions $s_1, \ldots, s_n$ are coordinates in the direction transversal to the torus and are functionally expressed in terms of integrals $f_1, \ldots, f_n$, i.e., $s_i = s_i(f_1, \ldots, f_n)$ for $1 \leqq i \leqq n$;

c) functions $\varphi_1, \ldots, \varphi_n$ are coordinates on the torus $T^n = S^1 \times \ldots \times S^1$, where φ_i is an angle coordinate on S^1 of number i, where $0 \leqq \varphi_i < 2\pi$;

d) each vector field v = grad F, where F is any of $f_1, \ldots, f_n$, being expressed in terms of coordinates $\varphi_1, \ldots, \varphi_n$ on T^n, are of the form $\dot{\varphi}_i = q_i(\xi_1, \ldots, \xi^n)$, i.e., components of this field are constant on the torus and integral trajectories on the field define the conditionally periodic motion of the system v, i.e., define the "straight winding" of the

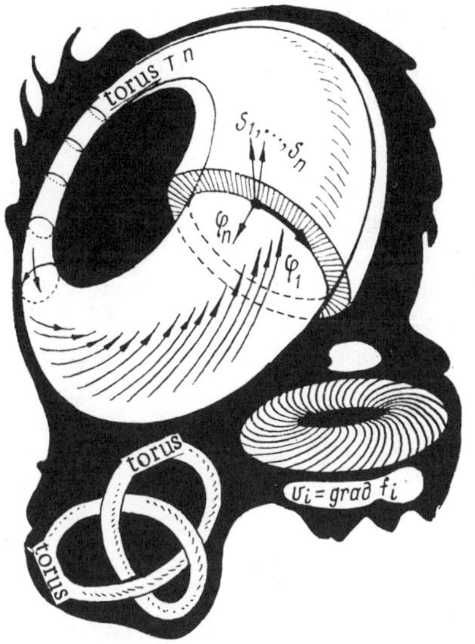

Fig. 214

torus T^n. (Here functions q_i, where $1 \leqq i \leqq n$, are defined in a neighborhood of a torus, and on nearby level surfaces we also have $\dot{\varphi}_i = q_i(s_1, \ldots, s_n)$.)

Thus, the initial system v = sgrad F is described in a neighborhood of T^n in coordinates $s_1, \ldots, \dot{\varphi}_n$ as $\dot{s}_i = 0$, $\dot{\varphi}_i = q_i(s_1, \ldots, s_n)$, $1 \leqq i \leqq n$ (see Fig. 214).

We see that if functions $f_1, \ldots, f_n$ are independent on M, then all nonsingular level surfaces M_ξ, i.e., such that the gradients of the functions are independent (in particular nonzero on them), are diffeomorphic to the same simple manifold, i.e., to an n-dimensional torus. Surely, the embedding of this torus into the enveloping manifold may be quite complicated (see Fig. 214). However, this "complicatedness" may be studied from the information on $f_1, \ldots, f_n$ which is known. Furthermore, on the torus the vector field grad f is as simple as possible since, with respect to angle coordinates $\varphi_1, \ldots, \varphi_n$, this field has constant coefficients, i.e., it is completely defined by the velocity vector at one point of the torus. In the general position, each trajectory of the field defines a dense winding of the torus.

The importance of this theorem is due to the fact that, as we will show below, various interesting mechanical systems and their analogs admit such a "Liouville integration." If we are given the concrete system v = sgrad, where F is a function on M, then we will say that this system is "completely integrable in a commutative (Liouville) sense" or "admits complete commutative integration" if there is a set of functions $f_1 = F$, f_2, ..., f_n satisfying the conditions of Theorem 19.1. In this case, the system v defines a conditionally periodic motion along tori of halved dimension. Below we will acquaint ourselves with the so-called "noncommutative integration." Thus, if the Hamiltonian F is fixed, then the first problem is the search for $n - 1$ more functions that generate, together with F, a functionally independent set of functions commuting with respect to the Poisson bracket. It is equivalent to the inclusion of F, considered here as an element of the Lie algebra F(M) (see above), into a commutative subalgebra of dimension n whose additive basis consists of functionally independent functions. We have encountered a similar problem earlier during the study of finite-dimensional Lie algebras and their commutative subalgebras. But now the situation is much more complicated because here we are dealing with an infinite-dimensional Lie algebra and there are no analogs of "finite-dimensional theorems" on some analogs of Cartan subalgebras. Therefore, the search for a full set of commuting functions that integrate the system is a nontrivial problem in each concrete case, defined by peculiarities of the considered system.

Proof of Theorem 19.1. Let $\xi = (\xi_1, \ldots, \xi_n)$ be a set of fixed values of functions $f_1, \ldots, f_n$. Since the set grad f_i, $1 \leqq i \leqq n$, is independent at each point on the level surface M_ξ, then by the implicit function theorem, this surface is a smooth n-dimensional submanifold in M. Consider smooth vector fields $\alpha(f_i)$ = sgrad f_i. Each of these is tangent to the level surface (see above); hence, we get the set of n tangent fields on M_ξ. We claim that these fields commute with each other and are linearly independent at each point. In fact, the symplectic structure ω defines the canonical identification ε of the tangent and the cotangent spaces, which transforms grad f_i into sgrad f_i (see Sec. 19.3). Since ε is an isomorphism of T_x and T_x^*, then linear independence of the covectors (grad f_i) is equivalent to the linear independence of the vectors (sgrad f_i).

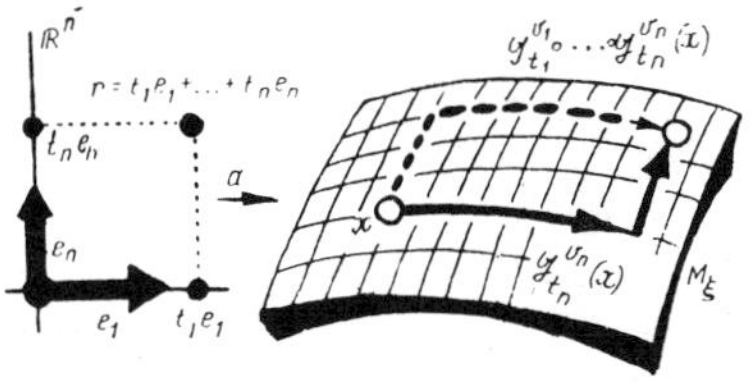

Fig. 215

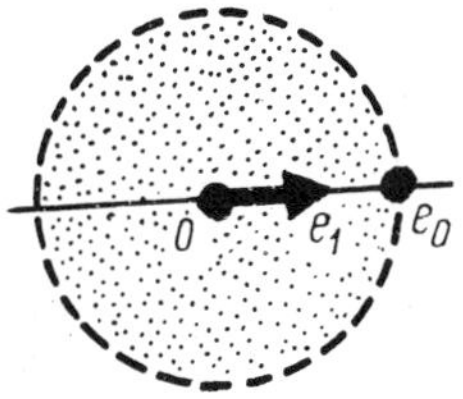

Fig. 216

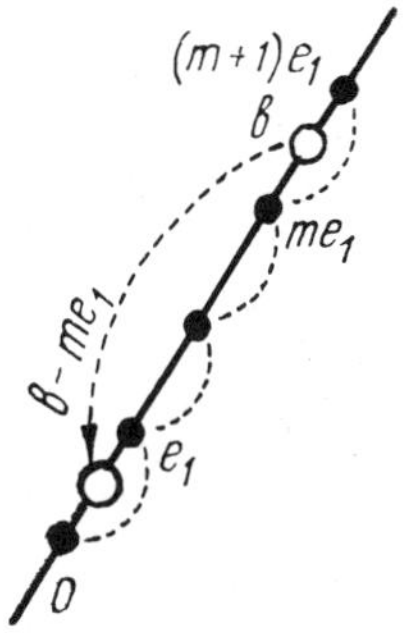

Fig. 217

Furthermore, Lemma 19.1 implies that $\alpha\{f_i, f_j\} = [\alpha f_i, \alpha f_j] \equiv 0$, since functions f_i, f_j are in involution. Hence, fields (sgrad f_i) commute with each other, as required. Thus, fields (sgrad f_i) constitute a basis in each tangent space to M_ξ. This, in particular, implies that the restriction of ω onto the tangent space $T_x M_\xi$ vanishes. Consider the Abelian group $\mathbf{R}^n$ and denote its generators by $e_1, \ldots, e_n$. Each e_i generates a one-dimensional subgroup $\mathbf{R}_i^1$ that will be presented as a one-parameter group $g_t^{v_i}$ of diffeomorphisms generated by the vector field v_i = sgrad f_i on M_ξ.

Thus, we have presented all generators $e_1, \ldots, e_n$ of $\mathbf{R}^n$ in terms of diffeomorphisms that preserve the symplectic structure in a neighborhood of M_ξ. Now let us define a smooth action of the whole $\mathbf{R}^n$ on M_ξ and in a neighborhood of M_ξ. For this, fix a point x on M_ξ and assign to the element $r = t_1 e_1 + \ldots + t_n e_n$ the smooth transformation $a(r): M_\xi \to M_\xi$, which is the composition $a(r)(x) = g_{t_1}^{v_1} \circ \ldots \circ g_{t_n}^{v_n}(x)$ (see Fig. 215). Since vector fields v_i commute on M_ξ, it follows that a smooth action of the commutative group $\mathbf{R}^n$ on M_ξ is well de-

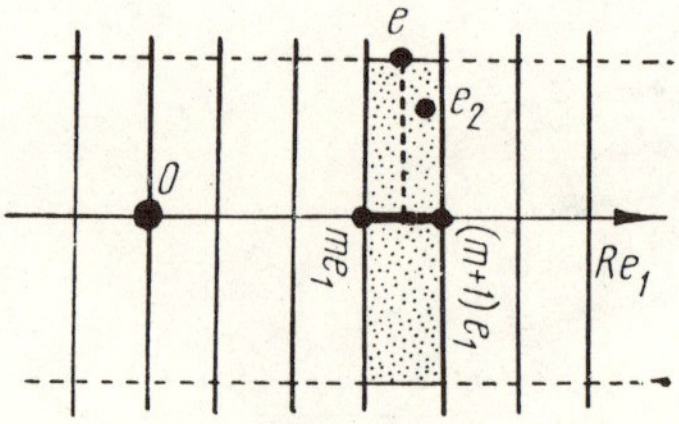

Fig. 218

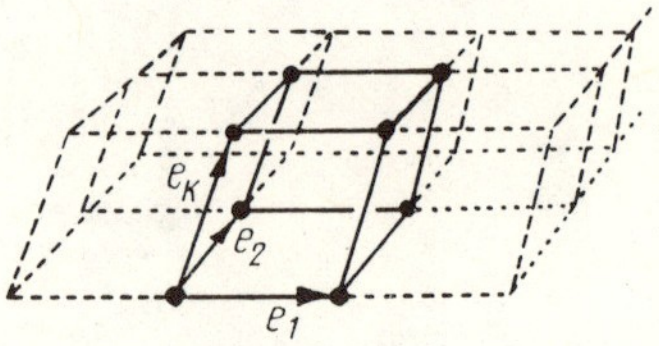

Fig. 219

fined when M_ξ is complete. It is also clear that this construction makes it possible to define this action in a neighborhood of a nonsingular surface M_ξ. Since fields v_i = = sgrad f_i are independent at each point of M_ξ, then $\mathbf{R}^n$ acts locally transitively on M_ξ and the mapping $a:\mathbf{R}^n \to M_\xi$ is onto. The definition of a also implies that M_ξ is an orbit of $\mathbf{R}^n$ that "grows" from $x_0 \in M_\xi$. Thus, n independent commuting vector fields are defined on M_ξ.

In $\mathbf{R}^n$, consider the stationary subgroup Γ of the point x_0, i.e., the set of all elements $r \in \mathbf{R}^n$ that preserve this point. The above implies that Γ is a discontinuous subgroup. Let us prove that there are k, where $0 \leqq k \leqq n$, independent vectors $e_1, \ldots, e_k \in \mathbf{R}^n$ such that Γ coincides with the set of all their integer linear combinations, i.e., $\Gamma = \mathbf{Z}(e_1, \ldots, e_k) = \mathbf{Z}^k$.

Let us assume that $a(0) = x_0$, where $0 \in \mathbf{R}^n$. If Γ is a nontrivial subgroup, we may consider the line $\mathbf{R}e_0$, where $e_0 \in \Gamma$, $e_0 \neq 0$. Let e_1 be a nonzero element of Γ that belongs to $\mathbf{R}e_0$ and is nearest to 0, i.e., $|e_1| \leqq |e_0|$, where $|\ \ |$ is the length of a vector in $\mathbf{R}^n$ (see Fig. 216). Such an element e_1 exists because in the ball of radius $|e_0|$ with center at 0, there are only a finite number of elements of Γ. The vector e_1 can coincide with e_0. All elements of Γ that belong to $\mathbf{R}e_0$ are necessarily of the form me_1, where $m \in \mathbf{Z}$, i.e., they are integer multiples of e_1. In fact, if inside of a segment $(me_1, (m + 1)e_1)$, some point b of Γ had appeared, then by moving it by the element $-me_1$ which belongs to Γ, we would have obtained on $\mathbf{R}e_0$ the new point $b - me_1$ situated closer to 0 than e_1 (see Fig. 217). This contradiction proves the statement.

Furthermore, if outside of $\mathbf{R}e_1$ there are no points of Γ, then the statement is proved. If there are such points

$e \in \Gamma$, then there is a point $e_2 \in \Gamma$ nearest to the line $\mathbf{R}e_1$ that does not belong to this line. In fact, the orthogonal projection of e onto $\mathbf{R}e_1$ belongs to a segment $I = (me_1, (m + 1)e_1)$. Consider the right circular cylinder with axis I and a circle of radius equal to the distance from I to e as the base (see Fig. 218). In this cylinder there are a finite number of elements of Γ. As e_2 let us take the element closest in this cylinder to I. Clearly, this point is the closest to the axis $\mathbf{R}e_1$, not only among points of Γ which belong to the cylinder, but among all points of Γ, because, moving the cylinder by integer multiples of e_1 along the axis $\mathbf{R}e_1$, we may drive into the cylinder with the axis I any element of Γ which belongs to this cylindrical band with the axis $\mathbf{R}e_1$ (see Fig. 219). The integer linear combinations of the form $ae_1 + be_2$ constitute a lattice in the plane $\mathbf{R}e_1 \oplus \mathbf{R}e_2$. Continuing this construction, we finally obtain the lattice $\mathbf{Z}^k = \mathbf{Z}(e_1, \ldots, e_k)$ of rank k, where $k \leq n$ (see Fig. 219). The lemma is proved.

Thus, the level surface M_ξ, being a homogeneous space of $\mathbf{R}^n$, is diffeomorphic to the quotient space $\mathbf{R}^n/\Gamma$. If M_ξ is compact, then the rank Γ is n and then M_ξ is an n-dimensional torus. If the level surface is noncompact, then it is diffeomorphic to the "cylinder" $\mathbf{R}^n/\Gamma$, where the rank Γ is less than n. We have constructed on M_ξ "angle coordinates" $\varphi_1, \ldots, \varphi_n$. Their role may be played by coordinates defined by coordinate lines – images of lines $\mathbf{R}e_1, \ldots, \mathbf{R}e_n \in \mathbf{R}^n$ with re-

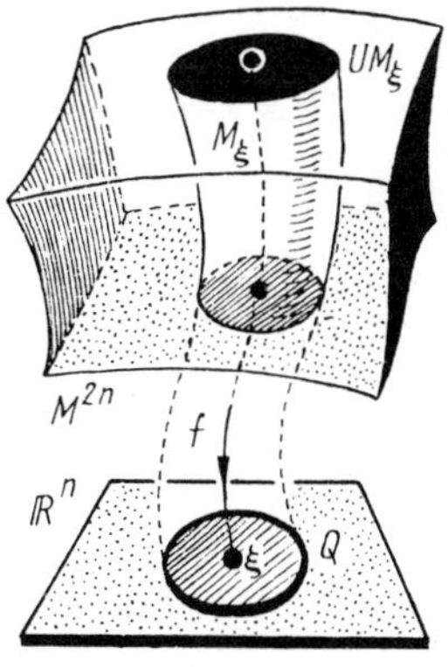

Fig. 220

spect to the projection $a : \mathbf{R}^n \to M_\xi = \mathbf{R}^n/\Gamma$. Furthermore, it is clear that integral trajectories of fields $v_i = \operatorname{sgrad} f_i$ define in coordinates $\varphi_1, \ldots, \varphi_n$ straight windings of T^n.

It remains to establish the existence of coordinates $s_1, \ldots, s_n$ "complementary" to $\varphi_1, \ldots, \varphi_n$, i.e., coordinates that parametrize a tubular neighborhood of a torus in directions orthogonal to this torus. Since ideas that constitute the background of this proof will not be used subsequently, we skip this proof, explaining briefly a geometrical structure of the fibration onto level surfaces.

On $T^n = M_\xi$, we have two families of vector fields. The first consists of $v_1, \ldots, v_n$; the second consists of $b_1, \ldots, b_n$, defined by $\varphi_1, \ldots, \varphi_n$, i.e., $b_i = \varphi_i$. At $x_0 \in M_\xi$ (as at any point of the torus), these two sets define two bases in the tangent space to the torus. Hence, there is a nondegenerate transformation defined by the matrix $D = (d_{ij})$ that transforms fields (v_i) into fields (b_i), i.e., $b_i = \sum_{1 \leq j \leq n} d_{ij} v_j$. The matrix D obviously depends on $\xi_1, \ldots, \xi_n$, i.e., on the set of values of integrals $f_1, \ldots, f_n$ that define this level surface. Thus, we may assume that $b_i = \sum_{1 \leq j \leq n} d_{ij}(f_1, \ldots, f_n) \times v_j$. The matrix D is therefore defined in a neighborhood that belongs to the set of values of integrals (f_i). In other words, we may consider the smooth mapping $f : M \to \mathbf{R}^n$, where $f(x) = (f_1(x), \ldots, f_n(x)) \in \mathbf{R}^n$. This mapping transforms each level surface into a point (see Fig. 220). Since integrals form a functionally independent set in a neighborhood of M_ξ, then f maps this neighborhood UM_ξ onto an n-dimensional domain Q in $\mathbf{R}^n$. The implicit function theorem implies that f defines the locally trivial bundle $f : UM_\xi \xrightarrow[M_\eta]{} Q$ whose fibers M_η are level surfaces close to M_ξ (see Fig. 220).

20. NONCOMMUTATIVE INTEGRATION OF HAMILTONIAN SYSTEMS

20.1. Noncommutative Lie Algebras of Integrals

In the Liouville theorem proved above, the main role is played by the fact that functions $f_1, \ldots, f_n$ commute. In other words, the linear space G of functions spanned by f_1, ..., f_n is a commutative Lie algebra of dimension n. The Hamiltonian of the integrated system enters this Lie algebra as one of its elements: $F = f_1$. However, in various concrete situations, Hamiltonian systems possess a set of integrals $f_1, \ldots, f_k$ that do not form a commutative Lie algebra, i.e., they are not in involution. Therefore, it would be quite useful to have a method that enables one to integrate such systems.

Suppose $f_1, \ldots, f_k$ are smooth functions on a symplectic manifold M^{2n} with the form ω functionally independent on an open dense subset in M. Consider the linear envelope G (over $\mathbf{R}$) of functions (f_i); then $\dim_{\mathbf{R}} G = k$. Suppose that the linear space G is closed with respect to the Poisson bracket, i.e., that the bracket $\{f_i, f_j\}$ of any pair is decomposable with respect to the basic functions $f_1, \ldots, f_k$ with constant (!) coefficients, i.e., $\{f_i, f_j\} = \sum_{q=1}^{k} c_{ij}^q f_q$, where $c_{ij}^q \in \mathbf{R}$.

This means that the linear space G is a finite-dimensional real Lie algebra with respect to the Poisson bracket. An important special case is the case where G is a commutative Lie algebra of dimension n, which falls under the action of the "commutative" Liouville theorem. Note that the Lie algebra G must not be compact. Even in the case of the Liouville theorem, it is a commutative algebra. Nevertheless, it turns out that the restriction of one more simple condition on G makes the system v = sgrad F, where $F = f_1$, completely integrable and enables one to describe trajectories of the system as simply as in the Liouville theorem case. Let us call the Lie algebra G, constructed above, the algebra of integrals. Let $\mathfrak{G}$ be the simply connected Lie group corresponding to G. Then $\mathfrak{G} = \exp G$. Each element of G is presentable as a Hamiltonian field on M. For this, we must consider the mapping $\alpha: f \to \operatorname{sgrad} f$. Hence, $\mathfrak{G}$ is presented as the group of diffeomorphisms of the manifold that preserves ω. Such transformations are called symplectic. Thus, the definition of the Lie algebra of integrals G defines a smooth symplectic action on the finite-dimensional group $\mathfrak{G}$ on M. Since in this section we will encounter problems where noncompact Lie algebras appear, let us give several useful definitions.

Let G^* be the space dual to G, i.e., the space of linear forms on G. The group $\mathfrak{G}$ is presented as the group of linear transformations of G^*, namely, $g \to Ad_g^*: G^* \to G^*$, where Ad_g^* is the transformation conjugate to $Ad_g: G \to G$. If $\xi \in G^*$ and $X \in G$, then let $\langle \xi, X \rangle$ be the value of the covector ξ on the vector X. Then $\langle Ad_g^* \xi, X \rangle = \langle \xi, Ad_g X \rangle$. Though we have used here the symbol $\langle\ ,\ \rangle$, used earlier for the scalar product, this will not lead to confusion since, if G is compact, then the value of a covector at a vector may be considered as the scalar product of two vectors, one of them being a covector with respect to the identification of G and G^*. Define also the transformation $ad_Y^*: G^* \to G^*$ due to the transformation $ad_Y: G \to G$. Then $\langle ad_Y^* \xi, X \rangle = \langle \xi, ad_Y X \rangle = \langle \xi, [Y, X] \rangle$. The representation $g \to Ad_g^*$ is called the coadjoint representation of $\mathfrak{G}$ and the correspondence $X \to ad_X^*$ is called the coadjoint representation of G. The group $\mathfrak{G}$ acting on G^* fibrates G^* into orbits $O^* = \mathfrak{G}(\xi)$ of the coadjoint representation. If $\xi \in O^*$, then the tangent space $T_\xi O^*$ to the orbit O^* consists of covectors of the form $ad_X^* \xi$, where X runs over G. This is proved similarly to Lemma 18.2.

Let $\xi \in G^*$ be a covector. In G consider the subspace $H_\xi = \operatorname{Ann} \xi$ consisting of all vectors X such that $ad_X^* \xi = 0$. The subspace H_ξ is called the annihilator of the covector ξ. Clearly, H_ξ is a Lie subalgebra since

$$\langle ad^*_{[X,Y]} \xi, Z \rangle = \langle \xi, [[X, Y], Z] \rangle = -\langle \xi, [[Z, X], Y] \rangle - \\ - \langle \xi, [[Y, Z], X] \rangle = \langle ad^*_Y \xi, [Z, X] \rangle + \langle ad^*_X \xi, [Y, Z] \rangle = 0.$$

If G is a compact Lie algebra, then identifying G and G^* with respect to the Killing form, we get the already familiar definition of the annihilator of a vector. Let us say that $\xi \in G^*$ is a generic covector if its annihilator has the minimum possible dimension. The rank of G is the dimension of the annihilator of a generic covector. If G is semisimple, then this definition coincides with the definition of the rank of G introduced earlier (verify!).

20.2. A Theorem on Noncommutative Integration

Here we will formulate a theorem that naturally generalizes the Liouville theorem and is proved in [21].

<u>Theorem 20.1</u>. Suppose that on the symplectic manifold M^{2n} the set of k smooth functions $f_1, \ldots, f_k$ is defined and

their linear combination is the Lie algebra G with respect to the Poisson bracket, i.e., $\{f_i, f_j\} = \sum_{q=1}^{k} c_{ij}^q f_q$, where C_{ij}^q are constants. Let M_ξ be the common level surface of the general position of functions (f_i), i.e., $M_\xi = \{x \in M : f_i(x) = \xi_i, 1 \leqq i \leqq n\}$. Suppose that on this level surface all k functions $f_1, \ldots, f_k$ are functionally independent. Suppose that the Lie algebra G satisfies dim G + rank G = dim M, i.e., k + + rank G = 2n. Then M_ξ is a smooth r-dimensional submanifold (where r = rank G), invariant with respect to each vector field v = sgrad h, where $h \in H_\xi$. Furthermore, let v be one of the following Hamiltonian fields on M:

a) v = sgrad h, where h is an element of the algebra of integrals G and belongs to the annihilator H_ξ of the covector ξ that defines the level surface M_ξ;

b) v = sgrad F is a Hamiltonian field on M such that all elements of G are integrals of v, i.e., $\{F, f\} \equiv 0$ for all $f \in G$.

Then, as in the "commutative" Liouville theorem, if M_ξ is connected and compact, then it is diffeomorphic to a small r-dimensional torus T^r and on this torus curvilinear coordinates $\varphi_1, \ldots, \varphi_r$ may be introduced such that the vector field v being expressed in terms of these coordinates becomes of the form $\dot{\varphi}_i = q_i(\xi_1, \ldots, \xi_k)$, i.e., components of this field are constant on the torus and integral trajectories of the field define the conditionally periodic movement of the system v, i.e., define the "straight winding" of T^r.

The proof will be given in the following section. In the special case when the Lie algebra of integrals G is commutative, the condition dim G + rank G = dim M becomes k + k = = 2n since here dim G = rank G = k. Thus, k = n, and we obtain the "commutative" Liouville theorem proved in the preceding section. In various concrete examples, the Lie algebra of integrals turns out to be compact and noncommutative. As we see from Theorem 20.1, the system moves along tori T^r, the dimension r being equal to the rank of G. As follows from Chapter 4, in the semisimple case, the rank r of G is less than its dimension and in all the main cases we may assume that $r \approx \sqrt{k}$, where k is the dimension of G. For instance, for A_{n-1} when $G = su_n$, we have $r = n - 1$, $k = \dim G = n^2 - 1$, i.e., rank $G \approx \sqrt{\dim G}$. This means that $r < k$ and since $r + k = 2n$, then $r < n = \frac{1}{2}\dim M$. In other words, the system $v = \operatorname{sgrad} F$ moves along tori whose dimension is less (and substantially less) than the halved dimension of the manifold. This shows that Hamiltonian systems with noncommutative symmetries, i.e., that possess a noncommutative algebra of integrals in the above sense, are "highly degenerate," i.e., their integral trajectories (in the general position) are dense on tori of a small dimension r. This constitutes the difference between these systems and those that satisfy the "commutative" Liouville theorem, where the motion occurs along tori of halved dimension, i.e., $r = n = \frac{1}{2}\dim M$. Thus the "noncommutative" Theorem 20.1 enables us to integrate highly degenerate systems, the degeneracy being greater when the rank of the algebra of the integrals of this system is smaller. Systems of this kind, being degenerate systems on the initial manifold, might appear to be illustrations of the "Liouville-type" systems on a submanifold.

Moreover, there is an interesting relationship between the commutative integration and the noncommutative one. For instance, if a Hamiltonian system possesses a noncommutative algebra of integrals (where $\dim G + \operatorname{rank} G = \dim M$), then in various cases it also possesses a commutative algebra of integrals of the halved dimension. Moreover, the following statement holds [22].

Theorem 20.2. Let $v = \operatorname{sgrad} F$ be a Hamiltonian system on the compact symplectic manifold M completely integrable in the noncommutative sense, i.e., possessing the Lie algebra of integrals G such that $\dim G + \operatorname{rank} G = \dim M$. Then this system is completely integrable in the usual commutative sense by Liouville, i.e., it possesses another commutative Lie algebra of integrals G' such that $\dim G' = \frac{1}{2}\dim M$.

Here we assume that additive generators of both algebras G and G' are functionally independent almost everywhere on M. Clearly, these algebras are not isomorphic if G is noncommutative. Thus, on a compact manifold a degenerate integrable Hamiltonian system possesses another commutative algebra of integrals of "general type." From a geometric viewpoint, the structure of such systems is extremely simple. Let $\{T^r\}$ be a family of r-dimensional tori, where $r < n$, along which trajectories of the system are moved, resulting in dense winding of these tori in the general position. Then (see Theorem 20.2) these tori of small dimension r can be

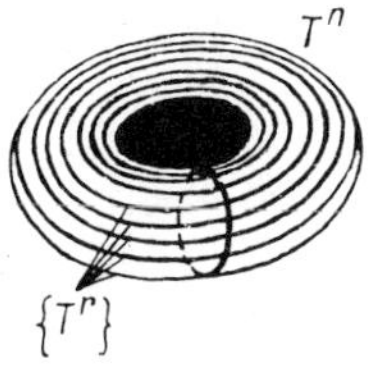

Fig. 221

organized into larger tori of dimension n, i.e., of the halved dimension along which trajectories of the system move. These big tori are level surfaces of another, now commutative, Lie algebra of integrals G' (see Fig. 221). Note that trajectories of such a system cannot be dense on the big torus T^n since the fibration of this torus by T^r of a small dimension is locally the direct product of T^r and a complementary submanifold of dimension $n - r$ (see Fig. 221). The proof of Theorem 20.2 is quite nontrivial; therefore, we skip it. For noncompact manifolds, a similar result has not been proved yet.

Conjecture. Any Hamiltonian system on any symplectic manifold completely integrable in the noncommutative sense is completely integrable in the commutative sense by Liouville.

The generalization of Theorem 20.2 onto noncompact manifolds is interesting because various concrete Hamiltonian systems are realized as flows on noncompact manifolds.

20.3. The Reduction of Hamiltonian Systems with Noncommutative Symmetries

We will describe a simple and beautiful construction that makes it possible to convert a Hamiltonian system with a group of symmetries into a Hamiltonian system on a symplectic manifold of smaller dimension. This procedure is called the reduction of a Hamiltonian system (J. Marsden, A. Weinstein). As an application we will prove Theorem 20.1.

Let, on the symplectic manifold (M^{2n}, ω), a Hamiltonian system $v = \operatorname{sgrad} F$ be defined with the Lie algebra of integrals G whose additive generators are k independent (almost everywhere) smooth functions $f_1, \ldots, f_k$. Let $\mathfrak{G}$ be the corresponding simply connected group that acts on M by symplectic diffeomorphisms (i.e., that preserve ω). For hereinafter

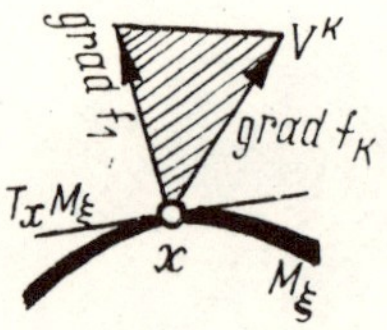

Fig. 222

it is convenient to consider the following mapping φ. Let us assign to each point $x \in M$ the linear functional (covector) φx on G. Let $\varphi x(f) = f(x)$, where $f \in G$. Thus, φx is an element of G^*. Hence, we have defined the smooth mapping $\varphi: M \to G^*$.

Lemma 20.1. Let $\xi \in G^*$ be an arbitrary covector. Then its pre-image $\varphi^{-1}\xi$ with respect to φ is a common level surface M_ξ of integrals $f_1, \ldots, f_k$ that generate G.

Proof. By definition, $\varphi^{-1}\xi = \{x \in M: f(x) = \xi(f)\}$, where $f \in G$. Since (f_i) is an additive basis of G, then $f = \sum_{i=1}^{k} a_i f_i$, i.e., $f(x) = \xi(f) = \sum_{i=1}^{k} a_i \xi(f_i)$. If $\xi(f_i) = \xi_i$, where $1 \leq i \leq k$, then $f_i(x) = \varphi x(f_i) = \xi(f_i) = \xi$ for $x \in M_\xi$. Thus, $M_\xi = \varphi^{-1}\xi = \{x \in M: f_i(x) = \xi_i$, where $\xi = (\xi_1, \ldots, \xi_k)\}$. The lemma is proved.

The mapping $\varphi: M \to G^*$ is not necessarily "onto." For example, if M is compact, then $\varphi(M)$ does not cover the whole G^*. Let us introduce a convenient notation. If $\xi \in G^*$, $f \in G$ then denote $\mathrm{ad}_f^*\xi$ by $a(\xi, f)$. Then $\langle \xi, [g, f] \rangle = \langle \mathrm{ad}_g^*\xi, f \rangle = \langle a(\xi, g), f \rangle$. We have $H_\xi \subset G$, $H_\xi = \{g \in G: a(\xi, g) = 0\}$, i.e., $H_\xi = \mathrm{Ann}\,\xi$.

Proposition 20.1. Let the element f of G belong to the annihilator H_ξ and $x \in M_\xi$, where M_ξ is a nonsingular surface. Then $\mathrm{sgrad}\, f(x) \in T_x M_\xi$.

This means that skew gradients of functions of the annihilator of ξ belong to the tangent space to the level surface M_ξ defined by this covector.

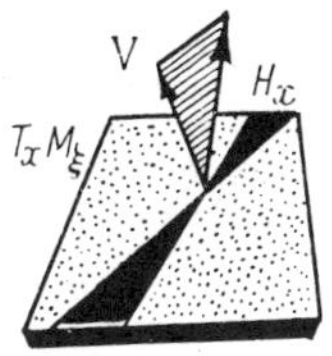

Fig. 223

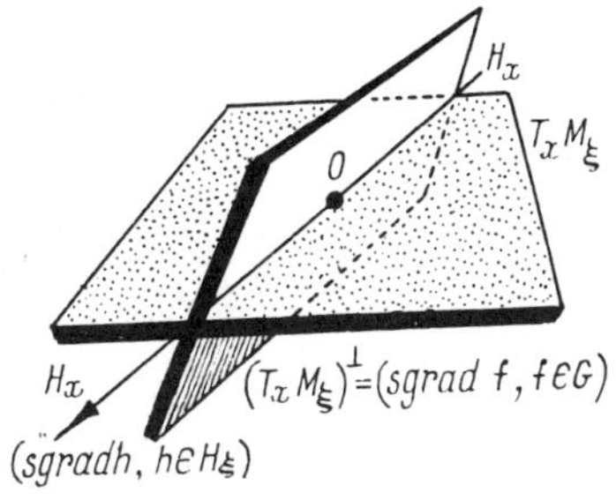

Fig. 224

Proof. Consider additive generators $f_1, \ldots, f_k$ of G and their gradients grad $f_1, \ldots,$ grad f_k. Since M_ξ is a nonsingular surface, then all these gradients are independent at all points of the surface and hence are transversal to M_ξ, i.e., a k-dimensional space spanned by (grad f_i) has only a zero intersection with T_xM_ξ and $V \oplus T_xM_\xi = T_xM$ (see Fig. 222). It is convenient to assume that M is endowed with a Riemannian metric. Then vectors (grad f_i) are orthogonal to the surface M_ξ. To prove that sgrad $f(x) \in T_xM_\xi$, it suffices to verify that (sgrad f)g $\equiv 0$ for any $g \in G$, i.e., that the derivative along sgrad f of any function g constant on M_ξ is zero. In fact, (sgrad f)g = (sgrad f, sgrad g), where (,) is the Riemannian metric on M. The vanishing of scalar products of sgrad f with all vectors (grad f_i) implies that sgrad f is orthogonal to V, i.e., sgrad $f \in T_xM_\xi$. Thus, (sgrad f)g = = {f, g}. Furthermore, $\{f, g\}(x) = \langle\xi, \{f, g\}\rangle = \langle a(\xi, f), g\rangle = 0$, since $a(\xi, f) = 0$, $f \in \mathrm{Ann}\ \xi$, as required.

Thus, all fields sgrad f generated by elements f of the annihilator of ξ are tangent to the corresponding level surface M_ξ.

Proposition 20.2. $(T_xM_\xi) \cap (\text{sgrad } f, f \in G) = H_x = (\text{sgrad } h, h \in H_\xi)$ (see Fig. 223).

Proof. We have proved above that $(\text{sgrad } h;\ h \in H_\xi) \subset T_xM_\xi \cap (\text{sgrad } f;\ f \in G)$. Let us prove the opposite statement. Let $X \in T_xM_\xi$ and $X = $ sgrad f, where $f \in G$. We must prove that $f \in H_\xi$. Since $\{f, g\}(x) = 0$, then $\langle a(\xi, f), g\rangle = \langle\xi, \{f, g\}\rangle \equiv 0$ for any $g \in G$. The first equality follows from $\{f, g\}(x) = (\text{sgrad } f)g|_x = X(g) = 0$ since $X \in T_xM_\xi$ and all $g \in G$ are constant on the level surface M_ξ. Thus, $\langle a(\xi, f), g\rangle \equiv 0$ for any $g \in G$. This means that $a(\xi, f) = 0$, i.e., $f \in \mathrm{Ann}\ \xi = H_\xi$. The statement is proved (see Fig. 224).

Consider the group $\mathfrak{H}_\xi$ with the Lie algebra H_ξ, i.e., $\mathfrak{H}_\xi = \exp H_\xi \subset \mathfrak{G}$.

Corollary 20.1. The level surface M_ξ is invariant with respect to the $\mathfrak{H}_\xi$-action on M.

Consider the form ω on M and let $\tilde{\omega} = \omega|M_\xi$ be the restriction onto M_ξ. The $\mathfrak{H}_\xi$-action on M_ξ generates at each point $x \in M_\xi$ the plane $H_x \subset T_xM_\xi$, generated by vectors sgrad f, where $f \in H_\xi$ (see Fig. 223). In other words, the space H_x is generated by the Lie subalgebra H_ξ.

Proposition 20.3. The kernel of $\tilde{\omega}$ (the restriction of ω onto M_ξ) coincides with the subspace $H_x \subset T_xM_\xi$.

Proof. Let us prove that $\operatorname{Ker} \tilde{\omega} \supset H_x$. Let X = sgrad h, where $h \in H_\xi$, $X \in H_x \subset T_xM_\xi$. We must prove that X belongs to the kernel of ω, i.e., $\omega(X, Y) = 0$ for any Y of T_xM_ξ. In fact, $\omega(X, Y) = \omega(\text{sgrad } h, Y) = Y(h) = 0$, since Y is tangent to the level surface and h, being an element of G, is constant on the level surface. Let us prove the oppositie inclusion, i.e., $\operatorname{Ker} \tilde{\omega} \subset H_x$. Let $\omega(X, Y) = 0$ for any $Y \in T_xM_\xi$. We must present X in the form X = sgrad h for some $h \in H_\xi$. Consider ω as a skew symmetric scalar product on the tangent space T_xM and let $(T_xM_\xi)^\perp$ be the orthogonal complement to T_xM_ξ with respect to ω in T_xM. Since ω is nondegenerate, then $\dim (T_xM_\xi)^\perp = \dim M - \dim T_xM_\xi = \dim V = k$. Recall that for the skew symmetric scalar product the space T_xM must not be presentable as the sum of T_xM_ξ and $(T_xM_\xi)^\perp$ since these subspaces can have a nonzero intersection. Clearly, $\operatorname{Ker} \tilde{\omega} = T_xM_\xi \cap \cap (T_xM_\xi)^\perp$. Let us prove that $(\text{sgrad } f;\ f \in G) = (T_xM_\xi)^\perp$. In fact, let $Y \in T_xM_\xi$; then $\omega(\text{sgrad } f, Y) = Y(f) = 0$ since f = const on M_ξ. Thus, $(\text{sgrad } f \mid f \in G) \subset (T_xM_\xi)^\perp$. Furthermore, $\dim (\text{sgrad } f \mid f \in G) = k = \dim G$. This equality follows from the fact that the dimension of the linear span of $(\text{grad } f \mid f \in G)$ is k (see the definition of G). The skew symmetric scalar product is nondegenerate and the dimension of the linear span of skew gradients is also k. Finally, we have proved that $\dim (T_xM_\xi)^\perp = k$, hence $(\text{sgrad } f \mid f \in G) = = (T_xM_\xi)^\perp$ (see Fig. 224). The statement is proved.

Let us collect all these facts together and study the geometry of the described submanifolds. The main objects of study are:

a) level surfaces M_ξ, where $\dim M_\xi = 2n - k$;

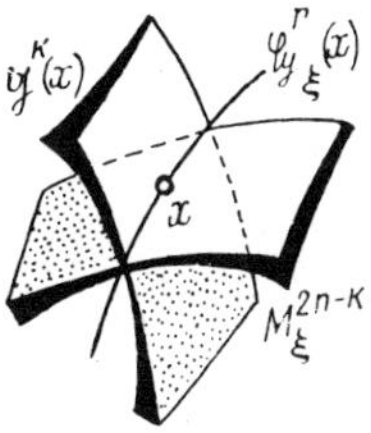

Fig. 225

b) the orbit $\mathfrak{G}(x)$ of the point x, where dim $\mathfrak{G}(x) = k$;

c) the orbit $\mathfrak{H}_\xi(x)$ of x under the action of $\mathfrak{H}_\xi = \exp H_\xi$.

Clearly, $T_x\mathfrak{G}(x) = (\text{sgrad } f |\ f \in G)$, $T_x\mathfrak{H}_\xi(x) = H_x = (\text{sgrad } h;\ h \in H_\xi)$. This implies that $\mathfrak{G}(x) \cap M_\xi = \mathfrak{H}_\xi(x)$ (see Fig. 225). Note that dim $\mathfrak{H}_\xi(x) = r$.

Consider the action of $\mathfrak{G}$ on M and suppose that in a small neighborhood of M_ξ this action has the same type of stationary subgroups, i.e., all orbits of $\mathfrak{G}$ close to $\mathfrak{G}(x)$ are diffeomorphic. Consider the projection $p:M \to M/\mathfrak{G}$ of M onto the space of orbits $M/\mathfrak{G} = N$. This space might appear nonsmooth and have singularities. It is important for us that the space $M/\mathfrak{G}$ is a smooth manifold of dimension $2n - k$ in a small neighborhood of the point $p\mathfrak{G}(x) \in M/\mathfrak{G}$. In fact, if $\mathfrak{G}$ is, for instance, a compact group that acts smoothly on M, then the union of generic orbits diffeomorphic to each other is an open dense subset in M, and hence N is a $(2n - k)$-dimensional manifold everywhere excluding a subset of zero measure. Note that the space (manifold) N must not be symplectic since, for example, it can be odd-dimensional. The projection p, being restricted onto M_ξ, maps it onto $Q_\xi = M_\xi/\mathfrak{H}_\xi$. Therefore, the space N is fibrated onto surfaces Q_ξ (see Fig. 226). Here we make use of Proposition 20.2.

Proposition 20.4. Manifolds Q_ξ, i.e., quotient manifolds of level surfaces M_ξ with respect to $\mathfrak{H}_\xi$, are symplectic manifolds with a nondegenerate closed form ρ, which is a projection of $\tilde{\omega}$ onto M_ξ with respect to $p:M_\xi \to Q_\xi$. Moreover, $p^*\rho = \tilde{\omega} = \omega|M_\xi$.

The proof follows from Proposition 20.3, since the kernel of $\tilde{\omega}$ on T_xM_ξ coincides with $H_x \subset T_xM_\xi$.

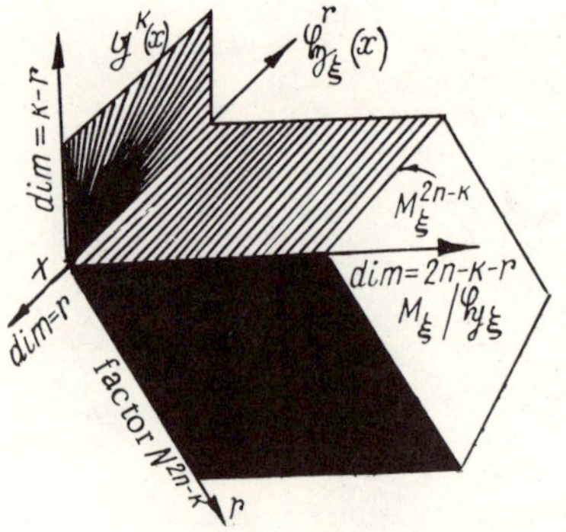

Fig. 226

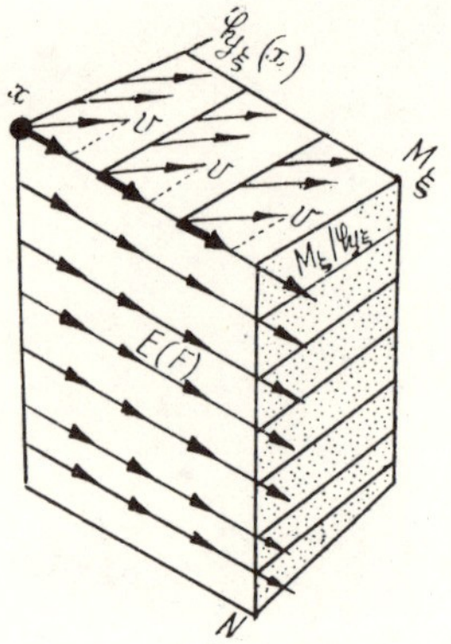

Fig. 227

Let us return to the study of the Hamiltonian systems on M. Let v = sgrad F be a system with the Lie algebra of integrals G, i.e., {F, G} = 0. Since F commutes (with respect to the Poisson bracket) with all elements of G, then F is invariant with respect to $\mathfrak{G}$. In fact, (sgrad f)(F) = = {f, F} = 0 for $f \in G$. In particular, the subgroup $\mathfrak{H}_\xi$ acting on M also transforms F onto itself. Thus, the natural projection of the vector field sgrad F onto $N = M/\mathfrak{G}$ is defined. Here sgrad F is tangent to M_ξ and is also mapped into a field E(F) on Q_ξ, since sgrad F is $\mathfrak{H}_\xi$-invariant. Thus, N is fibrated onto the symplectic manifolds Q_ξ, and the vector field E(F) on N is defined which is tangent to all Q_ξ (see Fig. 227). Finally, we have assigned to the triple (M^{2n}, sgrad F, ω) the new triple (Q_ξ, E(F), ρ).

Proposition 20.5. The vector field E(F) is a Hamiltonian field with respect to the symplectic form ρ on Q_ξ; its Hamiltonian is $\tilde{F}$, which is the projection of $F|M_\xi$ onto Q_ξ, i.e., $E(F) = \operatorname{sgrad}_\rho (p_* F|M_\xi)$.

The proof follows from Proposition 20.4 and the invariance of F with respect to the action of the group. The correspondence (M, sgrad F, ω) → (Q_ξ, E(F), ρ) constructed above is called the reduction of the initial Hamiltonian system on the manifold Q_ξ of dimension 2n − k − r which is less than the dimension 2n of the initial manifold M and $\dim Q_\xi < \dim M_\xi = 2n - k$. It may appear that the reduced system on Q_ξ is more simple than the initial system on M. Suppose that we are able to integrate the reduced system. This enables us

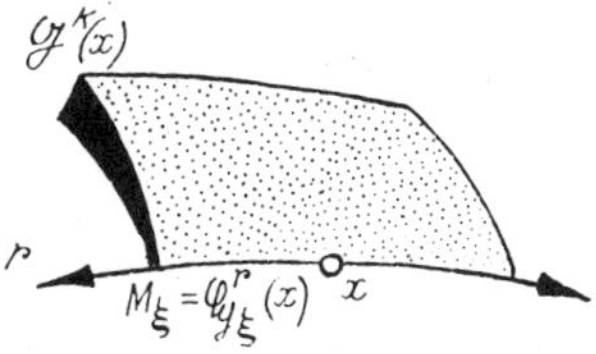

Fig. 228

to enlarge the number of integrals of the initial system sgrad F on M by "lifting" these integrals from N onto M.

Proposition 20.6. Let G be a finite-dimensional Lie algebra of integrals sgrad F on M satisfying all the conditions listed above, and let E(F) be the reduced system on $N = \bigcup_\xi Q_\xi$ which is Hamiltonian on each Q_ξ. Let G' be the linear space of functions on N such that their restriction onto Q_ξ constitutes the finite-dimensional Lie algebra of integrals of the flow E(F). Then the space of functions $G \oplus G''$, where $G'' = p^*G'$, i.e., $G'' = \{gp \mid g \in G', p: M \to N\}$, is the Lie algebra of integrals of the system sgrad F and $[G, G''] = 0$.

Proof. Let g be a function on N; then its pre-image gp with respect to the mapping $p: M \to N$ is a function on M which is evidently invariant with respect to the $\mathfrak{G}$-action on M. But this means that gp is in involution with the whole initial algebra of functions (integrals) G. Thus, any new function g which is an integral of the reduced flow E(F) on N gives an extra integral gp of the initial Hamiltonian flow sgrad F on M. These extra integrals are independent of elements on G as a result of the fact that their gradients are nonzero along submanifolds Q_ξ which belong (locally) to M_ξ. At the same time, gradients of elements of G are orthogonal to M_ξ. The proposition is proved.

Proof of Theorem 20.1. Let v be one of the systems mentioned in the formulation of the theorem, i.e., either v = = sgrad F, $\{F, Q\} = 0$ or v = sgrad h, where $h \in \mathrm{Ann}\ \xi = H_\xi$. Consider the above reduction technique. Since now the extra condition dim G + rank G = dim M is verified, i.e., $k + r =$ $= 2n$, then the dimension of M_ξ equals r. The dimension of $\mathfrak{H}_\xi(x)$ which is contained in M_ξ also equals r (by definition) (see Fig. 226). This immediately implies that $M_\xi = \mathfrak{H}_\xi(x)$, i.e., under the conditions of Theorem 20.1, the level surface

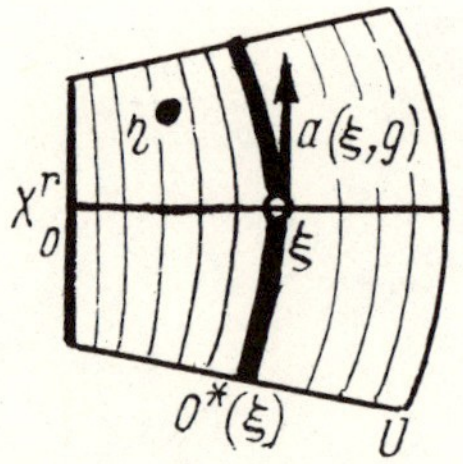

Fig. 229

M_ξ is the orbit of x under the action of $\mathfrak{H}_\xi$, whose Lie algebra is the annihilator of the covector ξ that defines this level surface. In particular, dim Q_ξ = 2n − k − 2 = 0. Therefore, in this case, the structure of the reduced system is extremely simple. Since Q_ξ is a point, then the flow E(F) is zero (see Fig. 228). Here dim N = n. Since M_ξ is a level surface of integrals of G for the flow v, then this flow is tangent to M_ξ in both cases (a) and (b) (see the formulation of the theorem), i.e., M_ξ is an r-dimensional submanifold invariant with respect to all fields of the form sgrad h, where $h \in$ Ann ξ, and sgrad F, where {F, G} = 0. It remains to prove that the level surface is a two-dimensional torus when M_ξ is compact and connected. For this we will need an auxiliary statement.

Proposition 20.7. Let $\xi \in G^*$ be a covector of general position. Then its annihilator Ann ξ is commutative; in particular, $\mathfrak{H}_\xi$ is commutative.

Proof. Consider the coadjoint action of $\mathfrak{G}$ = exp G in G*. Let O*(ξ) be the orbit passing through the point $\xi \in G^*$. Since dim H_ξ = r and dim G* = k, then dim O*(ξ) = k − r. Because ξ is a covector in general position, then orbits close to O*(ξ) are diffeomorphic to it and we may assume that a sufficiently small neighborhood U of ξ is fibrated onto homeomorphic fibers (see Fig. 229). Let X_0 be a local section of the fibration of U onto the orbits of $\mathfrak{G}$-action. We will make use of the fact that U is presentable as the direct product of the base and the fiber which is a part of the orbit (see Fig. 229). Let h(η) be a smooth function on U, constant on orbits. We claim that $a(\xi, dh(\xi)) = 0$, where dh (the differential of h) is interpreted as an element of the dual space (G*)*, i.e., $dh(\xi) \in G$. In other words, we claim that $dh(\xi) \in$ Ann ξ. We must verify that $a(\xi, dh(\xi))(g) = 0$ for any $g \in G$. We have

$$a(\xi, dh(\xi))(g) = \langle a(\xi, dh(\xi)), g\rangle =$$
$$= \langle \xi, [dh(\xi), g]\rangle = -\langle a(\xi, g), dh(\xi)\rangle = 0.$$

Since the covector $a(\xi, g) = \mathrm{ad}_g^*\xi$ belongs to the tangent space $T_\xi O^*$ to the orbit O^* at ξ and the function h is constant on orbits, it is, in particular, constant on O^* (see Fig. 229). On the section X_0, which is the smooth surface of dimension r transversal to orbits close to the orbit $O^*(\xi)$, let us consider the set of r independent functions $h_1, \ldots, h_r$ and extend them up to smooth functions on the whole U, assuming them to be constant along orbits O^*. We get $a(\xi, dh_i(\xi)) = 0$, where $1 \leqq i \leqq r$. Thus, $dh_i(\xi) \in \mathrm{Ann}\,\xi$ for $1 \leqq i \leqq r$. Since (h_i) were chosen to be independent, then all differentials $(dh_i(\xi))$ are independent in Ann ξ and their number is equal to r, i.e., coincides exactly with the dimension of Ann ξ. Thus, differentials $dh_i(\xi)$ constitute a basis in Ann ξ, and to prove the commutativity of the annihilator, it suffices to prove that commutators of any pair of these differentials is zero, i.e., $[dh_i(\xi), dh_j(\xi)] = 0$. Since $a(\xi, dh_i(\xi)) = 0$, we have $a(\xi, dh_i(\xi))(dh_j(\xi)) = 0$, implying $b(\xi) = = \langle a(\xi, dh_i(\xi)), dh_j(\xi)\rangle = \langle \xi, [dh_i(\xi), dh_j(\xi)]\rangle = 0$. Consider an arbitrary direction η in the neighborhood U and take the derivative of the function $b(\xi) \equiv 0$ along this direction, i.e., consider

$$0 = \frac{d}{d\eta} b(\xi) = \frac{d}{d\eta}\langle \xi, [dh_i(\xi), dh_j(\xi)]\rangle =$$
$$= \langle \eta, [dh_i(\xi), dh_j(\xi)]\rangle + \left\langle \xi, \left[\frac{d}{d\eta} dh_i(\xi), dh_j(\xi)\right]\right\rangle +$$
$$+ \left\langle \xi, \left[dh_i(\xi), \frac{d}{d\eta} dh_j(\xi)\right]\right\rangle = \langle \eta, [dh_i(\xi), dh_j(\xi)]\rangle -$$
$$- \left\langle a(\xi, dh_j(\xi)), \frac{d}{d\eta} dh_i(\xi)\right\rangle - \left\langle a(\xi, dh_i(\xi)), \frac{d}{d\eta} dh_j(\xi)\right\rangle =$$
$$= \langle \eta [dh_i(\xi), dh_j(\xi)]\rangle = 0,$$

since $a(\xi, dh_j(\xi)) = a(\xi, dh_i(\xi)) = 0$. Since $\eta \in U \subset G^*$ is arbitrary, then the equality $\langle r, [dh_i(\xi), dh_j(\xi)]\rangle = 0$ implies $[dh_i(\xi), dh_j(\xi)] = 0$. The statement is proved.

Let us return to the proof of Theorem 20.1. We have proved that M_ξ is an orbit of $\mathfrak{H}_\xi$, and since $\dim M_\xi = \dim \mathfrak{H}_\xi$, then M_ξ is a quotient group of $\mathfrak{H}_\xi$ with respect to the discontinuous lattice Γ. Since $\mathfrak{H}_\xi = \exp \mathrm{Ann}\,\xi$ is commutative, then if M_ξ is compact and connected, it is an r-dimensional torus. Other statements are proved as in the proof of the Liouville theorem.

Thus, one method of applying "noncommutative" Theorem 20.1 is the following. If v = sgrad f is a Hamiltonian system on M^{2n}, then we must find a noncommutative Lie algebra G such that the Hamiltonian f is contained in the annihilator of a covector $\xi \in G^*$ of general position. If there is such an algebra and dim G + rank G = 2n, then the flow v moves along the r-dimensional torus T^r, which coincides with the level surface M_ξ, where r = rank G.

20.4. Orbits of the (Co)adjoint Representation as Symplectic Manifolds

Let G be a finite-dimensional Lie algebra, G^* its dual space, and $Ad_{\mathfrak{G}}^*: G^* \to G^*$ the coadjoint representation of $\mathfrak{G} = \exp G$ on G^*. Then G^* is fibrated into orbits O^*. Each of the orbits is a smooth submanifold embedded into the linear space. It turns out that on each orbit a symplectic structure is naturally defined. This structure (the Kirillov form) is defined as follows. Let $x \in G^*$ be an arbitrary point and $\xi_1, \xi_2 \in T_xO^*$ be two arbitrary tangent vectors to the orbit. We must define the value of $\omega_x(\xi_1, \xi_2)$. Recall that each vector ξ tangent to the orbit is presentable in the form $ad_g^*x = a(x, g)$, since $T_xO^* = ad_G^*x$. Therefore, there are elements $g_1, g_2 \in G$ such that $\xi_i = a(x, g_i)$ for i = 1, 2. This presentation is nonunique, but it does not affect further construction. Define the value of the form at the point $x \in O^*$ at tangent vectors ξ_1, ξ_2 to the orbit O^* to be $\omega(\xi_1, \xi_2) = \langle x, [g_1, g_2]\rangle$, where $x \in G^*$, $g_1, g_2 \in G$.

Proposition 20.8. The form ω defined above has the following properties:

1) ω is bilinear and its value does not depend on the arbitrariness of the choice of representatives g_1, g_2;

2) ω is skew symmetric and defines the exterior 2-form on the orbit;

3) ω is nondegenerate on the orbit;

4) ω is closed on the orbit.

If G is compact, then it can be canonically identified with respect to the Killing form with G*, and then the form ω on orbits O of the adjoint representation is defined as $\omega_x(\xi_1, \xi_2) = \langle x, [g_1, g_2]\rangle$, where $\langle\ ,\ \rangle$ is the Killing form and $x, \xi_i, g_i \in G$. Note that, in the general case, the form may be written as follows:

$$\omega_x(\xi_1, \xi_2) = \langle a(x, g_1), g_2\rangle = \langle \xi_1, g_2\rangle = -\langle \xi_2, g_1\rangle.$$

Proof of Proposition 20.8. Let us apply the reduction technique developed above. Consider the cotangent bundle $M = T^*\mathfrak{G}$ to $\mathfrak{G}$ as the direct product $M = T^*\mathfrak{G} = T_E^*\mathfrak{G} \times \mathfrak{G} = G^* \times \mathfrak{G}$. Define the $\mathfrak{G}$-action on this direct product so that $\mathfrak{G}$ acts on the second multiple $\mathfrak{G}$ by left shifts and, trivially, on the first multiple G*. Consider the algebra of integrals V corresponding to the left $\mathfrak{G}$-action on the phase space $T^*\mathfrak{G}$. Clearly, this algebra of integrals is isomorphic to G. Any function $f \in V$ is right-invariant and takes on values according to the formula $f(\xi, g) = \langle Ad_{g^{-1}}^*(\xi), f\rangle$ for $f \in V = G$, $\xi \in G^*$, $g \in \mathfrak{G}$. Thus, the level surface M_ξ corresponding to $\xi \in V^*$ consists of all pairs (η, g) such that $f(\eta, g) = \langle \xi, f\rangle$, i.e., $Ad_{g^{-1}}^*(\eta) = \xi$ or $\eta = Ad_g^*\xi$. Since $(\xi, E) \in M_\xi$, then $M_\xi = \{(Ad_g^*\xi, g);\ g \in \mathfrak{G}\}$. From here we deduce that M_ξ is the bundle with the base $O^*(\xi)$ and the fiber $\mathfrak{H}_\xi$, where O^* are orbits of $\mathfrak{G}$ in G* and $\mathfrak{H}_\xi$ is the maximum torus corresponding to the Cartan subalgebra H_ξ that preserves the covector ξ. We have presented a generic orbit of the coadjoint representation as the quotient manifold $M_\xi/\mathfrak{H}_\xi$, where H_ξ is the annihilator of ξ. Proposition 20.4 implies that O^* is the symplectic manifold. Direct computation shows that a symplectic form that arises on O^* coincides with the canonical form described above.

Chapter 6

GEOMETRY AND MECHANICS

21. THE EMBEDDING OF HAMILTONIAN SYSTEMS INTO LIE ALGEBRAS

21.1. The Formulation of the Problem and Full Sets of Commutative Functions

Let us proceed to the study of several concrete mechanical systems whose complete integrability (in the Liouville sense) is proved. The most interesting are systems that describe

a) the motion of a multidimensional solid body with a fixed point (in the absence of gravity) and the motion of a "free solid body,"

b) the motion, from inertia, of a multidimensional solid body in an ideal fluid.

By the integrability of these (and similar) systems, we will understand their complete integrability in the Liouville sense, i.e., in our approach the main problem is the search for a sufficient number of commuting integrals. Usually, mechanical systems of this kind are described as systems of ordinary differential equations on a Euclidean space. For example, classical equations that describe the motion of a three-dimensional solid body with a fixed point (in the absence of gravity) are described in the three-dimensional Euclidean space $\mathbb{R}^3(x, y, z)$ as follows:

$$\dot{x} = \frac{\lambda_1 - \lambda_2}{\lambda_1 + \lambda_2} yz, \ \dot{y} = \frac{\lambda_3 - \lambda_1}{\lambda_3 + \lambda_1} xz, \ \dot{z} = \frac{\lambda_2 - \lambda_3}{\lambda_2 + \lambda_3} xy,$$

where λ_1, λ_2, λ_3 are real numbers. It turns out that several Hamiltonian systems in $\mathbf{R}^n$ have a hidden algebraic structure whose discovery enables one to integrate such systems. There are very many examples, but we will distinguish among them the structures related to Lie algebras. In the simplest form, this means that the studied system preserves the orbits of the (co)adjoint representation of a Lie group whose Lie algebra is identified with the Euclidean space where the system is defined; cf. [23-26].

Definition 21.2. We will say that the Hamiltonian system v on $\mathbf{R}^n$ admits an embedding into the Lie algebra G if $\mathbf{R}^n$ may be identified with the dual space G* of a Lie algebra of a Lie group $\mathfrak{G}$ so that

1) v is tangent (after this identification) to orbits O* of the coadjoint representation of $\mathfrak{G}$ in G* = $\mathbf{R}^n$, i.e., all orbits O* are invariant with respect to v;

2) v is Hamiltonian on orbits O* with respect to the canonical symplectic form ω described in Sec. 20, and is of the form v = sgrad f, where f is a function on O* (see Fig. 230).

The class of such systems contains important mechanical examples. Obviously, far from all systems admit an embedding into a Lie algebra because conditions 1 and 2 of Definition 21.1 place quite rigid restrictions on the structure of a system. At the same time, it is clear that the existence

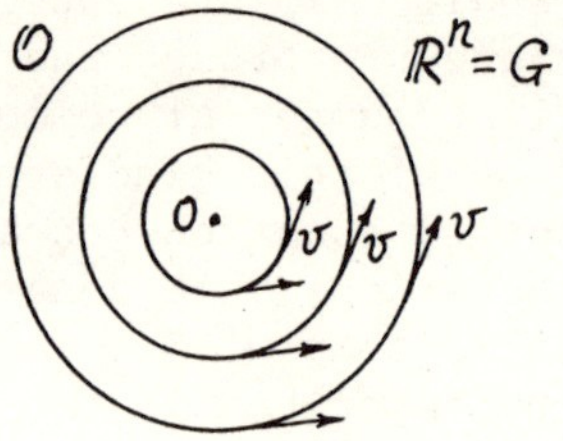

Fig. 230

of such an embedding enables one to apply the developed apparatus of the Lie algebra theory to the search for integrals of the system. If the system is defined on $\mathbf{R}^n$, then one of the methods of its integration is the search for its embeddings into an appropriate Lie algebra. It is worthwhile to mention the ambiguity in the realization of this problem. First, $\mathbf{R}^n$ may be identified with Lie algebras of different Lie groups; therefore, we must select from all the Lie algebras of the given dimension n. The number of algebras increases with increasing n. Second, in some cases, it turns out that it suffices to present the system v as a Hamiltonian system on one orbit O^* and not on all orbits. Since in a Lie algebra there are several orbits (generic orbits, singular orbits), this opens up the possibility of choosing while finding an embedding of the system into a Lie algebra.

In the above simple example of the system in $\mathbf{R}^3$, the required embedding into a Lie algebra exists and is quite simple. For this, it suffices to identify $\mathbf{R}^3 = G^* = G$ with the three-dimensional Lie algebra of the orthogonal group SO_3, i.e., with the space so_3 of skew symmetric real (3×3)-matrices. Here $x = x_{12}$, $y = x_{13}$, $z = x_{23}$, i.e., $(x, y, z) \to$

$\to \begin{pmatrix} 0 & x & y \\ -x & 0 & z \\ -y & -z & 0 \end{pmatrix} \in so_3$. As we already know, orbits of the adjoint representation of SO_3 on so_3 are standard two-dimensional spheres with center at the origin. The vector field $v = (v_x, v_y, v_z)$ that describes the motion of the solid body with a fixed point has the following two integrals: $P_\psi = (\lambda_1 + \lambda_2)x^2 + (\lambda_1 + \lambda_3)y^2 + (\lambda_2 + \lambda_3)z^2$ and $Q_\psi = (\lambda_1 + \lambda_2)^2x^2 + (\lambda_1 + \lambda_3)^2y^2 + (\lambda_2 + \lambda_3)^2z^2$. It is easy to verify that the derivative of these functions with respect to v is zero. The algebraic structure of the system and of its integrals becomes clear after the coordinate transform in $\mathbf{R}^3$: $x \to x/(\lambda_1 + \lambda_2)$, $y \to y/(\lambda_1 + \lambda_3)$, $z \to z/(\lambda_2 + \lambda_3)$. Then both integrals become $P_\varphi = x^2/(\lambda_1 + \lambda_2) + y^2/(\lambda_1 + \lambda_3) + z^2/(\lambda_2 + \lambda_3)$, and $Q_\varphi = x^2 + y^2 + z^2$, accordingly. After this transform, the field v becomes tangent to two-dimensional spheres; therefore, the system admits an embedding into a three-dimensional Lie algebra, since the canonical 2-form on orbits S^2 in so_3 coincides with the 2-form of the invariant Riemannian volume. The Euclidean scalar product coincides here with the Killing form.

Surely, the same system may have several embeddings into different Lie algebras. What is the advantage of the existence of an embedding of v into a Lie algebra from the point of view of the search for integrals? First, consider the following general problem. As is clear from the aforementioned, the following problem is quite natural: On the symplectic manifold M^{2n}, find n commutative independent functions. In other words, on M^{2n}, find a commutative (with respect to the Poisson bracket) n-dimensional Lie algebra of functions whose additive basis consists of smooth functions functionally independent almost everywhere on M. Such sets of functions H(M) will be briefly called full commutative sets. We are not interested yet in integration of Hamiltonian systems. If we have found a full commutative set on M, then we automatically obtain a series of completely integrable Hamiltonian systems in M: it suffices to consider fields sgrad f, where f is an element of H(M). Then the n-dimensional space of functions H(M) is a set of integrals for the system with the Hamiltonian f. Such an approach to problems of Hamiltonian mechanics considers as the main problem the construction on symplectic manifolds of the maximum possible number of different full commutative sets. The greater the supply, the more examples of completely integrable (in the commutative sense) systems we obtain. As we will see, there are important examples of manifolds where such full sets can be given explicitly. In particular, on any smooth symplectic manifold there always exists a full commutative set of smooth functions. The more precise question is whether there exists on any algebraic symplectic manifold a full commutative set consisting of algebraic (rational) functions. Perhaps there are topological obstructions that make it impossible to construct such an analytical (algebraic) set on an arbitrary symplectic manifold. If such manifolds (i.e., those that do not admit a full set) do exist, then any analytical (algebraic) Hamiltonian system on such a manifold is not completely integrable in the Liouville sense. The algebraic variant of the problem is of special interest since in concrete examples of already integrated systems the central role is played by polynomial (rational, algebraic) integrals. The most natural class of functions among which it is well worth searching for full commutative sets on algebraic varieties is the class of polynomial (rational or algebraic) functions.

Until now, full commutative sets were found on symplectic manifolds of the important class of orbits of the coadjoint representation of various Lie groups. In [21, 22], the following conjecture is formulated.

Conjecture A. Let G be an arbitrary finite-dimensional Lie algebra. Then there is a linear space of smooth functions on G^* whose restrictions onto generic orbits of the coadjoint representation of $\mathfrak{G} = \exp G$ constitute the full commutative set $H(O^*)$ on these orbits O^*, i.e., any pair of functions $f, g \in H(O^*)$ is in involution on orbits O^* with respect to the canonical form ω; the additive basis $f_1, \ldots, f_k$ in $H(O^*)$ consists of functions functionally independent almost everywhere in O^*, and the dimension of $H(O^*)$ is equal to the halved dimension of a generic orbit, i.e., $k = \frac{1}{2}\dim O^* = \frac{1}{2}(\dim G - - \operatorname{rank} G)$.

This conjecture is proved for all semisimple and reductive Lie algebras [27, 28], and also for various classes of noncompact real Lie algebras (see, for example, [29, 17]). It turns out that the full commutative sets found during the proof of the conjecture contain Hamiltonians of important mechanical systems which makes it possible to integrate them completely. We will study them in greater detail below. The meaning of Conjecture A is not restricted by the possibility of producing examples of integrable systems. Its validity would imply that Theorem 20.2 holds not only for compact but also for noncompact symplectic manifolds, i.e., in this case, any Hamiltonian system integrable in a noncommutative sense would automatically be integrable in the commutative sense. More precisely, the following statement holds (see [21]). Let the Lie algebra G of functionally independent functions be defined on the symplectic manifold M, where $\dim G + \operatorname{rank} G = = \dim M$ (see Theorem 20.1). Then if Conjecture A holds for G, there exists another now commutative Lie algebra of independent functions G' such that $\dim G' = \frac{1}{2}\dim G$. Let us now return to systems v that can be embedded into a finite-dimensional Lie algebra G (see Definition 21.1). If G satisfies Conjecture A, then on generic orbits $O^* \subset G^*$ there is a full commutative set of functions $H(O^*)$ (see above). If besides, the Hamiltonian f, where $v = \operatorname{sgrad} f$, belongs to $H(O^*)$, then we obtain a full commutative set of integrals for v. The knowledge of full commutative sets on orbits in G^* makes it possible to integrate in principle Hamiltonian systems that can be embedded in G^*.

Thus, one of the methods for integration of systems on $\mathbb{R}^n$ is the following:

a) let us present the system as a Hamiltonian system on orbits in G^* for an appropriate Lie algebra G;

b) if such a presentation exists, then find on $O^* \subset G^*$ a full commutative set of functions that contains the Hamiltonian of the system.

There are several tricks that enable one to construct full commutative sets on orbits. The method based on the idea of a shift of invariants of the coadjoint representation by a covector of general position is very effective (see the description below).

21.2. Equations of Motion of a Multidimensional Solid Body with a Fixed Point and Their Analogs on Semisimple Lie Algebras. Complex Series

Let G be a semisimple Lie algebra, $\langle\ ,\ \rangle$ the Killing form, and f a smooth function on G. To any such function assign a Hamiltonian system on the cotangent bundle $T^*\mathfrak{G}$ to $\mathfrak{G} = \exp G$, extending f up to a left-invariant function F defined on the whole $T^*\mathfrak{G}$. Since $T^*\mathfrak{G}$ is a symplectic manifold, then taking F as a Hamiltonian we obtain on $T^*\mathfrak{G}$ a Hamiltonian system.

This system is left-invariant and splits into two systems, one of which is defined on the cotangent space to the unit of the group isomorphic to G and is usually called the system of Euler equations. These equations admit a simple description. Let grad f be a field on G dual to df, i.e., $\langle \operatorname{grad} f, \xi\rangle = \xi(f)$. Then Euler equations are written as a commutator: $\dot{x} = [x, \operatorname{grad} f(x)]$. Cases of geodesic flows for a left-invariant matrix on $\mathfrak{G}$ are of special interest. Here f is a nondegenerate quadratic form on G and grad f(x) is defined by a self-conjugate linear operator in G, i.e., grad f(x) is of the form φX, where $\varphi: G \to G$ is self-adjoint. Now let us write equations that describe the motion of a multidimensional solid body with a fixed point (outside of gravity).

Let $G = so_n$ be the Lie algebra of the orthogonal group and I the diagonal real matrix $I = \begin{pmatrix} \lambda_1 & & 0 \\ & \ddots & \\ 0 & & \lambda_n \end{pmatrix}$, where $\lambda_i \neq \lambda_j$ for $i \neq j$. On so_n, consider the operator $\psi X = IX + XI$. Then equations $\psi\dot{X} = [X, \psi X]$ are called the equations of motions of an n-dimensional solid body. Let us write them explicitly, making use of the standard coordinates in so_n. Let us present so_n as the algebra of skew symmetric real matrices $X = (x_{ij})$; then $\psi X = ((\lambda_i + \lambda_j)x_{ij})$. Clearly, $\dot{x}_{ij} = \left(\frac{\lambda_j - \lambda_i}{\lambda_j + \lambda_i}\right) \sum_{q=1}^{n} x_{iq} x_{qj}$ (verify!). For n = 3, we get

$$\dot{x}_{12} = \frac{\lambda_2 - \lambda_1}{\lambda_2 + \lambda_1} x_{13} x_{32}, \quad \dot{x}_{13} = \frac{\lambda_3 - \lambda_1}{\lambda_3 + \lambda_1} x_{12} x_{23}, \quad \dot{x}_{23} = \frac{\lambda_3 - \lambda_2}{\lambda_3 + \lambda_2} x_{21} x_{13}.$$

Identifying so_3 with $\mathbf{R}^3$, we see that these equations coincide with the classical equations of the dynamic motion of a three-dimensional body (see above). This is the reason why equations $\psi\dot{X} = [X, \psi X]$ for an arbitrary n are called the equations of motion of a multidimensional solid body. Let I be chosen so that $\lambda_i + \lambda_j \neq 0$ for all i, j. Then the operator ψ is invertible on so_n and the inverse operator is of the form $(\varphi X)_{ij} = \frac{1}{\lambda_i + \lambda_j} x_{ij}$. In so_n, perform the coordinate transform $Y = \psi X$. Then the equation of motion of the solid body is presented in the form $\dot{Y} = [\psi^{-1} Y, Y]$. Multiplying it by −1 and redenoting Y by X, we get the Euler equation $\dot{X} = [X, \varphi X]$, where $\varphi: so_n \to so_n$ is a linear self-adjoint operator. In what follows we will study equations of this form. In coordinate notation, we have

$$\dot{x}_{ij} = (\lambda_i - \lambda_j) \sum_{q=1}^{n} \frac{x_{iq} x_{qj}}{(\lambda_j + \lambda_q)(\lambda_q + \lambda_i)} = \sum_{q=1}^{n} x_{iq} x_{qj} \left(\frac{1}{\lambda_j + \lambda_q} - \frac{1}{\lambda_i + \lambda_q} \right).$$

For n = 3, we get

$$\dot{x}_{12} = \left(\frac{1}{\lambda_2 + \lambda_3} - \frac{1}{\lambda_1 + \lambda_3} \right) x_{13} x_{32},$$

$$\dot{x}_{13} = \left(\frac{1}{\lambda_3 + \lambda_2} - \frac{1}{\lambda_1 + \lambda_2} \right) x_{12} x_{23},$$

$$\dot{x}_{23} = \left(\frac{1}{\lambda_3 + \lambda_1} - \frac{1}{\lambda_2 + \lambda_1} \right) x_{21} x_{13};$$

i.e.,

$$\dot{x} = \frac{-\lambda_1 + \lambda_2}{(\lambda_2 + \lambda_3)(\lambda_1 + \lambda_3)} yz, \quad \dot{y} = \frac{\lambda_1 - \lambda_3}{(\lambda_2 + \lambda_3)(\lambda_1 + \lambda_2)} xz,$$
$$\dot{z} = \frac{\lambda_3 - \lambda_2}{(\lambda_3 + \lambda_1)(\lambda_1 + \lambda_2)} xy.$$

The direct computation shows that the following two polynomials are functionally independent integrals of this system in $\mathbb{R}^3$:

$$x^2 + y^2 + z^2, \quad \frac{x^2}{\lambda_1 + \lambda_2} + \frac{y^2}{\lambda_1 + \lambda_3} + \frac{z^2}{\lambda_2 + \lambda_3}.$$

Evidently, these functions coincide with polynomials Q_φ and P_φ, obtained above from integrals Q_ψ and P_ψ of the flow $\psi\dot{X} = [X, \psi X]$ with respect to the form $Y = \psi X$, which defines mappings $P_\psi \to P_\varphi$, $Q_\psi \to Q_\varphi$ and changes the flow to the form $\dot{X} = [X, \varphi X]$. Integrals are defined nonuniquely. For the following, it is useful to note that they can be chosen as follows: $x^2 + y^2 + z^2$ and $x^2(\lambda_1{}^2 + \lambda_2{}^2) + y^2(\lambda_1{}^2 + \lambda_3{}^2) + z^2(\lambda_2{}^2 + \lambda_3{}^2)$. Verify that these functions are integrals of the flow $\dot{X} = [X, \varphi X]$.

Thus, for $n = 3$, we have shown an embedding of this system into the Lie algebra so_3. It turns out that a similar embedding exists for any n.

Proposition 21.1. The vector field $\dot{X} = [X, \varphi X]$, where $\varphi = \psi^{-1}$, $\psi X = IX + XI$, is tangent to all orbits of the adjoint representation of SO_n on its Lie algebra so_n. This field is Hamiltonian on orbits.

Proof. Let O be an orbit that passes through $X \in so_n = so_n^*$. Then $T_XO = \{[X, Y], Y \in so_n\}$. Therefore, $[X, \varphi X] \in T_XO$. Furthermore, $\dot{X} = \text{sgrad } F$, where $F(X) = \langle X, \varphi X\rangle$, as required.

Let us define analogs of the equations of motion of a solid body on an arbitrary semisimple Lie algebra. We will produce many-parameter families of operators $\varphi: G \to G$, not only for complex semisimple Lie algebras, but also for their compact and normal forms. It turns out that the systems $\dot{X} =$

$= [X, \varphi X]$ are completely integrable on generic orbits, and hence their integrals define full commutative sets of functions on semisimple Lie algebras and on several of their real forms.

Let G be a complex semisimple Lie algebra and $G = T \oplus V^+ \oplus V^-$ its root decomposition. Let a, $b \in T$, $a \neq b$ be two arbitrary regular elements of the Cartan subalgebra. Consider the operator $ad_a : G \to G$. Clearly, $ad_a|T \equiv 0$, $ad_a : V^+ \to V^+$, $ad_a : V^- \to V^-$, i.e., ad_a preserves the root decomposi- of G over $\mathbf{C}$. In fact, $ad_a E_\alpha = \alpha(a) E_\alpha$ for any $\alpha \in \Delta$. We assume that a and b are in general position, i.e., $\alpha(a) \neq 0$, $\alpha(b) \neq 0$. Then ad_a and ad_b are invertible on $V^+ \oplus V^- = V$, namely $ad_a^{-1} = E_\alpha = E_\alpha/\alpha(a)$. Define the linear operator $\varphi_{a,b,D} : G \to G$ by the formula $\varphi_{a,b,D} X = \varphi_{a,b} X' + D(t) = ad_a^{-1} \times ad_b X' + D(t)$, where $X = X' + t$ is the unique decomposition of X with respect to V and T, and $D : T \to T$ is an arbitrary linear operator which is symmetric on T with respect to the Killing form. The operator $\varphi_{a,b,D}$ is parametrized by a, b, D. Clearly, $\varphi_{a,b,D} E_\alpha = [\alpha(b)/\alpha(a)] E_\alpha$. In the Weyl basis $(E_\alpha, E_{-\alpha}, H_\alpha')$ the operator φ is defined by the matrix

$$\begin{array}{c} E_\alpha \\ \\ E_{-\alpha} \\ \\ H_\alpha' \end{array} \left(\begin{array}{c|c|c} \begin{matrix} \lambda_1 & & 0 \\ & \ddots & \\ 0 & & \lambda_q \end{matrix} & 0 & 0 \\ \hline 0 & \begin{matrix} \lambda_1 & & 0 \\ & \ddots & \\ 0 & & \lambda_q \end{matrix} & 0 \\ \hline 0 & 0 & D \end{array} \right) = \varphi_{a,b,D},$$

and $\varphi_{a,b}$ maps V onto V, where $\lambda_\alpha = \alpha(b)/\alpha(a)$, $q = \dim V^+ =$ = (number of roots $\alpha > 0$).

<u>Proposition 21.2</u>. $\varphi_{a,b,D}$ is symmetric with respect to the Killing form for any a, b, D that satisfy the mentioned restrictions.

<u>Proof</u>. Denote the Weyl basis in V by (e_i). It suffices to verify that $\langle \varphi e_i, e_j \rangle = \langle e_i, \varphi e_j \rangle$ for any i, j. We may assume that $i \neq j$. Recall that T is orthogonal to $V = V^+ \oplus V^-$. Since φ transforms V and T into themselves and D is symmetric on T, it suffices to verify the symmetry of $\varphi_{a,b}$ on V. Since E_α, where $\alpha \neq 0$ are eigenvectors of φ, then $\langle (\alpha(b)/\alpha(a)) E_\alpha, E_\beta \rangle = \langle E_\alpha, (\beta(b)/\beta(a)) E_\beta \rangle = 0$ for $\alpha + \beta \neq 0$,

since $\langle E_\alpha, E_\beta \rangle = 0$. If $\alpha + \beta = 0$, then $\alpha(b)/\alpha(a) = (-\alpha)(b)/(\alpha)(a)$. The proposition is proved.

The operator φ on V has in the general position q different eigenvalues of multiplicity 2. The operator $\varphi: V \to V$ is an isomorphism of V onto itself. Recall that V^+ is a nilpotent subalgebra; for instance, in our model example it is the subalgebra of upper-triangular matrices with zeros on the main diagonal. Since V^+ is generated by vectors E_α, where $\alpha > 0$, then $\varphi|V^+$ is symmetric with respect to the Killing form. In the general position, eigenvalues of this operator are different: $\lambda_1, \ldots, \lambda_q$. This series will be called the normal nilpotent series. By construction, one normal nilpotent series corresponds to each complex series. The operator $\varphi: G \to G$ transforms the subalgebra $V^+ \oplus T$ into itself and $\varphi|V^+ \oplus T$ is an automorphism of $V^+ \oplus T$. In our model example, $V^+ \oplus T$ is the subalgebra of upper-triangular matrices. As above, all eigenvalues of $\varphi|V^+ \oplus T$ are different and the operator is symmetric. Thus, to each complex series, a normal solvable series corresponds. In the Weyl basis, operators $\varphi|V^+$ and $\varphi|V^+ \oplus T$ are described by the matrices

$$\varphi|_{V^+} = \begin{pmatrix} \lambda_1 & & 0 \\ & \ddots & \\ 0 & & \lambda_q \end{pmatrix}, \quad \varphi|_{V^+ \oplus T} = \left(\begin{array}{ccc|c} \lambda_1 & & 0 & \\ & \ddots & & 0 \\ 0 & & \lambda_q & \\ \hline & 0 & & D \end{array}\right).$$

Thus, on each semisimple Lie algebra, we have constructed Hamiltonian systems $\dot{X} = [X, \varphi_{a,b,D} X]$ that are analogs of the equations of motion of a solid body and are completely integrable (see below). In a special case, we get equations on so_n.

21.3. Hamiltonian Systems of Compact and Normal Series

Here we will construct a similar family of Hamiltonian systems on an arbitrary simple compact real algebra making use of compact real forms of complex simple Lie algebras (see Sec. 17 and Theorem 17.1). Each semisimple complex Lie algebra G possesses a compact form G_u. We have constructed the canonical embedding of this compact subalgebra into the complex Lie algebra G. Recall that $G_u = \{E_\alpha + E_{-\alpha}, i(E_\alpha + E_{-\alpha}), iH_\alpha'\} = W^+ \oplus iT_0$. As in the preceding section, we will define the symmetric operator $\varphi: G_u \to G_u$ that defines the Hamiltonian system $\dot{X} = [X, \varphi X]$ on G_u, preserving the fibration of G_u into orbits of the adjoint representation. Let a, $b \in iT_0$ be elements in general position. Since $\mathrm{ad}_a E_\alpha = [a, E_\alpha] = i[a', E_\alpha]$, where $\alpha = ia'$, $a' \in T_0$, then $\mathrm{ad}_a E_\alpha = i\alpha(a')E_\alpha$, where $\alpha(a')$ is real. Hence,

$$\mathrm{ad}_a (E_\alpha + E_{-\alpha}) = \alpha(a')\,(i(E_\alpha - E_{-\alpha})),$$
$$\mathrm{ad}_a (i(E_\alpha - E_{-\alpha})) = -\alpha(a')\,(E_\alpha + E_{-\alpha}).$$

Thus, $\mathrm{ad}_a: W^+ \to W^-$ transforms $E_\alpha - E_{-\alpha}$ into a vector proportional to $i(E_\alpha - E_{-\alpha})$ and vice versa. The operator ad_b acts similarly and, due to the choice of $a \in iT_0$, the operator ad_a is invertible on W^+. Then vectors $E_\alpha + E_{-\alpha}$, $i(E_\alpha - E_{-\alpha})$ are eigenvectors of the operator $\varphi_{a,b} = \mathrm{ad}_a^{-1}\mathrm{ad}_b: W^+ \to W^+$ with eigenvalues $\alpha(b)/\alpha(a) = \alpha(b')/\alpha(a')$, where $a = ia'$, $b = ib'$, a', $b' \in T_0$. The case W^- is similar. Define the operator $\varphi_{a,b,D}: G_u \to G_u$ by the formula $\varphi X = \varphi(X' + t) = \varphi_{a,b}X' + D(t) = \mathrm{ad}_a^{-1} \cdot \mathrm{ad}_b X + D(t)$, where $X = X' + t$ is the unique decomposition of X in $G_u = W^+ \oplus iT_0$, where $X' \in W^+$, $t \in iT_0$ and $D: iT_0 \to iT_0$ is an arbitrary linear operator symmetric on iT_0. In the basis $(E_\alpha + E_{-\alpha},\ i(E_\alpha - E_{-\alpha}),\ iH_\alpha')$, the operator φ is defined by the matrix

$$\begin{array}{c} E_\alpha + E_{-\alpha} \\ \\ i(E_\alpha - E_{-\alpha}) \\ \\ iH_\alpha' \end{array} \left(\begin{array}{c|c|c} \begin{matrix} \lambda_1 & & 0 \\ & \ddots & \\ 0 & & \lambda_q \end{matrix} & 0 & \\ \hline 0 & \begin{matrix} \lambda_1 & & 0 \\ & \ddots & \\ 0 & & \lambda_q \end{matrix} & \\ \hline & & D \end{array}\right) = \varphi_{a,b,D},$$

where $\lambda_\alpha = \alpha(b)/\alpha(a)$ are real and $q = \dim W^+$.

Proposition 21.3. The operator $\varphi: G_u \to G_u$ is symmetric for any a, b, D satisfying the above restrictions.

Proof. Our considerations are similar to the proof of Proposition 21.2. We need only verify the orthogonality of the chosen basis in W^+. Recall that iT_0 is orthogonal to W^+. Furthermore, $\langle E_\alpha + E_{-\alpha}, i(E_\alpha - E_{-\alpha})\rangle = 0$ and the orthogonality of other vectors is proved above.

The operator $\varphi_{a,b}:W^+ \to W^+$ has in the general position q different eigenvalues each of multiplicity two.

Now let us construct the similar family of Hamiltonian systems on several simple compact real Lie algebras that correspond to classical normal compact subalgebras. In Sec. 17, we have shown in each compact form G_u the subalgebra G_n, spanned by the vectors $E_\alpha + E_{-\alpha}$, $\alpha \in \Delta$. Since all these vectors are eigenvectors of operators φ of the compact series, then by restricting them onto the subalgebra G_n we get the normal series. These operators just coincide with $\varphi_{a,b}:G_n \to G_n$, where $\varphi X = \mathrm{ad}_a^{-1}\mathrm{ad}_b X$ for $X \in G_n$, a, $b \in iT_0$, $\alpha(a) \neq 0$, $\alpha(b) \neq 0$. In the basis $(E_\alpha + E_{-\alpha})$, operators φ are defined by the matrices

$$\varphi_{ab} = \begin{pmatrix} \lambda_1 & & 0 \\ & \ddots & \\ 0 & & \lambda_q \end{pmatrix}, \quad q = \dim W^+.$$

Note here that a, $b \notin G_n$, i.e., to define operators of the normal series we need elements of a larger algebra. This is the difference between the normal series and the compact and complex ones for which elements a and b belong to the studied algebra. Not all compact semisimple Lie algebras can be presentable in the form G_n for a compact real form $G_u \subset G$. Below we list all such simple Lie algebras. As was shown, G_n coincides with fixed points of the automorphism $\tau:G \to G$, $\tau X = \bar{X}$ after its restriction onto G_u. Let $P \subset G_u$ be the subspace orthogonal to G_n in G_u, where $\tau = -1$. Then the following commutation relations are evident $[G_n, G_n] \subset G_n$, $[P, P] \subset G_n$, $[G_n, P] \subset P$. As we know, this defines a symmetric space $\mathfrak{G}_u/\mathfrak{G}_n$. Then P is identified with the tangent space to $\mathfrak{G}_u/\mathfrak{G}_n$, which is canonically embedded in $\mathfrak{G}_u$ as the Cartan model. Let us list all normal forms preserving the standard notation for the corresponding symmetric spaces.

Type AI. $G = sl(n, \mathbb{C})$, $G_u = su_n$, $G_n = so_n$, $\sigma X = \bar{X}$, $n > 1$. The algebra G_n is realized in G_u as the subalgebra of real skew symmetric matrices.

Type BDI. $G = so(p + q, \mathbb{C})$, $so(p, q)$ is the Lie algebra of the component of the unit of $SO(p, q)$. The Lie algebra $so(p, q)$ is realized in $sl(p + q, \mathbb{R})$ by matrices

$\begin{pmatrix} X_1 & X_2 \\ X_2^T & X_3 \end{pmatrix}$, where all X_i are real, X_1 and X_3 are skew symmetric of order p and q, respectively, and X_2 is arbitrary. Furthermore, $G_u = so_{p+q} \supset so_p \oplus so_q$, $p > 1$, $q < 1$, $p + q \neq 4$. For $p = q$ and $p = q + 1$ we get the normal forms, i.e., $G_n = so_q \oplus \oplus so_q$ and $G_n = so_q \oplus so_{q+1}$.

<u>Type CI</u>. $G = sp(n, \mathbf{C})$, $n \geqq 1$, where $sp(n, \mathbf{R})$ stands for the Lie algebra of matrices $\begin{pmatrix} X_1 & X_2 \\ X_3 & -X_1^T \end{pmatrix}$, where X_i are real matrices of order n, and X_2 and X_3 are symmetric. Furthermore, $G_u = sp_n$, $G_n = u_n$, and the embedding $G_n \to G_u$ is defined by the formula $A + iB \to \begin{pmatrix} A & B \\ -B & A \end{pmatrix}$, where $A + iB \in u_n$, A and B are real.

The above list exhausts all normal forms $G_n \subset G_u$ such that G_u is a classical simple Lie algebra, i.e., of the type A_n, B_n, C_n, or D_n. Besides these forms, there are several normal forms generated by exceptional Lie algebras whose description we will omit.

In conclusion, let us show that among Hamiltonian systems of normal series there are contained classical equations of motion of a multidimensional solid body with a fixed point (see Sec. 21.2). Consider the Lie algebra so_n and present it as a normal form in the Lie algebra su_n (see above). Consider the standard embedding of su_n in u_n and two regular elements a, b of the Cartan subalgebra iT_0 in u_n (not in su_n!). Let $a = \begin{pmatrix} ia_1 & & 0 \\ & \ddots & \\ 0 & & ia_n \end{pmatrix}$, $b = \begin{pmatrix} ib_1 & & 0 \\ & \ddots & \\ 0 & & ib_n \end{pmatrix}$, where $a_i, b_i \in \mathbf{R}$ and $a_i \neq \pm a_j$, $b_i \neq \pm b_j$ for $i \neq j$. Then the operator $\varphi_{a,b}: G_n \to \to G_n$ acts by the formula $\varphi_{a,b}(E_\alpha + E_{-\alpha}) = (\alpha(b)/\alpha(a))(E_\alpha + +E_{-\alpha})$. Since each root α is defined by a pair of indices (i, j), i.e., $\alpha = \alpha_{ij}$ (see above), then each eigenvector $E_\alpha + + E_{-\alpha}$ corresponding to the pair (i, j) is multiplied by the eigenvalue $\lambda_{ij} = (b_i - b_j)/(a_i - a_j)$. Thus, basic skew symmetric matrices $E_{ij} = T_{ij} - T_{ji} = \begin{pmatrix} \ddots & & 1 \\ & \ddots & \\ -1 & & \ddots \end{pmatrix}$ are multiplied under the action of φ by λ_{ij}. Therefore, the Hamiltonian system $\dot{X} = [X, \varphi X]$ is of the form

$$\dot{x}_{ij} = \sum_{q=1}^{n} x_{iq}x_{qj}(\lambda_{qj} - \lambda_{iq}) = \sum_{q=1}^{n} x_{iq}x_{qj}\left(\frac{b_q - b_j}{a_q - a_j} - \frac{b_i - b_q}{a_i - a_q}\right).$$

Now let $a = -ib^2$, i.e., $a_p = b_p{}^2$. Then

$$\dot{x}_{ij} = \sum_{q=1}^{n} x_{iq}x_{qj}\left(\frac{1}{a_j + a_q} - \frac{1}{a_i + a_q}\right).$$

Thus, for $a = -ib^2$, we get the familiar system (see Sec. 21.2) of equations of motion of a solid body with a fixed point. Moreover, among operators $\varphi_{a,b}$ of the normal series is contained the classical operator $\psi X = IX + XI$, where I is a real diagonal matrix. In fact, let $b = -ia^2$; then

$$\varphi_{ab}E_{ij} = \frac{b_i - b_j}{a_i - a_j}E_{ij} = (a_i + a_j)E_{ij},$$

i.e.,

$$\psi = \varphi_{a,-ia^2};\ \varphi_{ab}X = IX + XI,$$

where $I = -ia^2$. Thus, we have included the classical Hamiltonian system of the equations of motion of a solid body with a fixed point (without any potential) into a many-parameter family of similar Hamiltonian systems defined naturally on simple compact Lie algebras.

21.4. Sectional Operators and the Corresponding Dynamical Systems on Orbits

The above examples of dynamical systems are related to compact and semisimple groups. However, some Hamiltonian systems defined on $\mathbf{R}^n$ do not admit, in principle, an embedding into compact Lie algebras; therefore, to complete integration of such systems, it is necessary to consider noncompact Lie algebras. This situation arises, for instance, during the study of the equations of motion of a solid body from inertia in an ideal fluid. Thus, we arrive at the necessity of finding "noncompact analogs" of Hamiltonian systems of the form $\dot{X} = [X, \varphi_a bDX]$ described above. Since these systems are completely defined by the operator $\varphi_a bD$, i.e., by the Hamiltonian $F = \langle X, \varphi_a bDX\rangle$, where X = sgrad F, then, to find analogs of these systems in the noncompact case, we must give "noncompact Hamiltonians," defined by "noncompact operators" φ. This problem requires a wider approach since the operators φ described above essentially exploit the structure of semisimple Lie algebras. In particular, an important part of these operators is the operator $ad_a^{-1}ad_b$ which is well defined only in the semisimple case where there is a root decomposition and where ad_a is invertible on $V^+ \oplus V^-$. If the Lie algebra is noncompact, then there is no natural analog of the root decomposition and we need new considerations that enable us to include the noncompact case. It turns out that analogs of the "solid-body operators" $\varphi_a b,D$ exist also in the noncompact case (see [9, 27]). Let us describe them.

Let H be a Lie algebra, $\mathfrak{H}$ the corresponding group, $\rho:H \to \text{End } V$ the representation of H in the linear space V, and $\alpha: \mathfrak{H} \to \text{Aut } V$ the corresponding representation of the group. Let O(X) be an orbit of the $\mathfrak{H}$-action on V, where $X \in V$. If a linear operator $Q:H \to V$, which we will call the sectional operator, is defined, then a natural vector field $\dot{X}_Q = \rho(QX)X$ arises on orbits. The definition of such an operator Q also sometimes enables one to define a symplectic structure on orbits. In applications, an important role is played by the special class of sectional operators that form a many-parameter family, the main parameters being two elements $a \in V$, $b \in \text{Ker } \Phi_a$, where $\Phi_a h = (\rho h)a$. For instance, in the special case $H = so_n$ and $\rho = ad$, the field $\dot{X}_Q$ coincides with the equations of motion of a multidimensional solid body with a fixed point (in the absence of gravity). Thus, let a be an arbitrary generic point, i.e., the orbit that passes through a is of maximum dimension. Let K(H) be the annihilator of a and $K = \text{Ker } \Phi_a$, where Φ_a is defined above. If a is generic, then the dimension of K is the minimum possible. Let $b \in K$ be an arbitrary element. Consider the action ρb on V. Substitute M for $\text{Ker } (\rho b) \subset V$. Let K' be an arbitrary algebraic complement to K in H, i.e., $H = K + K'$ and $K \cap K' = 0$. The choice of K' is nonunique, and the possibility of varying this complement is responsible for the appearance of families of parameters in the construction. Clearly, $a \in M$. By the definition of K', the mapping $\Phi_a:H \to V$ transforms K' monomorphically into a subspace $\Phi_a K' \subset V$. Since $\Phi_a K' = \Phi_a H$, the subspace $\Phi_a K'$ does not depend on the choice of K' and is

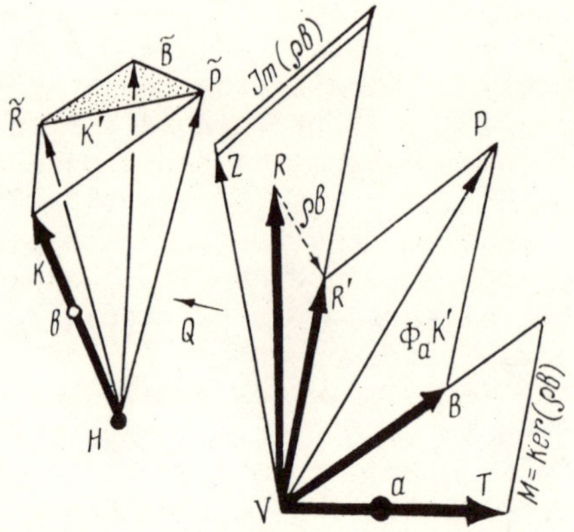

Fig. 231

uniquely defined by the choice of a and by ρ. Suppose there exists a b such that V splits into the sum of two subspaces M and Im (ρb), i.e., $V = M \oplus \mathrm{Im}\,(\rho b)$. For instance, for b we may take semisimple elements of K. The space $\Phi_a K'$ intersects with M and Im (ρb) in subspaces that will be denoted by B and R', respectively. We obtain the decomposition of $\Phi_a K'$ into the direct sum of three subspaces B + R' + P, where B and R' are uniquely defined and the complementary space P is chosen nonuniquely and has its own set of parameters. Consider the action ρb on Im (ρb); then ρb maps Im (ρb) onto itself isomorphically (see Fig. 231). In particular, ρb is invertible on Im (ρb). Let $(\rho b)^{-1}$ be the inverse of ρb on Im(ρb). Let $R = (\rho b)^{-1}R'$; then $\rho b: R \to R'$. The subspace R is uniquely defined. In Im (ρb) consider an algebraic complement Z to R; then Im (ρb) = Z + R' and R ~ R'. Let T be the complement to B in M. We have constructed the decomposition of V into the direct sum of four subspaces V = T + B + R + Z. Here R, B, M, Im (ρb) are defined uniquely, and Z, T nonuniquely, and have their own set of parameters. If in V a scalar product is defined, then Z, T are uniquely defined as orthogonal complements. Since K' is isomorphic to $\Phi_a K'$, then $K' = \tilde{B} + \tilde{R} + \tilde{P}$, where $\tilde{B} = \Phi_a^{-1}B$, $\tilde{R} = \Phi_a^{-1}R$, $\tilde{P} = \Phi_a^{-1}P$. Thus, a many-parameter decomposition of H into the direct sum of four subspaces $K + \tilde{B} + \tilde{R} + \tilde{P}$ is defined.

Let us define the sectional operator $Q: V \to H$, $Q: T + B + R + Z \to K + \tilde{B} + \tilde{R} + \tilde{P}$, setting

$$Q = \begin{pmatrix} D & 0 & 0 & 0 \\ 0 & \Phi_a^{-1} & 0 & 0 \\ 0 & 0 & \Phi_a^{-1}\rho b & 0 \\ 0 & 0 & 0 & D' \end{pmatrix},$$

where $D:T \to K$ is an arbitrary linear operator, $\Phi_a{}^{-1}:B \to \tilde{B}$ the inverse to Φ on B, and

$$\Phi_a^{-1}\rho\, b: R \to \tilde{R},\ \rho\, b: R \to R',\ \Phi_a^{-1}: R' \to \tilde{R},\ D': Z \to \tilde{P}$$

(see Fig. 231). Thus, Q is of the form $Q(a, b, D, D')$. Now let us construct the dynamical system $\dot{X}_Q = \rho(QX)X$, where $X \in V$. We took a in the general position in V because, in this case, the dimension of K' is maximum, i.e., operators $\Phi_a{}^{-1}\rho b$ and $\Phi_a{}^{-1}$ have the largest domain of definition. Note some important special cases of the above construction.

If $V = H^*$, $\rho = \mathrm{ad}^*:H \to \mathrm{End}\, H^*$, then $\Phi_a{}^{-1}\rho b = \Phi_a{}^{-1}\mathrm{ad}_b^*$. Take as an example of H the noncompact Lie algebra $so_n \oplus \mathbf{R}^n$, i.e., the Lie algebra of the group of motions of the Euclidean space. We get (see below) that the system $\dot{X}_Q = \mathrm{ad}_{Q(X)}^* X$ turns into the equations of motion of a solid body from inertia in an ideal fluid. Here we have $K = K^*$ under the natural identification of H and H^* and $Z = \tilde{Z} = 0$, $R = \tilde{R}$.

Let $\mathfrak{G}/\mathfrak{H}$ be a compact symmetric space; then G splits into the sum H + V, where H is the stationary subalgebra and V is the tangent space to $\mathfrak{G}/\mathfrak{H}$. The subalgebra H acts on V according to the adjoint representation. Then the decomposition $V = T + B + R + Z$ that defines the sectional operator is of the form: T is a maximum commutative subspace in V, $a \in T$, $R = R'$, $Z = 0$, $b \in K$, $\Phi_a K' + T = V = T + B + R$. If $C:V \to H$ is a sectional operator, then on orbits $O(X) \subset V$, an exterior 2-form $F_C(X, \xi, \eta) = \langle CX, [\xi, \eta]\rangle$ arises, where $\xi, \eta \in T_X O$.

There is an abundance of symmetric spaces and sectional operators for which this form defines (almost everywhere on the orbit) a symplectic structure that is noninvariant under the action of the group. We return again to the general case.

Let $\xi, \eta \in T_X O$ be tangent vectors; then there exist uniquely defined vectors $\eta', \xi' \in K'(X)$ such that $\rho\xi' = \xi$, $\rho\eta' = \eta$. Suppose $C:V \to H$ is the sectional operator. Define the bilinear form $\tilde{F}_C = \langle CX, [\xi', \eta'] \rangle$, where $[\xi', \eta'] \in H$, $CX \in H$. This form is defined on orbits and is skew symmetric. On orbits the flow $\dot{X}_Q$ is also defined.

<u>Question A</u>. For which operators C is the form $\tilde{F}_C$ closed on orbits and nondegenerate?

<u>Question B</u>. For which C and Q is the flow $\dot{X}_Q$ Hamiltonian with respect to $\tilde{F}_C$?

It turns out that, for symmetric spaces, we can give quite complete answers to these questions.

For example, consider the symmetric space SU_3/SO_3. Then equations $\dot{X}_Q$ on B coincide with the Euler equations of motion of a three-dimensional solid body with a fixed point and an arbitrary tensor of inertia. Note that among the systems $\dot{X}_Q = \mathrm{ad}_{QX}^* X$ constructed above are contained Hamiltonian equations of motion of a solid body with a fixed point (for any n, not only for n = 3). To verify this, it suffices to take as a symmetric space a semisimple group $\mathfrak{H}$. It is presentable in the form $\mathfrak{H}\times\mathfrak{H}/\mathfrak{H}$, where the involution $\sigma: \mathfrak{H}\times\mathfrak{H}\to\mathfrak{H}\times\mathfrak{H}$ is defined by the formula $\sigma(x, y) = (y, x)$. The corresponding decomposition in the Lie algebra $G = H + V$ is of the form $V = (X, -X)$, where $X \in H$ and $H = (X, X)$, $X \in H$ (here H and its realization in G are denoted by one letter); $\sigma V = -V$, $\sigma H = H$. It is easy to verify that the form F_C constructed above becomes the canonical symplectic structure on orbits of the coadjoint representation, and the field $\dot{X}_Q$ for $D' = 0$ becomes the required "solid-body equations." Thus, we have found a "multidimensional" series of dynamical systems that contain the equations studied earlier and are interesting because they are also defined on noncompact Lie algebras, being at the same time analogs of systems of the "solid-body" type.

21.5. Equations of Motion from Inertia of a Multidimensional Solid Body in an Ideal Fluid

Here we will present explicitly the embedding of the system of equations mentioned in the title into a noncompact algebra of the group of motions of the Euclidean space. The system will turn out to be Hamiltonian on generic orbits and completely integrable (see [30]).

Let E(n) be the group of eigenmotions of $\mathbf{R}^n$. It is known that E(n) is a semidirect product of SO_n and the commutative group R of parallel translations which is the normal subgroup in E(n) isomorphic to the Euclidean space of dimension n. The matrix realization E(n) is of the form

$$\left(\begin{array}{c|c} SO_n & \begin{matrix} x_1 \\ \vdots \\ x_n \end{matrix} \\ \hline 0\ldots 0 & 1 \end{array}\right),$$

where $(x_1, \ldots, x_n) \in R$. The action of SO_n on the normal subgroup R coincides with the standard representation of SO_n in $\mathbf{R}^n$. Hence, the Lie algebra e(n) of E(n) is the semidirect sum $so_n \oplus_\varphi \mathbf{R}^n$, where $\varphi: so_n \to \mathrm{End}\, \mathbf{R}^n$ is the differential of the standard representation of SO_n in $\mathbf{R}^n$ and $\mathbf{R}^n$ is considered as the commutative Lie algebra. The matrix presentation of e(n) is of the form

$$\left(\begin{array}{c|c} so_n & \begin{matrix} y_1 \\ \vdots \\ y_n \end{matrix} \\ \hline 0\ldots 0 & 0 \end{array}\right).$$

Commutation in e(n) is produced according to the formula $[x + \xi, y + \eta] = [x, y] + x(\eta) - y(\xi)$, where $x(\eta)$ and $y(\xi)$ are the results of the action of matrices x and y on vectors η and ξ, respectively, under the standard so_n-action on $\mathbf{R}^n$. The space $e(n)^*$ dual to e(n) will be identified with e(n). For this, we define a nondegenerate (noninvariant) scalar product in e(n). We have $e(n) = so_n + \mathbf{R}^n$ as linear spaces. Let $\langle\ ,\ \rangle$ be the Killing form and $(\ ,\ )_e$ the Euclidean product on $\mathbf{R}^n$. Then set $((x_1, y_1), (x_2, y_2)) = \langle x_1, x_2\rangle + \langle y_1, y_2\rangle_e$, where $x_1, x_2 \in so_n$, $y_1, y_2 \in \mathbf{R}^n$. Let us present all subspaces of $e(n)^*$ as subspaces of e(n) with respect to the above identification. Let us compute explicitly the image of ad^* under the isomorphism $e^*(n) = e(n)$. Recall that we have denoted $ad_\xi^* x$, where $\xi \in G$, $x \in G^*$, by $a(x, \xi)$.

Proposition 21.4. Let $\xi \in so_n$, $x \in \mathbf{R}^n$, $S \in so_n^* = so_n$, $M \in \mathbf{R}^{n*} = \mathbf{R}^n$, $\xi + x \in so_n \oplus_\varphi \mathbf{R}^n = e(n)$, $S + M \in e(n)^* = (so_n \oplus \mathbf{R}^n)^* = so_n \oplus \mathbf{R}^n$. Then

$$a(S + M, \xi + x)|_{so_n} = [S, \xi] + {}^1/_2(Mx^T - xM^T),$$
$$a(S + M, \xi + x)|_{\mathbf{R}^n} = -\xi M$$

Here M, $a \in \mathbf{R}^n$ are columns of coordinates and T is the transposition.

The proof follows from the definition of the operation $a(,)$.

Let G be an arbitrary Lie algebra. The Euler equations on G* are the system of differential equations $\dot{x} = a(x, C(x))$ on G*, where $C:G^* \to G$ is a linear operator and $a(x, \xi)$ is the linear functional described above. Then $\dot{x} = a(x, C(x))$ flows along orbits of the coadjoint representation Ad* of $\mathfrak{G}$. On these orbits Euler equations are Hamiltonian with respect to the canonical symplectic structure (see [31]).

Let us apply the construction used in Sec. 21.4 to define the sectional operator for the coadjoint representation of E(n). We get a many-parameter family of sectional operators $Q:e(n)^* \to e(n)$ for which the Euler equations are completely integrable Hamiltonian systems on orbits of the coadjoint representation of E(n). In e(n)*, hence in e(n), we consider the subspace

$$K = \bigoplus_{k=0}^{\left[\frac{n+1}{2}\right]-2} \mathbf{R}(E_{2k+1,\,2k+2}) \oplus \mathbf{R}e_n \subset e(n),$$

where E_{ij} is an elementary skew symmetric matrix and e_i is the standard orthobasis in $\mathbf{R}^n$. Let K* be the corresponding subspace in e(n)*.

Proposition 21.5. There is an invariant description of K and K*, i.e., $K = \mathrm{Ann}\,(x_1) = \{\xi \in G \,|\, a(x_1, \xi) = 0\}$, where $G = e(n)$ and $K^* = \{\xi' \in G^* \,|\, a(\xi', x_2) = 0\}$, where $x_1 \in G^*$ and $x_2 \in K \subset G$ are generic elements.

The proof follows from Proposition 21.4.

The orthogonal complement to W in e(n) or in e(n)* with respect to the scalar product $\langle\, , \rangle + (,)_e$ will be denoted by $W^\perp$.

Lemma 21.2. Let $a \in K^*$. Consider the operator $\Phi_a: e(n) \to e(n)^*$, $x \to a(a, x) \in e(n)^*$. Then $\Phi_a K^\perp \subset K^{*\perp}$, where $K^\perp$ is the orthogonal complement to K, and $K^{*\perp}$ the complement to K*.

The proof is evident. If a is generic, then $K = \mathrm{Ker}\,\Phi_a$, and $\Phi_a: K^\perp \to K^{*\perp}$ is an isomorphism; hence, the inverse mapping $\Phi_a^{-1}: K^{*\perp} \to K$ is defined. We have the direct sum decompositions $e(n) = K^{\perp} \oplus K$ and $e(n)^* = K^{*\perp} \oplus K^*$ (as linear spaces). The general method of Sec. 21.4 states that if $a \in K^*$, $b \in K$, where a is generic and $z = x + y \in e(n)^*$, where $x \in K^{*\perp}$, $y \in K^*$, then $Q(a, b, D)z = \Phi_a^{-1} ad_b^* x + D(y)$, where $D: K^* \to K$ is an arbitrary operator.

Now we may express the main equations $\dot{X}_Q = ad_{QX}^* X$ on $G^* = (so_n \oplus R^n)^* \cong so_n \oplus R^n$, where $Q(a, b, D)$ is the sectional operator constructed in Sec. 21.4 and which is a noncompact analog of operators $\varphi_{a,b,D}$ that describe the motion of a solid body. In our case, $Q(a, b, D): e(n)^* \to e(n)$; hence $\dot{X}_Q$ is defined on G^*. It is possible to write the equations $\dot{X}_Q = ad_{QX}^* X$ explicitly in the form

$$\dot{S} = [S, \xi] + {}^1/_2 (Mx^T - xM^T), \qquad \dot{M} = -\xi M, \tag{Γ}$$

where $\xi \in so_n$, $x \in R^n$ are functions in elements S, M, and this dependence is defined by $Q(a, b, D)$, i.e., $\xi + x = Q(a, b, D) \times (S + M)$. Here $S + M \in e(n)^*$ and $\xi + x \in e(n)$. Since $Q(a, b, D)$ is explicitly defined, it is not difficult to compute explicitly how ξ and x depend on S, M, a, b, D.

<u>Proposition 21.6</u>. The system of differential equations $\dot{X} = ad_{Q(a,b,D)X}^* X$ on $e^*(n)$ [whose explicit form is (Γ)] is a Hamiltonian system on generic orbits. Moreover, for n = 3, this system coincides with the equations of motion from inertia of a solid body in an ideal fluid. Therefore, this system admits an embedding into a noncompact Lie algebra e(n) in the sense of Definition 21.1.

This proposition is the reason why we will say that equations (Γ) for an arbitrary n describe the motion from inertia of a multidimensional solid body in an ideal fluid. Before we come to the proof of Proposition 21.6, recall the classical equations of motion from inertia of a three-dimensional solid body in an ideal fluid (for details, see, for example, [32]). Let us connect the coordinate system with the moving body, and let u_i be components of the velocity of the translational movement of the origin and ω_i components of the angular velocity of rotation of the solid body. Then the kinetic energy of the system fluid–solid body is of the form $T = \frac{1}{2}(A_{ij}\omega_i\omega_j + B_{ij}u_iu_j) + C_{ij}\omega_i\omega_j$, where A_{ij}, B_{ij}, C_{ij}

are constants depending on the shape of the body and on the densities of the body and the fluid (we sum indices that appear twice from 1 to 3). Let $N = (y_1, y_2, y_3)$, where $y_i = \partial T/\partial \omega_i$ and $K = (x_1, x_2, x_3)$, where $x_i = \partial T/\partial u_i$. Then the motion from inertia of a solid body in an ideal fluid is described by the equations

$$\begin{aligned} dN/dt &= N\times\omega + K\times U, \\ dK/dt &= K\times\omega, \end{aligned} \qquad (*)$$

where $V = (u_1, u_2, u_3)$, $\omega = (\omega_1, \omega_2, \omega_3)$.

The kinetic energy of the solid body is an arbitrary positive-definite homogeneous quadratic form in six variables u_i, ω_i; therefore,it is defined by 21 coefficients A_{ij}, B_{ij}, C_{ij}. Equations (*), in the general case, have three classical Kirchhoff integrals; therefore, for complete integrability of (*), we need one extra integral which is functionally independent of them. Since generic orbits have dimension 4, then four independent integrals, two of which define the orbit and two of which are no longer constant on the orbit, suffice to complete the integrability of the system. Recall three classical cases when this fourth extra integral exists (for a complete review, see, for example, [33]).

The first general solution of the equations of motion of a solid body in an ideal fluid was given by Kirchhoff for a body of revolution. In 1871, Klebsch gave two more forms of the function T (kinetic energy) that enable one to add to three Kirchhoff integrals the fourth one, hence to integrate the problem in quadratures. The solution, in Klebsch's first case, when the fourth integral is, generally speaking, a linear homogeneous function in x_i, y_i, was given by Halphen. In 1878, Weber studied Klebsch's second case, when the fourth integral is the homogeneous quadratic function in x_i, y_i, with some technical assumptions on arbitrary constants. The general solution of the last problem was given by Kötter. The third kind of kinetic energy that makes (*) explicitly integrable was discovered by V. A. Steklov.

1. Klebsch's first case. The kinetic energy in variables x_i, y_i is of the form

$$\begin{aligned} T = {}^1/_2\, b_{11}(x_1^2 + x_2^2) + {}^1/_2\, b_{33} x_3^2 + b_{14}(x_1y_1 + x_2y_2) + \\ + b_{36} x_3 y_3 + {}^1/_2\, b_{44}(y_1^2 + y_2^2) + {}^1/_2\, b_{66} y_3^2, \end{aligned}$$

and the equations of motion (*) admit, in this case, a fourth linear integral in x_i, y_i. The body with this energy preserves its shape after rotation around the 0_z axis by $\pi/2$. For $b_{14} = b_{36} = 0$, we get the body of rotation.

2. Klebsch's second case. The kinetic energy is of the form

$$T = b_{11}x_1^2 + b_{22}x_2^2 + b_{33}x_3^2 + b_{44}y_1^2 + b_{55}y_2^2 + b_{66}y_3^2,$$

where $(b_{22} - b_{33})/b_{44} + (b_{33} - b_{11})/b_{55} + (b_{11} - b_{22})/b_{66} = 0$. The equations of motion admit a fourth integral, i.e., the homogeneous quadratic form. The solid body considered is symmetric with respect to three pairwise perpendicular planes.

3. The Steklov case. Equations (*) are defined by the kinetic energy function $2T = \Sigma b_{11}x_1{}^2 + 2\sigma\Sigma b_{55}b_{66}x_1y_1 + \Sigma b_{44} \times y_1{}^2$, where b_{ii} are defined from the relations $b_{11} = \sigma^2 b_{44} \times (b_{55}{}^2 + b_{66}{}^2)$, $b_{22} = \sigma^2 b_{55}(b_{66}{}^2 + b_{44}{}^2)$, $b_{33} = \sigma^2 b_{66}(b_{44}{}^2 + b_{55}{}^2)$. Equations (*) admit, besides three classical integrals, a fourth one which is the integer homogeneous polynomial of degree 2 in x_i, y_i. Here σ is an arbitrary constant, and Σ stands for the sum of expressions obtained from the written ones by cyclic permutation of groups of indices 1, 2, 3 and 4, 5, 6.

Now let us prove Proposition 21.6.

Lemma 21.2. Consider the mapping $\psi: so_3 \to \mathbf{R}^3$ that transforms $X = xE_{12} + yE_{13} + zE_{23}$ into $(z, -y, x)$. Then $\psi[X, Y] = -\psi X \times \psi Y$, where $\psi X \times \psi Y$ is the vector product in $\mathbf{R}^3$. Furthermore, $\psi(Mx^T - xM^T) = M \times x$, $M, x \in \mathbf{R}^3$ and $\xi M = -\psi(\xi) \times M$, where $\xi \in so_3$.

Proof. If $z = \xi + x \in e(3)$, $z = S + M \in e(3)^*$, then $a(z, z) = (y, x)$, as shown in Proposition 21.4, where $y = [S, \xi] + \frac{1}{2}(Mx^T - xM^T) \in so_3$ and $X = -\xi M$; the vectors M and x are expressed as columns and so_3 is realized by skew symmetric matrices. The statement is proved.

Let us write the operators $Q(a, b, D)$ constructed above in the simplest three-dimensional case for $e(3) = so_3 \oplus \mathbf{R}^3$ when several multidimensional effects vanish, thus simplifying the explicit notation. In this case,

$$K = K^* = \begin{pmatrix} 0 & a_1 & 0 \\ -a_1 & 0 & 0 \\ 0 & 0 & 0 \end{pmatrix} \oplus \begin{pmatrix} 0 \\ 0 \\ a_2 \end{pmatrix},$$

$$K^{\perp} = K^{*\perp} = \begin{pmatrix} 0 & 0 & x_2 \\ 0 & 0 & x_3 \\ -x_2 & -x_3 & 0 \end{pmatrix} \oplus \begin{pmatrix} u_1 \\ u_2 \\ 0 \end{pmatrix}.$$

Let

$$f = \begin{pmatrix} 0 & 0 & f_2 \\ 0 & 0 & f_3 \\ -f_2 & -f_3 & 0 \end{pmatrix} \oplus \begin{pmatrix} u_1 \\ u_2 \\ 0 \end{pmatrix} \in K^{*\perp},$$

then

$$\mathrm{ad}_a^* f = \begin{pmatrix} 0 & 0 & -a_1 f_3 + {}^1/_2\, u_1 a_2 \\ 0 & 0 & a_1 f_2 + {}^1/_2\, u_2 a_2 \\ a_1 f_3 - {}^1/_2\, u_1 a_2 & -f_2 a_1 - {}^1/_2\, u_2 a_2 & 0 \end{pmatrix} \oplus \begin{pmatrix} -a_1 u_2 \\ a_1 u_1 \\ 0 \end{pmatrix}.$$

Let

$$b = \begin{pmatrix} 0 & b_1 & 0 \\ -b_1 & 0 & 0 \\ 0 & 0 & 0 \end{pmatrix} \oplus \begin{pmatrix} 0 \\ 0 \\ b_2 \end{pmatrix}, \quad x = \begin{pmatrix} 0 & 0 & x_2 \\ 0 & 0 & x_3 \\ -x_2 & -x_3 & 0 \end{pmatrix} \oplus \begin{pmatrix} y_1 \\ y_2 \\ 0 \end{pmatrix},$$

then

$$\Phi_b(x) = \begin{pmatrix} 0 & 0 & b_1 x_3 - {}^1/_2\, b_2 y_1 \\ 0 & 0 & -b_1 x_2 - {}^1/_2\, b_2 y_2 \\ x_3 b_1 - {}^1/_2\, b_2 y_1 & -x_2 b_1 - {}^1/_2\, b_2 y_2 & 0 \end{pmatrix} \oplus \begin{pmatrix} -b_2 x_2 \\ -b_2 x_3 \\ 0 \end{pmatrix}.$$

Hence,

$$\Phi_b^{-1}\, \mathrm{ad}_a^* \left(\begin{pmatrix} 0 & 0 & f_2 \\ 0 & 0 & f_3 \\ -f_2 & -f_3 & 0 \end{pmatrix} \oplus \begin{pmatrix} u_1 \\ u_2 \\ 0 \end{pmatrix} \right) =$$

$$= \begin{pmatrix} 0 & 0 & \frac{a_1}{b_2} u_2 \\ 0 & 0 & \frac{-a_1}{b_2} u_1 \\ \frac{-a_1}{b_2} u_2 & \frac{a_1}{b_2} u_1 & 0 \end{pmatrix} \oplus \begin{pmatrix} z_1 \\ z_2 \\ 0 \end{pmatrix},$$

$$z_1 = 2\frac{a_1}{b_2} f_3 + u_1 \frac{-b_2a_2 - 2b_1a_1}{b_2^2},$$

$$z_2 = -2\frac{a_1}{b_2} f_2 - u_2 \frac{b_2a_2 + 2b_1a_1}{b_2^2}.$$

Finally,

$$Q(a, b, D)\left(\begin{pmatrix} 0 & f_1 & f_2 \\ -f_1 & 0 & f_3 \\ -f_2 & -f_3 & 0 \end{pmatrix} \oplus \begin{pmatrix} u_1 \\ u_2 \\ u_3 \end{pmatrix}\right) =$$

$$= \begin{pmatrix} 0 & \alpha f_1 + \beta u_3 & \frac{a_1}{b_2} u_2 \\ -\alpha f_1 - \beta u_3 & 0 & \frac{-a_1}{b_2} u_1 \\ \frac{-a_1}{b_2} u_2 & \frac{a_1}{b_2} u_2 & 0 \end{pmatrix} \oplus \begin{pmatrix} 2\frac{a_1}{b_2} f_3 - u_1 \frac{b_2a_2 + 2b_1a_1}{b_2^2} \\ -2\frac{a_1}{b_2} f_2 - u_2 \frac{b_2a_2 + 2b_1a_1}{b_2^2} \\ \gamma f_1 + \delta u_3 \end{pmatrix},$$

where α, β, γ, δ are constants that define the operator $D: K^* \to K$. The kinetic energy is of the form $\langle X, Q(a, b, D)X \rangle$. The matrix of this quadratic form is

$$A = \begin{pmatrix} -2\alpha & 0 & 0 & 0 & 0 & \frac{\gamma}{2} - \beta \\ 0 & 0 & 0 & 0 & -\frac{2a_1}{b_2} & 0 \\ 0 & 0 & 0 & \frac{a_1}{b_2} & \frac{a_1}{b_2} & 0 \\ 0 & 0 & \frac{a_1}{b_2} & b_2^{-2}(-b_2a_2 + 2b_1a_1) & 0 & 0 \\ 0 & -\frac{2a_1}{b_2} & \frac{a_1}{b_2} & 0 & b_2^{-2}(-b_2a_2 + 2b_1a_1) & 0 \\ \frac{\gamma}{2} - \beta & 0 & 0 & 0 & 0 & \delta \end{pmatrix}.$$

Clearly,

$$\det A = -a_1^4 b_2^{-4}\left(\left(\frac{\gamma}{2}-\beta\right)^2 + 2\delta\alpha\right),$$

i.e., the sign of det A can be chosen to be arbitrary by varying, for instance, D. Setting $\lambda = b_2^{-2}(b_{2a2} + 2b_{1a1})$, we get that T is reduced to the following diagonal form:

$$T = -2\alpha\left(f_1 - \frac{\gamma-2\beta}{4\alpha}u_3\right)^2 + \left[\frac{(\gamma-2\beta)^2}{8\alpha}+\delta\right]u_3^2 -$$
$$-\lambda\left(u_1 - \frac{a_1}{b_2\lambda}f_3\right)^2 - \lambda\left(u_2 + \frac{2a_1}{b_2\lambda}f_2 - \frac{a_1}{b_2\lambda}f_3\right)^2 +$$
$$+\frac{(2a_1)^2}{b_2\lambda}\left(f_2 - \frac{f_3}{2b_2}\right)^2 + \frac{2a_1^2(b_2-1)}{b_2^3\lambda}f_3^2,$$

which implies that T is indefinite.

In dimensions greater than 3, explicit formulas become much more cumbersome and, therefore, we will not analyze them here.

22. COMPLETE INTEGRABILITY OF SEVERAL HAMILTONIAN SYSTEMS ON LIE ALGEBRAS

22.1. The Shift of the Argument Method and the Construction of Commutative Algebras of Integrals on Orbits in Lie Algebras

It turns out that the Hamiltonian systems listed above not only admit embeddings into Lie algebras, but are completely integrable in the commutative Liouville sense (see [31, 28, 30]). In particular, we obtain in these cases a positive answer to Conjecture A (see Sec. 21), since we have produced full commutative sets of functions on generic orbits in semisimple and compact Lie algebras. Integrals of these Hamiltonian systems are quite simple: to construct them, it suffices to know invariants of the Lie algebra, i.e., the set of functions constant on generic orbits. Roughly speaking, the process of construction of integrals is as follows. Let f be an invariant of the Lie algebra which is a function on G* (or on G in the compact and semisimple case). Let $a \in G^*$

be a covector in general position. Let us shift the argument of the function f, i.e., consider the function $f(x + \lambda a)$, where $\lambda \in \mathbf{C}$ or $\mathbf{R}$. Since f is a polynomial for all the cases in which we are interested, we may expand $f(x + \lambda a)$ in degrees of the formal variable λ which gives us the decomposition of the form $f(x + \lambda a) = \sum_k P_k(x, a)\lambda^k$. The remarkable fact is that these polynomials $P_k(x, a)$ [or, similarly, the functions $f(x + \lambda\ )$] constitute full commutative sets of functions (integrals) in all the cases listed above (see [31, 28, 30]). This method for construction of integrals will be called the shift of the argument method. This general scheme is the development of the idea proposed in [10] in the case so_n.

It turns out that the shift of the argument method gives positive results for various noncompact Lie algebras. Full commutative sets of functions on generic orbits may be obtained by applying the shift of the argument method not only to invariants of the algebra (sometimes there are not a sufficient quantity of them to get the full set), but also to the so-called semiinvariants, i.e., to functions on orbits that under the (co)adjoint action of the group are multiplied by the character of this representation. Invariants are obviously special cases of semiinvariants since, under the (co)adjoint action, invariants are fixed points in the space of functions. There are other general methods of construction of full commutative sets of integrals which we are unable to discuss here (see, for example, [22, 29]).

Now let us pass on to the construction of commutative sets of integrals on generic orbits. In what follows we will prove that these sets are full. We will restrict ourselves mainly to consideration of a complex semisimple Lie algebra and Euler equations of the form $\dot{X} = [X, \varphi X]$, where operators $\varphi_{a,b,D}$ define Hamiltonians of the complex series (see Sec. 21.2). Consider the adjoint action of the complex semisimple Lie group $\mathfrak{G}$ on its Lie algebra G (here we may assume $G = G^*$). The group $\mathfrak{G}$ fibrates G into orbits and we will assume that $\mathrm{Ad}_g X = gXg^{-1}$ for $g \in \mathfrak{G}$.

<u>Lemma 22.1</u>. Any smooth function f(X), where $X \in G$, invariant with respect to the adjoint action, i.e., constant on orbits, is an integral of the Euler equation $\dot{X} = [X, \varphi X]$, where $\varphi: G \to G$ is an arbitrary self-adjoint operator.

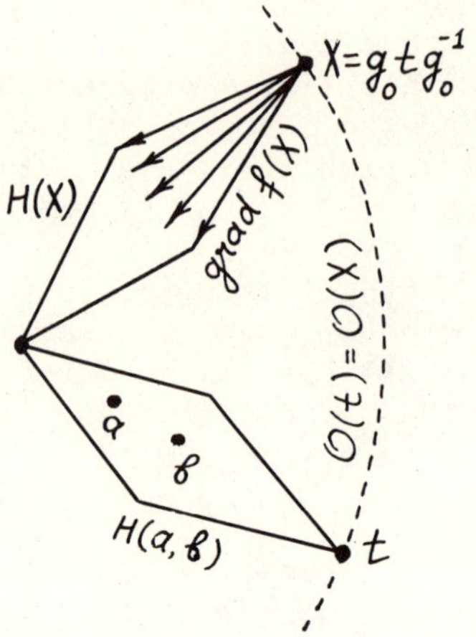

Fig. 232

The proof follows straightforwardly from the fact that $T_XO = \{[X, y]\}$, where y runs over the whole algebra.

Note that in the complex case not all elements of the algebra belong to the orbit O(t) for a certain $t \in H$, where H is a fixed Cartan subalgebra.

Consider the set of all complex vectors grad f(x), where $f \in IG$ and IG stands for the ring of invariant polynomials on G. Let H(X) be the subspace in G consisting of all elements that commute with X. If $X \in \text{Reg } G$, then H(X) is a Cartan subalgebra. Recall that in a semisimple Lie algebra any two Cartan subalgebras are conjugate. In particular, if $X \in \text{Reg } G$, then $H(X) = g_0 H(a, b) g_0^{-1}$ for some $g_0 \in \mathfrak{G}$ and $H(a, b)$ is the Cartan subalgebra that contains a, b. Clearly, H(X) is contained in the subspace generated by grad f(x), where $f \in IG$; and if $X \in \text{Reg } G$, then $H(X) = \{\text{grad } f(x), f \in IG\}$. This follows from the nondegeneracy of the Killing form and the orthogonality of H(X) to the tangent space to the orbit (see Fig. 232).

Lemma 22.2. A smooth function f is constant on orbits of the algebra iff $[X, \text{grad } f(X)] = 0$ holds for any $X \in G$.

Let grad f(X) be the value of grad f at X.

Proof. Recall that $T_X O = \{[X, \xi]\}$, where ξ runs over G. This implies <grad f(X), [X, ξ]> = 0 for any ξ since $[X, \xi] \times f(X) = 0$. Since ad_X is skew symmetric, then <[grad f(X), X], ξ> = 0; this, in view of the nondegeneracy of the Killing form, means that [grad f(X), X] = 0. The converse statement is similarly proved.

Proposition 22.1. Let $f \in IG$, i.e., f is an invariant and constant on orbits of the algebra. Then complex functions $h_\lambda(X) = f(X + \lambda a)$ are (for any λ) integrals of the equation $\dot{X} = [X, \varphi_{a\,b D}X]$, where φ is an operator of the complex series (see above). The function $F(X) = \langle X, \varphi X\rangle$ is also an integral.

Proof. Let us verify the identity $0 = (d/d\tau)h_\lambda(X)$, where τ is a parameter along trajectories of the flow $\dot{X}$. This identity is equivalent to <grad $h_\lambda(X)$, $\dot{X}$> = 0. We have

$$\langle \operatorname{grad} h_\lambda(X), \dot{X}\rangle = \langle \operatorname{grad} f(X + \lambda a), [X, \varphi X]\rangle =$$
$$= \langle \operatorname{grad} f(X + \lambda a), [X + \lambda a, \varphi X]\rangle - \lambda\langle \operatorname{grad} f(X + \lambda a),$$
$$[a, \varphi X]\rangle = \langle [\operatorname{grad} f(X + \lambda a), X + \lambda a], \varphi X\rangle -$$
$$- \lambda\langle \operatorname{grad} f(X + \lambda a), [a, \operatorname{ad}_a^{-1}\operatorname{ad}_b X' + D(t)]\rangle =$$
$$= \langle [\operatorname{grad} f(X + \lambda a), X + \lambda a], \varphi X\rangle - \lambda\langle \operatorname{grad} f(X + \lambda a),$$
$$[b, X']\rangle - \lambda\langle \operatorname{grad} f(X + \lambda a), [a, D(t)]\rangle.$$

We have made use of the definition of φ from Sec. 21; here $t \in H(a, b)$, $X' \in V$. The first summand in the obtained sum vanishes by Lemma 22.2 applied at the point $X + \lambda a$. The third summand vanishes since $D(t) \in H(a, b)$. Let us transform the second summand:

$$\lambda\langle \operatorname{grad} f(X + \lambda a), [b, X']\rangle = \lambda\langle \operatorname{grad} f(X + \lambda a),$$
$$[b, X' + t + \lambda a]\rangle = \lambda\langle \operatorname{grad} f(X + \lambda a), [b, X + \lambda a]\rangle =$$
$$= -\lambda\langle \operatorname{grad} f(X + \lambda a), X + \lambda a], b\rangle = 0$$

by Lemma 22.2, and in view of $[b, t + \lambda a] = 0$. Thus, $(d/dt)h_\lambda(X) = 0$ along $\dot{X}$. Furthermore, $(d/dt)F(X) = \langle \dot{X}, \varphi X\rangle + \langle X, \varphi\dot{X}\rangle = 2\langle [X, \varphi X], \varphi X\rangle = 0$ by the symmetry of < , > and the skew symmetry of ad.

Consider the model example sl(n, **C**). Clearly, the standard symmetric polynomials in eigenvalues of X are integrals constant on orbits of the algebra. Let us transform the equation, writing it in the form $(X + \lambda a)' = [X + \lambda a, \varphi X + \lambda b]$. In fact, after simplifications, we get the initial equation $\dot{X} = [X, \varphi X]$. We have made use of $[a, b] = 0$, $[X, b] + [a, \varphi X] = 0$, and the definition of $\varphi_{a b, D}$. Thus, the equation was not affected but we now obtain a new series of integrals: symmetric polynomials in eigenvalues of the matrix $X + \lambda a$. These integrals can be expressed by two methods:

1) consider the expansion of the polynomial $\det (X + \lambda a - \mu E) = \sum_{\alpha,\beta} P_{\alpha\beta} \lambda^\alpha \mu^\beta$ in λ and μ; then all polynomials $P_{\alpha\beta}(X, a)$ are integrals of the equation;

2) consider functions $S_k = \operatorname{Tr} (X + \lambda a)^k$ and their expansions in λ, i.e., $S_k = \sum_\alpha Q_\alpha^{(k)}(X, a)\lambda^\alpha$.

The relationship of Newton polynomials with symmetric polynomials σ_i defines the relationship of $Q_\alpha^{(k)}$ with $P_{\alpha\beta}$.

Now let us proceed to the construction of integrals for the compact series. Let G_u be the compact form of G. Let $x \in G_u$, a, $b \in H_u$, $X + \lambda a \in G_u$ and λ be real. Consider the $\mathfrak{G}_u$-action on G_u. Unlike the complex case, the union of orbits which grow from the Cartan subalgebra $H_u = H_u(a, b)$ coincides with G_u. Let $\varphi: G_u \to G_u$ be an operator of the compact series.

<u>Lemma 22.3</u>. Any smooth function f, invariant with respect to the adjoint action of $\mathfrak{G}_u$ (i.e., constant on orbits), is an integral of the Euler equation $\dot{X} = [X, \varphi X]$, where $\varphi: G_u \to G_u$ is an arbitrary self-adjoint operator.

The proof is obvious. Let IG_u be the ring of invariant polynomials on G_u. Let us produce explicitly multiplicative generators of the ring IG_u. Let N be the normalizer of H_u in G_u; then $N/\mathfrak{H}_u = \Phi$ is the Weyl group (see above). Let $t \in H_u$; then the orbit $O(t)$ is orthogonal to H_u and this orbit returns on H_u, piercing it in a finite number of points which are images of t with respect to the action of the Weyl group. The ring IG_u is identified with the ring of polynomials on H_u, invariant with respect to the Weyl group. This ring admits a simple description: if $\mathfrak{G}_u$ is connected, then IG_u is a free algebra in r variables, where $r = \operatorname{rank} G_u$; their role can be played by homogeneous algebraically independent polynomials P_{k_1}, ..., P_{k_r}, where $k_i = \deg P_{k_i}$. For simple Lie algebras, the numbers k_i, i.e., the degrees of the polynomials, are of the following form:

A_n: 2, 3, 4, ..., n, $n+1$;
B_n: 2, 4, 6, ..., $2n$;
C_n: 2, 4, 6, ..., $2n$;
D_n: 2, 4, 6, ..., $2n-2$, n;
G_2: 2, 6;
F_4: 2, 6, 8, 12;
E_6: 2, 5, 6, 8, 9, 12;
E_7: 2, 6, 8, 10, 12, 14, 18;
E_8: 2, 8, 12, 14, 18, 20, 24, 30.

Polynomials P_{k_i} can be written explicitly. Consider the linear representation of G_u of the minimum possible dimension of matrices of size $m \times m$; let Λ_1, ..., Λ_m be weights of the representation, i.e., linear functionals on H_u corresponding to eigenvectors of operators of H_u in the space of the representation. Coordinates Λ_1, ..., Λ_m on H_u can be linearly dependent. Polynomials P_{k_i} are of the form

$$A_n: \sum_{j=1}^{n+1} \Lambda_j^{k_i};$$

$$B_n: \sum_{j=1}^{n} \Lambda_j^{k_i};$$

$$C_n: \sum_{j=1}^{n} \Lambda_j^{k_i};$$

$$D_n: \sum_{j=1}^{n} \Lambda_j^{k_i}, \text{ where } k_i = 2, 4, 6, \ldots, 2n-2, \text{ and } P_n' = \Lambda_1 \cdot \Lambda_2 \cdot \ldots \cdot \Lambda_n.$$

If G_u is an exceptional simple Lie algebra, then $P_{k_i} = \sum_{j=1}^{m} \Lambda_j^{k_i}$. Clearly, all rings IG_u are subrings of the ring of symmetric polynomials $S(\Lambda_1, \ldots, \Lambda_m)$. All the above functions are of the form Tr $X^{k_i} = \sum_{j=1}^{m} \Lambda_j^{k_i}$, excluding the case of series D_n, where there is one more polynomial $\sqrt{\det X}$.

Proposition 22.2. Let $f \in IG_u$, i.e., f is constant on orbits of G_u. Then functions $h_\lambda(X) = f(X + \lambda a)$ are (for any λ) integrals of the equation $\dot{X} = [X, \varphi X]$, where φ is an operator of the compact series $X + \lambda a \in G_u$, $\lambda \in \mathbf{R}$. The function $F(X) = \langle X, \varphi X\rangle$ is also an integral.

The proof follows the scheme of the proof of Proposition 22.1. For details, see [28].

Now consider integrals of the normal series. Consider the embedding of G_n into G_u. Operators $\varphi_{a,b}: G_n \to G_n$ are generated by vectors $a, b \in H_u$; in particular, $a, b \notin G_n$; hence $X + \lambda a \notin G_n$ if $X \in G_n$, $\lambda \in \mathbf{R}$.

Proposition 22.3. Let $f \in IG_u$, i.e., f is constant on orbits of G_u. Consider functions $q_\lambda(X)$, where $\lambda \in \mathbf{R}$, $X \in G_n \subset$ $\subset G_u$, which are restrictions of $h_\lambda(X) = f(X + \lambda a)$ onto $G_n \subset G_u$. Then q_λ are integrals of the equation $\dot{X} = [X, \varphi_{ab} X]$, where φ_{ab} is an operator of the normal series. The function $F(X) =$ $= \langle X, \varphi X\rangle$ is also an integral.

22.2. Examples of Lie Algebras so_3 and so_4

Let us illustrate the series of integrals constructed above by the simplest Lie algebras. In particular, we will see that, among these integrals, there are known classical integrals. Let $G_u = so_3$. Let us present so_3 in the form su_2, making use of the known isomorphism. Let us embed su_2 as a compact real form G_u into $G = sl(2, \mathbf{C})$; then su_2 coincides with the set of fixed points of the involution $\sigma X = -\bar{X}^T$. Clearly, G_u splits into the sum of three one-dimensional subspaces generated by vectors $E_+ = E_\alpha + E_{-\alpha}$, $E_- = i(E_\alpha - E_{-\alpha})$, E_0, where $E_0 \in H_u = iH_0$,

$$E_0 = \begin{pmatrix} i & 0 \\ 0 & -i \end{pmatrix}, \quad E_+ = \begin{pmatrix} 0 & 1 \\ -1 & 0 \end{pmatrix}, \quad E_- = \begin{pmatrix} 0 & i \\ i & 0 \end{pmatrix}.$$

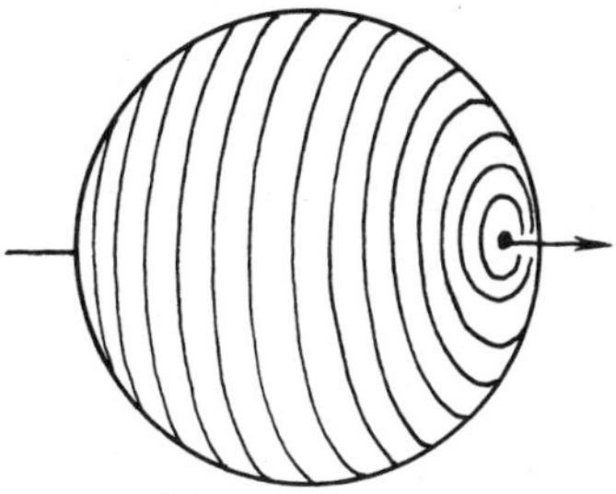

Fig. 233

The operator of the compact series $\varphi: su_2 \to su_2$ acts according to the formula

$$\varphi E_+ = \frac{\alpha(b)}{\alpha(a)} E_+, \ \varphi E_- = \frac{\alpha(b)}{\alpha(a)} E_-, \ \varphi E_0 = \lambda_0 E_0,$$

where $\lambda_0 \neq 0$ is an arbitrary real number,

$$b = \lambda_+ a, \ \lambda_+ \neq 0, \ \alpha(a) \neq 0, \text{ i.e., } \lambda_+ = \frac{\alpha(b)}{\alpha(a)}.$$

Finally, $\varphi E_+ = \lambda_+ E_+$, $\varphi E_- = \lambda_+ E_-$, $\varphi E_0 = \lambda_0 E_0$ and are nonzero. For φ in the general position, we have $\lambda_+ \neq \lambda_0$. In the case $G_u = su_2$, operators φ of the compact series form a two-parameter family (λ_+, λ_0). If

$$X = \begin{pmatrix} iz & x+iy \\ -x+iy & -iz \end{pmatrix} \in su_2,$$

then

$$\varphi X = \begin{pmatrix} i\lambda_0 z & \lambda_+(x+iy) \\ \lambda_+(-x+iy) & -i\lambda_0 z \end{pmatrix}.$$

Clearly, $\langle X, \dot{X}\rangle = 0$, i.e., the velocity vector $\dot{X}$ is tangent to orbits of the adjoint action of SU_2 in su_2. Orbits are two-dimensional spheres with center at the origin 0 and the point 0 itself. All orbits, except 0, are generic. Let us fix an arbitrary generic orbit. Then integral trajectories of the flow $\dot{X}$ on a sphere coincide with the trajectories of points of the sphere with respect to its rotations around the axis E_0 (see Fig. 233). The functions Tr $(X + \lambda a)^k$ must be integrals of the flow. We have

$$X + \lambda a = \begin{pmatrix} i(z + q\lambda) & x + iy \\ -x + iy & -i(z + q\lambda) \end{pmatrix},$$

where $a = qE_0$, $q \neq 0$, implying $S_1 = 0$, $S_2 = -2(x^2 + y^2 + z^2 + 2zq\lambda + q^2\lambda^2)$. The coefficients of degrees of λ, i.e., $Q_1(x, a) = x^2 + y^2 + z^2$, $Q_2(X, a) = zq$, $Q_3(X, a) = q^2$ are integrals, i.e., the functions z, $x^2 + y^2$ are actually integrals also. Integral trajectories are intersections of spheres with the planes z = const. We have obtained one of the simplest classical cases of the motion of a solid body: the integral $Q_1(X, a) = Q_1(X)$ is the kinetic moment, and the integral z = const is equivalent to the integral of the energy in the special case $I_1 = I_2$ (ellipsoid of rotation).

Let us consider again so_3 and realize it now as a normal form, i.e., let us study integrals of the normal series for so_3. Let $G = sl(3, \mathbf{C})$, $G_u = su_3$, $G_n = so_3$, $\sigma X = -\bar{X}^T$, $\tau X = \bar{X}$, and G_n be the set of fixed points of involutions σ and τ. The subalgebra $G_n = so_3$ is generated by three vectors $E_{ij} = E_\alpha + E_{-\alpha}$:

$$E_{12} = \begin{pmatrix} 0 & 1 & 0 \\ -1 & 0 & 0 \\ 0 & 0 & 0 \end{pmatrix}, \quad E_{13} = \begin{pmatrix} 0 & 0 & 1 \\ 0 & 0 & 0 \\ -1 & 0 & 0 \end{pmatrix}, \quad E_{23} = \begin{pmatrix} 0 & 0 & 0 \\ 0 & 0 & 1 \\ 0 & -1 & 0 \end{pmatrix}.$$

Let $a, b \in H_u$ (diagonal purely imaginary 3×3 matrices with trace zero). Then the operators $\varphi_{a,b}: G_n \to G_n$ are of the form

$$\varphi E_{12} = \frac{b_1 - b_2}{a_1 - a_2} E_{12}, \quad \varphi E_{13} = \frac{b_1 - b_3}{a_1 - a_3} E_{13},$$

$$\varphi E_{23} = \frac{b_2 - b_3}{a_2 - a_3} E_{23}.$$

The set $\{\varphi_{ab}\}$ for the normal series constitutes a three-parameter family unlike the two-parameter family for the compact series. There is no compact operator which is normal (verify!). Set $\lambda_{ij} = (b_i - b_j)/(a_i - a_j)$; then $\dot{X} = \gamma\beta(\lambda_{13} - \lambda_{23})E_{12} + \alpha\gamma(\lambda_{23} - \lambda_{12})E_{13} + \alpha\beta(\lambda_{12} - \lambda_{13})E_{23}$, where $X = \alpha E_{12} + \beta E_{13} +$

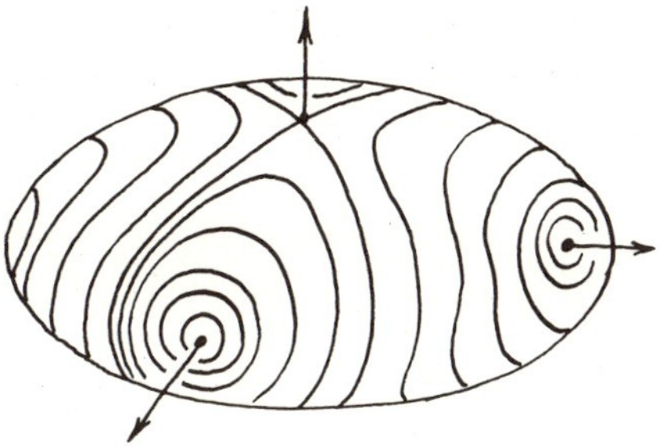

Fig. 234

$+ \gamma E_{23}$. Clearly, $\langle X, \dot{X}\rangle = 0$, i.e., the vectors $\dot{X}$ are tangent to spheres with the center at 0. Recall that for $so_3 = su_2$ the Killing form coincides with the Euclidean scalar product. We have

$$X + \lambda a = \begin{pmatrix} i\lambda a_1 & \alpha & \beta \\ -\alpha & i\lambda a_2 & \gamma \\ -\beta & -\gamma & i\lambda a_3 \end{pmatrix}.$$

Integrals are defined by functions $\mathrm{Tr}\,(X + \lambda a)^k$, where $1 \leqq k \leqq 3$. The computation gives us $P(X) = \alpha^2 + \beta^2 + \gamma^2$, $Q(x, a) = \alpha^2(a_1 + a_2) + \beta^2(a_1 + a_3) + \gamma^2(a_2 + a_3)$. These integrals coincide with classical ones: $P = M^2$ is the integral of kinetic moment, and $Q = E$ is the energy. Integral trajectories are described in Fig. 234. Unlike in the previous case, here we describe an ellipsoid of energy level E = const and its intersections with spheres M^2 = const. Euler equations are completely integrable for any $a, b \in H_u$. The flow $\dot{X}$ for the compact series is obtained as the limit of the flow of the normal series (verify!).

As the next example, consider flows of the normal series for $G_n = so_4 \subset G_u = su_4 \subset G = sl(4, \mathbf{C})$. The Lie algebra so_4 is realized in G as skew symmetric matrices and is spanned by vectors $E_{ij} = E_\alpha + E_{-\alpha}$ of the standard form. Let us present $X \in so_4$ in the form $X = \alpha E_{12} + \beta E_{13} + \gamma E_{14} + \delta E_{23} + \rho E_{24} + \varepsilon E_{34}$, where all coefficients are real. Recall that rank $so_4 = 2$ and generic orbits are four-dimensional manifolds $S^2 \times S^2$. Let $a, b \in H_u \subset su_4$; then,

$$\varphi_{ab} X = \alpha \frac{b_1 - b_2}{a_1 - a_2} E_{12} + \beta \frac{b_1 - b_3}{a_1 - a_3} E_{13} + \gamma \frac{b_1 - b_4}{a_1 - a_4} E_{14} +$$

$$+ \delta \frac{b_2 - b_3}{a_2 - a_3} E_{23} + \rho \frac{b_2 - b_4}{a_2 - a_4} E_{24} + \varepsilon \frac{b_3 - b_4}{a_3 - a_4} E_{34}.$$

For each pair a, b in general position, we obtain the flow $\dot{X}$ on $S^2 \times S^2$. Functions $\mathrm{Tr}\,(X + \lambda a)^k$, where $1 \leqq k \leqq 4$, and

$$X + \lambda a = \begin{pmatrix} \lambda a_1 & \alpha & \beta & \gamma \\ -\alpha & \lambda a_2 & \delta & \rho \\ -\beta & -\delta & \lambda a_3 & \varepsilon \\ -\gamma & -\rho & -\varepsilon & \lambda a_4 \end{pmatrix}$$

are integrals. The computation gives the following four integrals: $h_1 = \mathrm{Tr}\, X^2$, $h_2 = \mathrm{Tr}\, X^4$, $h_3 = \mathrm{Tr}\, X^2 a$, $h_4 = \mathrm{Tr}\,(X^2 a^2 + XaXa)$. Integrals h_1 and h_2 are constant on orbits and are of the form

$$h_1 = \alpha^2 + \beta^2 + \gamma^2 + \delta^2 + \rho^2 + \varepsilon^2,$$

$$h_2 = h_1^2 + 4(\beta\delta\gamma\rho - \alpha\delta\gamma\varepsilon + \alpha\rho\beta\varepsilon) - 2(\alpha^2\varepsilon^2 + \beta^2\rho^2 + \gamma^2\delta^2).$$

In fact, h_2 is the squared integral q of degree 2 (after subtracting h_2 from h_1^2), where $q = \alpha\varepsilon - \beta\rho + \gamma\delta$. Thus, two quadratic integrals h_1 and q are generators of Iso_4, i.e., any polynomial constant on orbits depends only on h_1 and q. It is easy to verify that h_1 and q are independent. Equations $h_1 = p$ and $q = t$, where p, t are constants, define generic orbits. These integrals, in particular q, were considered in [24]. Integrals h_3 and h_4 are not constant on orbits and are of the form $h_3 = \alpha^2(a_1 + a_2) + \beta^2(a_1 + a_3) + \gamma^2(a_1 + a_4) + \delta^2(a_2 + a_3) + \rho^2(a_2 + a_4) + \varepsilon^2(a_3 + a_4)$, $h_4 = \alpha^2(a_1^2 + a_1 a_2 + a_2^2) + \beta^2(a_1^2 + a_1 a_3 + a_3^2) + \gamma^2(a_1^2 + a_1 a_4 + a_4^2) + \delta^2(a_2^2 + a_2 a_3 + a_3^2) + \rho^2(a_2^2 + a_2 a_4 + a_4^2) + \varepsilon^2(a_3^2 + a_3 a_4 + a_4^2)$. It is easy to verify that the integrals h_1, q, h_3, h_4 are functionally independent and the integrals h_3 and h_4 are in involution on orbits.

22.3. Cases of the Complete Integrability of the Equations of Motion of a Multidimensional Solid Body with a Fixed Point in the Absence of Gravity and the Complete Integrability of Their Analogs on Semisimple Lie Algebras

Here we give a brief outline of the proof; for technical details, see [31, 28].

Theorem 22.1. 1) Let G be a complex semisimple Lie algebra, and let $\dot{X} = [X, \varphi_{a,b,D}X]$ be Euler equations with an operator of the complex series. Then this system is completely integrable (by Liouville) on generic orbits. Let f be any invariant function on G. Then all functions $h_\lambda(X, a) = f(X + \lambda a)$ are integrals of the flow $\dot{X}$ for any λ. Any two integrals $h_\lambda(X, a)$ and $\rho_\mu(X, a)$ constructed from f, $g \in IG$ are in involution in orbits. The Hamiltonian $F = \langle X, \varphi X\rangle$ of $\dot{X}$ also commutes with all integrals of the form $h_\lambda(X, a)$. It is possible to choose from the set of these integrals the halved dimension of the orbit of integrals functionally independent on generic orbits. The integral F is functionally expressed in terms of integrals of the form $h_\lambda(X, a)$.

2) Let G_u be the compact real form of a semisimple Lie algebra, and let $\dot{X} = [X, \varphi X]$ be the Hamiltonian system defined by an operator φ of the compact series. Then the set of functions of the form $f(X + \lambda a)$, where $f \in IG_u$, constitute the full commutative set on generic orbits in G_u.

3) Let G_n be the normal compact subalgebra in the compact Lie algebra G_u, and let $\dot{X} = [X, \varphi X]$ be the Hamiltonian system of the normal series. Then the set of functions of the form $f(X + \lambda a)$, where $f \in IG_n$, constitutes the full commutative set of functions on generic orbits.

First, let us prove that integrals of the complex series are in involution. For the background, let us compute explicitly sgrad f for any smooth function f on G, expressing sgrad f in terms of grad f.

Lemma 22.4. sgrad $f(X) = [\text{grad } f(X), X]$ for any smooth function f on G.

Proof. Let ξ be a vector of T_XO; then,

$$\omega(\text{sgrad} f, \xi) = \xi f(X) = \langle \text{grad} f, \xi\rangle;$$

By definition of ω, we set $\omega(\mathrm{sgrad}\, f, \xi) = \langle \mathrm{sgrad}\, f, y\rangle$, where $\xi = [X, y]$, whence $\langle \mathrm{grad}\, f, [X, y]\rangle = \langle \mathrm{sgrad}\, f, y\rangle$, i.e., $\langle [\mathrm{grad}\, f, X], y\rangle = \langle \mathrm{sgrad}\, f, y\rangle$. Since this identity holds for any y, then $\mathrm{sgrad}\, f = [\mathrm{grad}\, f, X]$.

If $F = \langle X, \varphi X\rangle$, then $\varphi X = \mathrm{grad}\, f(X)$, whence $-\dot{X} = [\varphi X, X] = \mathrm{sgrad}\, f$, proving that $\dot{X}$ is Hamiltonian. Thus, if f and g are two functions on G, then $\{f, g\} = \langle [X, \mathrm{grad}\, f], \mathrm{grad}\, g\rangle$. Finally, $\{f, g\} = \langle X, [\mathrm{grad}\, f, \mathrm{grad}\, g]\rangle$. We have proved the following statement.

Lemma 22.5. $\{f, g\} = \langle X, [\mathrm{grad}\, f, \mathrm{grad}\, g]\rangle$ for any smooth functions f and g on G.

Lemma 22.6. Let f and g be smooth functions on G, constant on orbits. Then $[\mathrm{grad}\, f, \mathrm{grad}\, g] = 0$.

Proof. First, let $X \in \mathrm{Reg}\, G$. Since f and g are constant on orbits, then their gradients are orthogonal to the orbit, i.e., both of them belong to H(X); hence, they commute. Since regular elements are dense, the lemma is proved.

Proposition 22.4. Let f and g be smooth functions on G, constant on orbits. Consider functions $h_\lambda(X, a) = f(X + \lambda a)$, $d_\mu(X, a) = g(X + \mu a)$, where $a \in H(a, b)$. Then integrals h_λ and d_μ commute. Moreover, $\{F, h_\lambda\} = 0$ for any $f \in IG$.

Proof. Recall that functions h_λ and d_μ are integrals of the flow $\dot{X} = [X, \varphi X]$ in view of Proposition 22.1. By Lemma 22.5, it suffices to prove that $\langle X, [\mathrm{grad}\, h_\lambda, \mathrm{grad}\, d_\mu]\rangle = 0$, i.e., $\langle X, [\mathrm{grad}\, f(X + \lambda a), \mathrm{grad}\, g(X + \mu a)]\rangle = 0$. Let $Y = X + \lambda a$; then $X = Y - \lambda a$ and $X + \mu a = Y + \nu a$, where $\nu = \mu - \lambda$. First, suppose that $\nu \neq 0$. Then

$$\begin{aligned} Z &= \langle Y - \lambda a, [\mathrm{grad}\, f(Y), \mathrm{grad}\, g(Y + \nu a)]\rangle = \\ &= \langle [Y, \mathrm{grad}\, f(Y)], \mathrm{grad}\, g(Y + \nu a)\rangle - \\ &- \langle [\lambda a, \mathrm{grad}\, g(Y + \lambda a), \mathrm{grad}\, f(Y)]\rangle. \end{aligned}$$

Since $f \in IG$, then $[Y, \mathrm{grad}\, f(Y)] = 0$ by Lemma 22.2. Since $g \in IG$, then by the same lemma $[Y + \nu a, \mathrm{grad}\, g(Y + \nu a)] = 0$; hence $[Y, \mathrm{grad}\, g(Y + \nu a)] = -\nu [a, \mathrm{grad}\, g(Y + \nu a)]$. Substituting this in Z, we get

$$Z = -\frac{\lambda}{\nu} \langle [Y, \mathrm{grad}\, g(Y + \nu a)], \mathrm{grad}\, f(Y)\rangle =$$

$$= \frac{\lambda}{\nu} \langle \operatorname{grad} g(Y + \nu a), [Y, \operatorname{grad} f(Y)] \rangle = 0,$$

since $f \in IG$. The statement is proved for $\lambda \neq \mu$. If $\lambda = \mu$, then $0 = \langle X, [\operatorname{grad} f(X + \lambda a), \operatorname{grad} g(X + \lambda a)] \rangle$ by Lemma 22.6. It remains to prove that $\{F, h_\lambda\} = 0$, i.e., to compute $L = \langle X, [\varphi X, \operatorname{grad} f(X + \lambda a)] \rangle$, since $\operatorname{grad} F(X) = \varphi X$. Let $Y = X + \lambda a$. Then

$$L = \langle [Y - \lambda a, \varphi Y - \lambda \varphi a], \operatorname{grad} f(Y) \rangle =$$
$$= \langle [Y, \varphi Y], \operatorname{grad} f(Y) \rangle - \lambda \langle [Y, \varphi a], \operatorname{grad} f(Y) \rangle -$$
$$- \lambda \langle [a, \varphi Y], \operatorname{grad} f(Y) \rangle + \lambda^2 \langle [a, \varphi a], \operatorname{grad} f(Y) \rangle = 0,$$

since in the first summand $[Y, \operatorname{grad} f(Y)] = 0$ and similarly in the second one; in the fourth one, $\varphi a = Da \in H(a, b)$, i.e., $[a, \varphi a] = 0$, while in the third one $[a, \varphi Y] = a d_b Y = [b, Y]$, and again $[Y, \operatorname{grad} f(Y)] = 0$.

The involutive property of integrals of the compact series is similarly proved.

Proposition 22.5. Let $f, g \in IG_u$. Set $h_\lambda(X, a) = f(X + \lambda a)$, $d_\mu(X, a) = g(X + \mu a)$, where $a, b \in H_u(a, b)$. Then integrals h_λ and d_μ commute and $\{F, h_\lambda\} = 0$ for any $f \in IG_u$.

The same scheme underlies the proof of the involutive property of integrals of the normal series. Let us now prove that this commutative set of functions is full. This last part of the proof is technically more delicate; therefore, we restrict ourselves to the scheme of its construction.

Let G be a complex semisimple Lie algebra, $X \in \operatorname{Reg} G$. Let $f_1, \ldots, f_r \in IG$ be a full set of invariants of the algebra. At the point X, a set of complex vectors $\operatorname{grad} h_{\lambda,k}$ arises, where $h_{\lambda,k}(X, a) = f_k(X + \lambda a)$. Let $V(x, a)$ be a subspace in G generated by vectors $\operatorname{grad} h_{\lambda,k}(X, a)$. Our goal is to estimate $\dim V(x, a)$ from below. Consider decompositions

$h_{\lambda,k} = \sum_{i=0}^{q_k+1} P_k^i \lambda^i$, where $q_k + 1 = \deg f_k$. Let f_k be ordered with respect to the growth of their degree. Let $N + 1 = q_r + 1 = \deg f_r$ be the largest degree among generators f_k. All polynomials $h_{\lambda,k}$ can be considered polynomials of degree N + + 1 with only some coefficients of large powers of λ equal

to zero. We have $\operatorname{grad} h_{\lambda,k} = \sum_{i=0}^{q_k} U_k^i \lambda^i$, where $U_k^i(X, a) =$

$= \operatorname{grad} P_k^i(X, a)$ are polynomials of degree i in X and a. Clearly, $U_k^{q_k}(a)$ does not depend on X since $P_k^{q_k}$ is linear in X. All vectors $U_k^{q_k}$ generate the Cartan subalgebra H(X) which does not depend on the choice of X. Clearly, V(X, a) is generated by all vectors U_k^i.

Lemma 22.7. For any k, the following recursive relations in U_k^i hold:

$$
\begin{aligned}
&[U_k^0, X] = 0,\\
&[U_k^1, X] + [U_k^0, a] = 0,\\
&\cdots\cdots\cdots\cdots\cdots\\
&[U_k^i, X] + [U_k^{i-1}, a] = 0,\\
&\cdots\cdots\cdots\cdots\cdots\\
&[U_k^N, X] + [U_k^{N-1}, a] = 0,\\
&[U_k^N, a] = 0.
\end{aligned}
$$

Proof. By Lemma 22.2, $[X, \operatorname{grad} f_k(X)] = 0$. Applying this identity to functions $h_{\lambda,k}$, we get $[X + \lambda a, \operatorname{grad} h_{\lambda,k}(X, a)] = 0$, i.e., $\left[X + \lambda a, \sum_{i=0}^{N} U_k^i \lambda^i\right] = 0$, implying the following statement.

Since H(X) is the Cartan subalgebra, we may construct the root decomposition of G with respect to H(X) and choose the Weyl basis.

Lemma 22.8. If $a \in H(X) \oplus V^+(X)$, then $\operatorname{grad} h_{\lambda,k}(X, a) \in H(X) \oplus V^+(X)$, i.e., $U_k^i \in H(X) \oplus V^+(X)$.

In what follows, we assume that $a \in H(X) \oplus V^+(X)$. Consider simple roots $\alpha_1, \ldots, \alpha_r$; then each positive root can be expressed in the form $\alpha = \sum_{i=1}^{r} m_i \alpha_i$, where $m_i \geq 0$ and m_i are integers. The height of α is the integer $k = k(\alpha) = \sum_{i=1}^{r} m_i$. Let

$V_k^+(X)$ be the subspace in $V^+(X)$ generated by vectors X_α such that $k(\alpha) = k$. Then evidently $V^+(X) = V_1^+ \oplus \ldots \oplus V_s^+$ and V_1^+ is generated by $X_{\alpha_1}, \ldots, X_{\alpha_r}$, i.e., by simple roots. Let us choose $a \in H(X) \oplus V^+(X)$ more precisely. Let $a \in V_1^+$

and $a = \sum_{i=1}^{r} \nu_i X_{\alpha_i}$, where $\nu_i \neq 0$ for $1 \leqq i \leqq r$. Then $[V_k^+,$

$a] \subset V_{k+1}^+$, $[H(X), a] \subset V_1^+$.

<u>Lemma 22.9</u>. Let X, a be chosen as above. Then $V_1^+ = [H(X), a]$ and $ad_a: H(X) \to V_1^+$ is an isomorphism.

<u>Lemma 22.10</u>. Let X, a be the vectors mentioned above. Then $[V_k^+, a] = V_{k+1}^+$, i.e., $ad_a: V_k^+ \to V_{k+1}^+$ is an epimorphism.

<u>Lemma 22.11</u>. $U_k^0 \in H(X)$ and $U_k^i \in H(X) \oplus V_1^+ \oplus \ldots \oplus V_i^+$ for any $1 \leqq k \leqq r$.

<u>Lemma 22.12</u>. Vectors U_k^j, where $j \leqq i$, generate the subspace $H(X) \oplus V_1^+ \oplus \ldots \oplus V_i^+$.

<u>Lemma 22.13</u>. Let a be an element in general position. Then $\dim_{\mathbb{C}} V(X, a) \geqq \dim H(X) \oplus V^+(X) = \frac{1}{2}(\dim G + \operatorname{rank} G)$, where $V(X)$ is the complex subspace generated by all vectors of the form $\operatorname{grad} h_{\lambda,k}$ for all X of the open dense subset in G.

To complete the proof of the theorem, it suffices to note that there is a symmetry between X and a in all the above statements since $f(X + \lambda a) = \lambda^q f((1/\lambda)X + a)$, where $f \in IG$ is a polynomial of degree q.

The proof of fullness of the above commutative sets for compact and normal series is obtained from the above scheme after involutions that define these two series are taken into account.

22.4. Cases of the Complete Integrability of the Equations of Motion from Inertia of a Multidimensional Solid Body in an Ideal Fluid

Consider the embedding of the system mentioned in the title into the noncompact Lie algebra of the group of motions of a Euclidean space. It turns out that, in this case, the shift of the argument method makes it possible to construct the full commutative set of integrals on generic orbits.

Lemma 22.14. Let f be an invariant of the coadjoint representation of the group of motions of the Euclidean space $\mathbb{R}^n$. Then functions $f(X + \lambda a)$ for any real λ are integrals of the Euler equations $\dot{X} = \mathrm{ad}^*_{QX}X$, where $Q(a, b, D)$ are the above sectional operators $Q: e^*(n) \to e(n)$.

Proof. It suffices to verify the equality $\langle a(X, QX), df(X + \lambda a)\rangle = 0$, where $\langle X, \xi\rangle$ is the value of the functional X at the vector ξ. Evidently, we have

$$A = \langle a(X, QX),\ df(X + \lambda a)\rangle = -\langle QX, a(X + \lambda a,\ df(X + \lambda a))\rangle - \lambda\langle a(QX, a), df(X + \lambda a)\rangle.$$

Since f is invariant, then the first summand is zero. From the definition of $Q(a, b, D)$, we get

$$-\frac{1}{\lambda}A = \langle a(\Phi_a^{-1}\mathrm{ad}_b^* X_1, a),\ df(X + \lambda a)\rangle + \langle a(DX_2, a),\ df(X + \lambda a)\rangle,$$

where $X_1 \in K^{*\perp}$, $X_2 \in K^*$. The second summand is zero since $DX_2 \in K$, $a \in K^*$. The first summand equals $\langle a(X_1 + X_2 + \lambda a, b), df(X + \lambda a)\rangle$. Since $X_2, b \in \mathrm{Ann}(a)$, then

$$\langle a(X_1 + X_2 + \lambda a, b),\ df(X + \lambda a)\rangle = \langle a(X + \lambda a,\ df(X + \lambda a)), b\rangle = 0$$

because f is an invariant. The lemma is proved.

Theorem 22.2. 1) The system of differential equations $\dot{X} = \mathrm{ad}^*_{QX}X$, where $Q = Q(a, b, D)$ on $e^*(n)$, is completely integrable on generic orbits.

2) Let f be an invariant function $e^*(n)$. Then $h_\lambda(X) = f(X + \lambda a)$ are integrals of motion for any λ. Any two integrals h_λ and g_μ are in involution on all orbits of the representation Ad* of E(n) and the number of independent integrals of the above-mentioned form is equal to the halved dimension of a generic orbit. If O is a (generic) orbit of maximum dimension in the coadjoint representation, then codim O = [(n + 1)/2].

Proof. In Lemma 22.14, we have verified that these functions are integrals. Their involutive property is proved as in Sec. 22.3. It remains to verify that shifts of invariants $f(X + \lambda a)$ constitute a full commutative set on a generic orbit. codim O = [(n + 1)/2] is verified by standard methods. Let us generate the full set of invariants of the algebra. For this let us present $e(n)^*$ in the matrix form

$$e(n)^* = \left(\begin{array}{c|c} so_n & \begin{matrix}0\\ \vdots \\ 0\end{matrix} \\ \hline y_1 \ldots y_n & 0 \end{array}\right).$$

The minor of the matrix X that consists of intersections of rows with numbers $i_1, \ldots, i_s$ and columns with numbers $j_1, \ldots, j_s$ will be denoted by $M^{i_1,\ldots,i_s}_{j_1,\ldots,j_s}(X)$, where $1 \leqq i_1 < \ldots < i_s \leqq n$ and $1 \leqq j_1 < \ldots < j_s \leqq n$. Then the functions

$$J_s(X) = \sum_{1 \leqslant i_1 < \ldots < i_s \leqslant n} M^{i_1\ldots i_s,\, n+1}_{i_1\ldots i_s,\, n+1}\left(\frac{1}{2}(X - X^T)\right)$$

are invariants of the algebra. Functions with even numbers vanish identically, and those with odd numbers constitute the full set of invariants; their verification is straightforward. Let (f_i) be a full set of polynomial invariants; then

$$f_i(X + \lambda a) = \sum_{s=0}^{N_i} P_{is}(X, a)\lambda^s, \quad \text{where } df_i \in e(n)^{**} = e(n).$$

Set $df_i(X+\lambda a) = \sum_{s=0}^{N_i} U_{is}(X,a)\lambda^s$, where $U_{is} \in e(n)$.

Lemma 22.15. The following recursive relations hold:

$$a(X, U_{i0}) = 0,$$
$$a(X, U_{i1}) + a(a, U_{i0}) = 0,$$

$$a(X, U_{i,N_i}) + a(a, U_{i,N_i-1}) = 0,$$
$$a(a, U_{i,N_i}) = 0.$$

Let $n = 2s + 1$. Consider the complexification $Ce(n)$. The Lie algebra $so(n, \mathbf{C})$ is simple. Let $so(n, \mathbf{C}) = H \oplus \sum_{i\geqslant 1} G_i^+ \oplus \sum_{i\geqslant 1} G_i^-$, where subspaces $G_i^{\pm}$ are spanned by root vectors e_α such that the height of α equals $\pm i$ and $H = \bigoplus_{k=0}^{[n/2]-1} \mathbf{C}E_{2k+1,\,2k+2}$.

In $e(n)$ consider the graded subspace $e(n)^+ = (H \oplus \mathbf{C}e_n) \oplus (G_1^+ \oplus B_1) \oplus \ldots \oplus (G_s^+ \oplus B_s) \oplus \sum_{k\geqslant s+1} G_k^+ = \bigoplus_{i\geqslant 0} H_i$, where $B_{s+1-j} \in \mathbf{C}(e_{2j-1} + ie_{2j}) \subset \mathbf{C}^n$ for $j = 1, \ldots, s$. Choose X, $a \in Ce(n)^*$ so that $X \in K^*$ is in general position; $a \in G_1^+ \oplus B_1$ is such that all components in the root decomposition of G_1^+ and the coordinate with respect to the basic vector $e_{n-2} + ie_{n-1} \in \mathbf{C}^n$ are nonzero.

Lemma 22.16. Let $a \in G_1^+ \oplus B_1 \subset e(n)^*$ be the element constructed above. Then $a(a, H_i) \subset H_{i+1} \subset e(n)^*$ for $H_i \subset e(n)$ and $i \geqq 0$.

Lemma 22.17. Let X, $a \in e(n)^*$ be chosen as above; then the mapping $H_i \to H_{i+1} \subset e(n)^*$ defined by the formula $y \to a(a, y)$, $y \in H_i \subset e(n)$ is an epimorphism.

Lemma 22.18. $U_{j0} \in H_0$ and $U_{jk} \in H_k$ for any j.

Lemma 22.19. Vectors U_{jk} generate the whole subspace H_k.

Finally, we obtain that the dimension of the subspace generated by $df(X + \lambda a)$ is no less than $\dim e(n)^{+} = s^2 + 2s + 1$. For complete integrability, we must have

$$\operatorname{codim} O + 1/2\,(\dim e(n)^* - \operatorname{codim} O) = s^2 + 2s + 1$$

functionally independent integrals on G^*. Hence, the theorem is proved when $n = 2s + 1$. The case $n = 2s$ is considered according to a similar scheme, and we will omit here the technical details.

22.5. Equations of Magnetohydrodynamics and Cases of Their Complete Integrability

Let G be an arbitrary Lie algebra. Then we can construct the new Lie algebra ΩG defined as follows. Consider the semidirect sum of two copies of G, i.e., consider the algebra $G \oplus_\rho G$, where G acts on the second summand with respect to the adjoint representation $\rho = \mathrm{ad}$ and the second summand is considered as an Abelian Lie algebra. This semidirect sum will be denoted by ΩG. Algebras ΩG, ΩG_u, ΩG_n, where G is a complex semisimple Lie algebra, and G_u and G_n, which are its compact form and normal compact subalgebra, respectively, are of special interest. Let ΩG^*, $\Omega G_u{}^*$, $\Omega G_n{}^*$ be corresponding dual spaces. All these algebras are noncompact. It turns out (V. V. Trofimov) that on each of these algebras there are full commutative sets of functions (integrals) on generic orbits and these sets are analogs of the "solid-body-like" Hamiltonian systems constructed above with the use of sectional operators. Hamiltonian systems of the form $\dot{X}_Q$ constructed from sectional operators of the type $Q(a, b, D)$ are, as can be verified, analogs of systems of magnetohydrodynamics on arbitrary semisimple Lie algebras. If for G we take the orthogonal Lie algebra so_n, then these equations become the equations of magnetohydrodynamics studied in [34]. Thus, these equations admit an embedding into a Lie algebra in the sense of Definition 21.1. So, for instance, in the case $G = so_n$, these equations are of the form $\dot{M} = [\Omega, M] - [H, J]$, $\dot{H} = [\Omega, H]$, where $\Omega = (R_{g^{-1}})_* \dot{g}$ is the right shift onto the unit of the group of the velocity vector $\dot{g} \in T_g SO_n$; $J = \mathrm{Ad}_{g^{-1}}{}^* j$, where j is the density of the current in the body; $H = \mathrm{Ad}_g\, h$, where h is the strength of the magnetic field in the body and M is the kinetic momentum in the space.

It turns out that, as in previous examples in Secs. 21.3 and 21.4, these multidimensional analogs of the equations of magnetohydrodynamics of an arbitrary semisimple Lie algebra and compact and normal subalgebras are completely integrable Hamiltonian systems for many-parameter series of sectional operators Q(a, b, D). Moreover, the full set of commuting integrals also exists in the case so_n.

In conclusion, let us return once more to Conjecture A (see Sec. 21) regarding the existence of a full commutative set of functions (integrals) on generic orbits for an arbitrary finite-dimensional Lie algebra. This conjecture holds and can be easily verified for all Lie algebras of small dimensions, namely, for Lie algebras of dimension no greater than 5 and for nilpotent Lie algebras of dimension no greater than 6.

The reader may obtain additional information concerning the subject of this book in [35-40].

REFERENCES

1. A. S. Mishchenko and A. T. Fomenko, A Course in Differential Topology, Mosk. Gos. Univ., Moscow (1980).
2. B. A. Dubrovin, S. P. Novikov, and A. T. Fomenko, Modern Geometry, Nauka, Moscow (1979).
3. D. B. Fuks, A. T. Fomenko, and V. L. Gutenmacher, Homotopic Topology, Mosk. Gos. Univ., Moscow (1969).
4. Yu. G. Borisovich, N. M. Blisnyakov, Ya. A. Israilevich, and T. N. Fomenko, Introduction to Topology, Vysshaya Shkola, Moscow (1980).
5. J. Milnor, Singular Points of Complex Hypersurfaces [Russian translation], Mir, Moscow (1971).
6. S. P. Novikov, "Hamiltonian formalism and a multidimensional analog of the Morse theory," Usp. Mat. Nauk, 37, No. 5, 3-49 (1982).
7. J. Milnor, Lectures on the h-Cobordism Theorem, Princeton Univ. Press, Princeton (1965).
8. R. C. Kirby and M. G. Scharlemann, "Eight faces of the Poincaré homology 3-sphere," in: Geometric Topology, Academic Press, New York—San Francisco—London (1979), pp. 113-146.
9. A. T. Fomenko, "Group symplectic structures on homogeneous spaces," Dokl. Akad. Nauk SSSR, 253, No. 5, 1062-1067 (1980).
10. S. V. Manakov, "A remark on integration of Euler equations of n-dimensional solid body dynamics," Funkt. Anal. Ego Prilozhen., 10, No. 4 (1976).
11. I. A. Volodin and A. T. Fomenko, "Manifolds, knots, algorithms," Proceedings of the Seminar on Vector and Tensor Analysis, Vol. 18, Mosk. Gos. Univ., Moscow (1978), pp. 94-128.

12. I. A. Volodin, V. E. Kuznetsov, and A. T. Fomenko, "On the problem of algorithmic recognition of the standard three-dimensional sphere," Usp. Mat. Nauk, 24, No. 5, 71-168 (1974).
13. J. H. C. Whitehead, "On certain sets of elements in a free group," Proc. London Math. Soc., Ser. 2, 41, 48-56 (1936).
14. M. Ochiai, "A counterexample to a conjecture of Whitehead and Volodin—Kuznetsov—Fomenko," J. Math. Soc. Jpn., 31, 87-691 (1979).
15. O. Ya. Viro and V. L. Kobelski, "The Volodin—Kuznetsov—Fomenko conjecture on Heegaard diagrams of a three-dimensional sphere is false," Usp. Mat. Nauk, 32, No. 5, 175-176 (1977).
16. T. Homma, M. Ochiai, and M. Takahashi, "An algorithm for recognizing S^3 in 3-manifolds with Heegaard splittings of genus 2," Osaka J. Math., 17, 625-648 (1980).
17. A. A. Markov, "Insolvability of the homeomorphicity problem," Dokl. Akad. Nauk SSSR, 121, No. 2, 218-220 (1958); Proc. Int. Congr. Math. (1960), pp. 300-306.
18. S. I. Adyan, "Insolvability of some algorithmic problems of group theory," Tr. Mosk. Mat. Ova., 6, 231-298 (1957).
19. Seminar "Sophus Lie," 1954/1955, Paris (1955).
20. V. I. Arnold, Mathematical Methods of Classical Mechanics, Nauka, Moscow (1974).
21. A. S. Mishchenko and A. T. Fomenko, "A generalized Liouville method of integration of Hamiltonian systems," Funkt. Anal. Ego Prilozhen., 12, No. 2, 46-56 (1978).
22. V. V. Trofimov, "Euler equations on Borel subgroups of semisimple Lie algebras," Izv. Akad. Nauk SSSR, 43, No. 3, 714-732 (1979).
23. M. Adler and P. van Moerbeke, "Completely integrable systems, Euclidean Lie algebras and curves," Adv. Math., 38, No. 3, 267-317 (1980).
24. M. Langlois, "Contribution à l'étude du mouvement du corps rigide a N dimensions autour d'un point fixe," in: Thèse Présentée a la Faculté des Sciences de l'Université de Besancon, Besancon (1971).
25. A. G. Reiman, "Integrable Hamiltonian systems connected with graded Lie algebras," Zap. Nauchn. Sem. Leningr. Otd. Mat. Inst. im. V. A. Steklova, Akad. Nauk SSSR, 95, 3-54 (1980).
26. A. G. Reiman and M. A. Semjonov-Tian-Shanski, "Reduction of Hamiltonian systems, affine Lie algebras, and Lax equations," Invest. Math., 54, 81-100 (1979).

27. A. T. Fomenko, "On symplectic structures and integrable systems on symmetric spaces," Mat. Sb., 115, No. 2, 263-280 (1981).
28. A. S. Mishchenko and A. T. Fomenko, "Integration of Hamiltonian systems with noncommuting symmetries," Proceedings of the Seminar on Vector and Tensor Analysis, Vol. 20, Mosk. Gos. Univ., Moscow (1980), pp. 5-54.
29. V. V. Trofimov, "Finite-dimensional representations of Lie algebras and completely integrable systems," Mat. Sb., 111, No. 4, 610-621 (1980).
30. V. V. Trofimov and A. T. Fomenko, "A method of construction of Hamiltonian flows on symmetric spaces and integrability of hydrodynamical systems," Dokl. Akad. Nauk SSSR, 254, No. 6, 1349-1353 (1980).
31. A. S. Mishchenko and A. T. Fomenko, "The Euler equations on finite-dimensional Lie groups," Izv. Akad. Nauk SSSR, 42, No. 2, 396-415 (1978).
32. I. E. Kochin, I. A. Kibel, and N. V. Rose, Theoretical Hydrodynamics, Part I, Fizmatgiz, Moscow (1963).
33. G. B. Gorr, L. B. Kudryashova, and L. A. Stepanova, Classical Problems of Solid Body Theory. Development and Present State, Naukova Dumka, Kiev (1978).
34. S. M. Vishik and F. V. Dolzhanski, "Analogs of Euler–Poisson equations and of equations of magnetohydrodynamics connected with Lie groups," Dokl. Akad. Nauk SSSR, 238, No. 5, 1032-1035 (1978).
35. J. S. Birman and H. M. Hilden, "Heegaard splittings of branched coverings of S^3," Trans. Am. Math. Soc., 213, 315-352 (1975).
36. B. A. Dubrovin, V. B. Matveev, and S. P. Novikov, "Nonlinear Korteweg–de Vries-type equations, finite-zonal linear operations, and Abelian varieties," Usp. Mat. Nauk, 26, No. 1 (1976).
37. S. Helgasson, Differential Geometry, Lie Groups, and Symmetric Spaces, Academic Press, New York–London (1972)
38. D. Kazhdan, B. Kostant, and S. Sternberg, "Hamiltonian systems of Calogero type," Commun. Pure Appl. Math., 31, No. 4, 481-507 (1978).
39. J. Moser, "Various aspects of integrable Hamiltonian systems," in: Proc. CIME Conf., Bressanone, Italy (June 1978).
40. F. Waldhausen, "Heegaard–Zerlegungen der 3-sphere," Topology, 7, No. 2, 195-203 (1968).